Intelligent Infrastructure and Smart Materials

Shray Pathak • Anoop Kumar Shukla •
Shreya Sharma • Vijay P. Singh
Editors

Intelligent Infrastructure and Smart Materials

Sustainable Technologies for a Greener Future

Springer

Editors
Shray Pathak
Department of Civil Engineering
Indian Institute of Technology Ropar
Rupnagar, Punjab, India

Shreya Sharma
AWaDH
Indian Institute of Technology Ropar
Rupnagar, Punjab, India

Anoop Kumar Shukla
Manipal School of Architecture and Planning
Manipal Academy of Higher Education
Manipal, Karnataka, India

Vijay P. Singh
Department of Biological and Agricultural Engineering and Zachry Department of Civil and Environmental Engineering
Texas A&M University
College Station, TX, USA

ISBN 978-3-031-92420-0 ISBN 978-3-031-92421-7 (eBook)
https://doi.org/10.1007/978-3-031-92421-7

This Springer imprint is published by the registered company Springer Nature Switzerland AG
The registered company address is: Gewerbestrasse 11, 6330 Cham, Switzerland

Preface

The twenty-first century is marked by rapid urbanization, growing environmental concerns, and an urgent need for sustainable solutions in infrastructure development. Traditional materials and construction methods, while effective in the past, are no longer sufficient to address the challenges of climate change, resource depletion, and increasing demands for resilient infrastructure. In response, the integration of *intelligent infrastructure* and *smart materials* has emerged as a transformative approach to building a greener and more sustainable future.

This book, *Intelligent Infrastructure and Smart Materials: Sustainable Technologies for a Greener Future*, explores the latest advancements in the field, bridging the gap between cutting-edge engineering, material science, and environmental sustainability. It provides an in-depth analysis of how intelligent infrastructure—enabled by data-driven technologies, automation, and sensor networks—can enhance the efficiency, resilience, and adaptability of modern cities. Additionally, the book delves into the evolution of smart materials, including self-healing concrete, shape-memory alloys, intelligent water resources, and bio-inspired composites, which are revolutionizing the way we construct and maintain infrastructure.

Aimed at researchers, engineers, policymakers, and sustainability advocates, this book offers a multidisciplinary perspective on the role of intelligent and adaptive materials in creating eco-friendly infrastructure. By integrating case studies, recent research, and real-world applications, we present a comprehensive guide to sustainable technologies that will shape the future of urban and industrial landscapes.

As we embark on this journey towards a more sustainable world, we hope that the insights shared in this book inspire innovation and collaboration in the development of intelligent infrastructure systems. The transition to greener technologies is not just an option—it is an imperative for the well-being of future generations and the planet.

Rupnagar, Punjab, India — Shray Pathak
Manipal, Karnataka, India — Anoop Kumar Shukla
Rupnagar, Punjab, India — Shreya Sharma
College Station, TX, USA — Vijay P. Singh

Contents

About the Editors

Shray Pathak is working as an Assistant Professor in the Department of Civil Engineering, Indian Institute of Technology Ropar, Rupnagar, Punjab, India. Dr. Pathak has received the Bachelor's degree in Civil Engineering from the National Institute of Technology (NIT), Kurukshetra, India in June 2012 and completed his Master's and Ph.D. degree in Civil Engineering from the Indian Institute of Technology (IIT), Roorkee, Uttarakhand, India in June 2014 and February 2019, respectively. His research interests include hydrology, remote sensing, GIS, urban sustainability, and climate change. Earlier, he was employed as an Assistant Professor in the Department of Civil Engineering, Netaji Subhas University of Technology, New Delhi from Dec 2021 to Sept 2023. He has a post-doctoral experience of 2.5 years in School of Geographic Sciences, East China Normal University (ECNU), Shanghai, China (June 2019 to Dec 2021) and worked on the risk assessment analysis of urban flooding in coastal and non-coastal regions under extreme climate scenarios. Till date he has amassed over 45 publications, spanning peer-reviewed journals, conferences, and book chapters. He has received the 2022 Best Practice Oriented Paper for the Journal paper published in ASCE Journal of Hazardous, Toxic and Radioactive Waste awarded by the American Society of Civil Engineers (ASCE).

Anoop Kumar Shukla is an Assistant Professor at Manipal School of Architecture and Planning (MSAP), Manipal Academy of Higher Education, Manipal, Karnataka, India. He holds 8 years of research experience in Research and Development in Geomatics Engineering, Water and Wastewater Management, Remote Sensing, and Geographical Information Systems (GIS), as well as in Water Quality, Climate Change, Crop Yield Modeling, and related subjects. Dr. Shukla has published 35 high-impact journal papers, 45 International/National Conference papers, and 23 book chapters across ASCE, Wiley, and Springer publications. He is a member of various national and international academic, research, and administrative bodies including IAH, INCA, IEEE-GRSS, AGU, and ASCE.

He earned his M.Tech in Geomatics Engineering and Ph.D. in Remote Sensing and GIS applications in water resources planning and management from the Indian

Institute of Technology Roorkee, Uttarakhand, India. Following this, he pursued a Post-Doctoral Fellowship (PDF) at the Institute of Geographic Sciences and Natural Resources Research (IGSNRR), Chinese Academy of Sciences (CAS), Beijing, China. Dr. Shukla serves as an Associate Editor for the Journal of Hazardous, Toxic, and Radioactive Waste (ASCE) and as a Review Editor on the Editorial Board of *Water and Wastewater Management and Organic Pollutants Frontiers in Environmental Science* journal. Dr. Shukla is a reviewer for more than 90 peer-reviewed international SCI journals.

Shreya Sharma serves as the Research and Development (R&D) and Corporate Manager at the Agriculture and Water Technology Development Hub (AWaDH), a Technology Innovation Foundation (TIF) at the Indian Institute of Technology (IIT) Ropar in Rupnagar, Punjab, India. In this capacity, she plays a pivotal role in advancing research initiatives and fostering corporate collaborations within the hub. Before her current role, Dr. Shreya Sharma served as an Assistant Professor at Chitkara Business School, Chitkara University, and also held a similar position at Amity University. She earned her Bachelor's degree in Commerce (B.Com) in 2015, followed by a Master's degree in Business Administration (MBA) with a specialization in Finance in 2017. In 2023, she completed her Ph.D. in Management from Devi Ahilya University, Indore. Her research interests span various domains, including capital market finance, corporate finance, green finance, environmental, social, and governance (ESG) factors, as well as sustainability and the Sustainable Development Goals (SDGs). To date, she has published five research articles in reputed journals, contributed three book chapters, edited one book, and presented her work at multiple national and international conferences.

Vijay P. Singh is a Distinguished Professor, a Regents Professor, and the inaugural holder of the Caroline and William N. Lehrer Distinguished Chair in Water Engineering at Texas A&M University, USA. He received his B.Sc. and Tech., M.S., Ph.D., and D.Sc. degrees, all in engineering. He is a registered professional engineer, a registered professional hydrologist, and an Honorary Diplomate of ASCE-AAWRE. He is a member of National Academy of Engineering (NAE), an Academician of Georgia Fazisi Academy, a member of European Academy of Sciences and Arts, and a fellow/member of 12 other international science/engineering academies. He is a Distinguished Member of ASCE, an Honorary Distinguished Member of IWRA, a Distinguished Fellow of AGGS, and an Honorary Member of AWRA, and a Fellow of AGU, EWRI, ASCE, IAH, ISAE, IWRS, and IASWC. He has published extensively in the areas of hydrology, irrigation engineering, hydraulics, groundwater, hydraulic geometry, water quality, food-water-energy-environment nexus, water resources, entropy theory, copula theory, and climate change impacts with more than 1570 refereed journal articles; 40 books; 95 edited reference books, as well as the Handbook of Applied Hydrology and Encyclopedia of Snow, Ice and Glaciers; 130 book chapters; and 330 conference papers. His Google Scholar citations are more than 105,231; h-index of 134; and I10-index of 1226.

He has received four honorary doctorates from Italy and Canada, and more than 115 national and international awards, including Norman Medal (twice), Chow Award, Torrens Award, Arid Lands Hydraulic Engineering Award, and Outstanding Projects and Leader (OPAL) Award, Lifetime Achievement Award, Best paper Awards (twice), and Roy C. Tipton Award, all given by American Society of Civil Engineers; Hancor Award and Lalit and Aruna Verma Award for Global excellence, both given by the American Society of Agricultural and Biological Engineers; Merriam Improved Irrigation Award, given by International Commission on Irrigation and Drainage; Linsley Award, Wetzel Award, and Founders Award, given by the American Institute of Hydrology; Crystal Drop Award and Chow Memorial Award, given by the International Water Resources Association; Outstanding Scientist Award given by Sigma Xi; Distinguished Scientist Award, given by the Chinese Academy of Science; Distinguished Professor Award, given by the Mexican Academy of Science; Professor Gajendra Singh Gold Medal for Education, given by the Indian Society of Agricultural Engineers; to name but a few. He has served as President of the American Institute of Hydrology (AIH); President of the American Academy of Water Resources Engineers; President of the International Association for Water, Environment, Energy, and Society; and Chair of the Watershed Council of American Society of Civil Engineers. He has served as editor-in-chief of five journals and two book series and has served/serves on editorial boards of more than 40 journals and three book series. He has given more than 180 keynote lectures and 300 invited seminars all over the world and has organized more than 25 international conferences.

Professor Singh has long been engaged in philanthropic activities. He founded G.B. School in 1994 in his native village Naglavishnu in District Agra, U.P., India, in memory of his parents. The school imparts quality education to children in rural Agra. He has been bearing most of the expenditures involved in the operation of school, including the cost of building construction, staff salaries, furniture, maintenance, management of the school, and so on. The school now has four campuses: (1) Primary School (Grade 1–5), (2) Inter College (Grade 6–12), (3) Degree College (B.A. and B.Sc.), and (4) Industrial Training Center (Fitter and Electronics).

Factors Responsible for Sustainable Habitat: An End-User Satisfaction

Manish Bharadwaj, Manoj Patwardhan, and Charu Goyal

1 Introduction

The construction industry, comprising government agencies and private firms, is pivotal in shaping urban environments' physical and social fabric. Infrastructure development, housing projects, and urban renewal initiatives are all critical components of this sector's contributions to urbanization and sustainable development [1]. Sustainable habitat includes three dimensions as essential realms that manifest their significance in group housing societies. It encompasses various facets of communal life, including social cohesion, equity, inclusivity, and the quality of resident relationships. Unlike environmental sustainability, which often involves tangible metrics like carbon emissions or resource conservation [2], social sustainability is the intangible aspect of human interaction and community development [3], and the urban shift holds the promise of economic growth, innovation, and improved living standards [4]. There is a growing movement among various stakeholders, including architects, planners, developers, engineers, and local authorities, advocating for a comprehensive approach to urban planning and construction. Practical efforts are needed to develop a sustainability measurement framework that considers end-user satisfaction, explores community strength and quality of life pertaining to new housing developments, and highlights a shift toward more comprehensive strategies to cope with urban development challenges [5]. The Sustainable Development

M. Bharadwaj (✉) · M. Patwardhan
Department of Management Studies, Atal Bihari Vajpayee-Indian Institute of Information Technology and Management (ABV-IIITM), Gwalior, India
e-mail: manishb@iiitm.ac.in; manojp@iiitm.ac.in

C. Goyal
Centre for Computer Science and Business Management, MITS-DU, Gwalior, India
e-mail: goyal.charu09@mitsgwalior.in

S. Pathak et al. (eds.), *Intelligent Infrastructure and Smart Materials*,
https://doi.org/10.1007/978-3-031-92421-7_1

Goals (SDGs) set forth by the United Nations recognize the urgency of addressing these challenges, with SDG 11, "Make cities and human settlements inclusive, safe, resilient, and sustainable," being a linchpin in this effort [6–8]. SDG 11 shapes our cities' physical and social infrastructure, influencing their functionality, resilience, and inclusivity, whether through constructing affordable housing, developing green spaces, or enhancing public transportation to achieve its targets. The individuals who drive its operations are at the heart of the construction sector. The workforce within government agencies and private construction firms represents the industry's lifeblood. Their skills, motivation, and commitment are instrumental in delivering quality projects that fulfill the needs of end users [9]. Since the adoption of the 2030 Agenda, studies have focused on implementing the SDGs, exploring synergies between goals and targets, identifying trade-offs [10], and improving the capacity for progress reporting [11, 12]. This chapter measures end-user satisfaction by considering various group factors centered on sustainable habitats and achieving SDG-related objectives.

2 Dimensions of Sustainable Habitat

Sustainability encompasses many practices and principles that minimize construction projects' negative impact in terms of environmental, social, and economic aspects along with promotion of long-term viability and responsible resource management. It underpins seven dimensions necessary to achieve a sustainable habitat, as shown in Fig. 1.

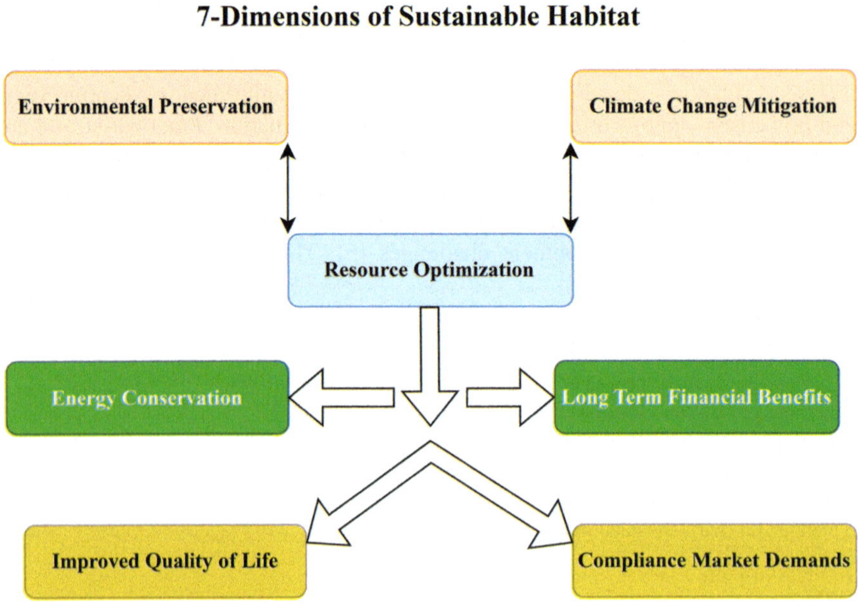

Fig. 1 Dimensions of sustainable habitat

2.1 Environmental Preservation

Sustainability practices in construction are pivotal for diminishing the environmental footprint of projects. A sustainable habitat involves utilizing environmentally friendly materials, minimizing waste generation, and integrating energy-efficient designs and technologies [13]. These efforts are instrumental in curbing resource consumption, mitigating greenhouse gas emissions, and safeguarding natural ecosystems.

2.2 Resource Optimization

Construction is notorious for its intensive use of natural resources, such as water, energy, and raw materials. Sustainable construction prioritizes resource efficiency by incorporating recycled materials, embracing renewable energy sources, and streamlining construction processes to curtail waste generation [14].

2.3 Climate Change Mitigation

The construction sector bears a significant responsibility for global greenhouse gas emissions. Sustainable construction practices mitigate climate change by advocating for energy-efficient building designs, using low-carbon construction materials, and reducing carbon emissions associated with transportation and construction procedures.

2.4 Energy Conservation

Sustainable construction emphasizes the use of energy-efficient building designs and technologies. The implementation of high-performance building envelopes, efficient HVAC systems, and renewable energy sources such as solar panels can significantly reduce a building's energy consumption and long-term operating costs.

2.5 Improved Quality of Life

Sustainable construction goes beyond environmental concerns; it also enriches the quality of life for building occupants. Features like abundant natural lighting, effective indoor air quality management, and noise reduction measures contribute to healthier and more comfortable living and working environments.

2.6 Long-Term Financial Benefits

While sustainable construction may involve an initial investment, it frequently leads to substantial financial savings over the long haul. Energy-efficient buildings, for instance, typically exhibit lower operational expenses, while durable materials and construction practices minimize maintenance and replacement costs over time.

2.7 Compliance and Market Demand

Governments and local authorities have implemented regulations and incentives that endorse sustainable construction practices. Moreover, there is an escalating market demand for environmentally friendly and sustainable buildings [15]. Construction firms that prioritize sustainability are not only in compliance with regulations but also align with the evolving expectations of their clientele.

3 Group Factors and Methodology

Wieser et al. (2019) emphasize the interconnected nature of the SDGs, especially in the context of urban sustainability, and introduce the use of the iMODELER tool for visualizing and analyzing these interdependencies. Johnsson et al. (2020) reported that the construction industry must align with the Sustainable Development Goal (SDG) 11 (Sustainable Cities and Communities) while focusing primarily on SDG 13 (Climate Action). The study underscores the need for a holistic approach to the SDGs, increased pressure on construction companies for sustainability reporting, and the necessity of transformative changes in business operations to achieve SDG 11. Sarkar and Bharat (2021) analyze the government agencies' policies and programs aimed toward achieving Targets 6.1 and 6.2 of SDG 6, which focus on safe drinking water and sanitation. The alignment of the policies and programs impacts these targets of SDG 6. The construction industry contributes significantly to decent work and economic conditions. Firstly, it generates employment opportunities through schemes like Pradhan Mantri Awas Yojana (PMAY), which focuses on affordable housing and boosts demand for construction materials, creating jobs for both skilled and unskilled laborers. PMAY also encourages innovative and green technology, further enhancing employment prospects. Secondly, the construction sector aids economic growth by addressing housing shortages in urban areas, benefiting economically weaker sections and low-income groups. Several benefits not only improve living conditions but also stimulate economic activity within the construction industry. Financial assistance and subsidies provided by PMAY facilitate housing market growth. Moreover, the construction industry plays a crucial role in infrastructure development, creating jobs and fostering economic growth. Schemes like PMAY also support local economies by promoting local materials, skills, and technology, reducing construction costs, and enhancing overall economic conditions. In summary, the construction industry, mainly through initiatives like PMAY, contributes to decent work and improved economic circumstances through employment generation, affordable housing solutions, and infrastructure development. Forte and Russo (2017) proposed several interventions that will enhance the quality of the housing complex and the satisfaction of its residents. These interventions include improvements to the safety of the complex, upgrades to the quality of the green areas, and increases in the availability of social services. The authors conclude by arguing that the customer satisfaction method is a valuable tool for evaluating the quality of public housing and identifying areas where improvement is needed. They also emphasize the importance of involving residents in the evaluation process so that their needs and preferences can be taken into account when developing interventions. Although the benefits are high, as mentioned, the end-user satisfaction is to be measured. It requires a selection of factors that create boundaries for the study. Based on the literature review and SDG 11 targets, six main factors are selected, which include sub-factors (attributes), as shown in Table 1. In the pursuit of a comprehensive understanding of the research landscape, a comparative review of relevant literature was conducted, guiding the formulation of a structured questionnaire designed to gather quantitative data against these factors to measure satisfaction weights.

Table 1 Factors responsible for sustainable habitat

Factors	Attributes	Category	References
Environmental aspect	1. Distant for noisy environment 2. Distant from any hazardous chemical industry 3. Air Quality Index	Environment	[2, 3, 13]
Safe and secure environment	1. Adequate lighting 2. Acoustics 3. Emergency services provision		[4, 6, 11]
Access to public services	1. Water and Sanitation Facilities 2. Reliable electricity supply 3. Accessible health-care facilities	Social	[10, 16, 17]
Inclusive community spaces	1. Diverse and Inclusive Community spaces 2. Facilities for social interactions 3. Regional places		[12, 15]
Affordable housing	4. Cost-effective housing options 5. Accessibility Affordability	Economic	[1, 5, 9],
Accessibility and mobility	1. Accessible and efficient transportation options 2. Public transit 3. Walkable infrastructure		[7, 14]

3.1 End-User Satisfaction Measurement

Respondents from a spectrum of construction industry end users are selected to ensure a diverse and knowledgeable participant pool. The sampling strategy employed a combination of non-probability sampling approaches, specifically purposive sampling. These approaches were deemed most suitable for targeting individuals possessing specialized insights. The sample population encompasses academics, construction industry professionals well-versed in sustainable construction practices, and end users. This whole procedure aims to devise a means to evaluate user satisfaction, obtained regularly through the questionnaire. The questionnaire captures responses for garnering user satisfaction on various aspects directly related to building and external factors that can influence user satisfaction. The questionnaire was sent to 150 persons, of whom 126 (84%) responses were received. After analysis, 114 data (76%) were found valid. After gathering data, an analysis is performed to generate the User Satisfaction Index in two phases.

1. Calculation of average normalized average score (SSA).
2. Evaluation of User Satisfaction Index (USI).

The following section shows the results derived by applying the above techniques.

4 Results and Discussion

The average normalized score is calculated by following these steps:

1. Let n be the total number of respondents, k be the total division on the Likert scale, and t be the index for Likert values.
2. Let c_{ijt} be the total count of t value on the Likert scale for sub-attribute j of attribute i.
3. The *average normalized scores* SSA_{ij} of attribute j for factors i based on responses from the user survey are derived using the following equation.

$$\mathrm{SSA}_{ij} = \frac{1}{n * k} \sum_{t=1}^{k} t * c_{ijt} \tag{1}$$

Here, i is the index for factors that vary from 1 to 6, and the value of j varies for different attributes as per their number of attributes.

Thus, SSA_{ij}, $i = 1,\ldots,6$ *and* $j = 1,\ldots,4$ *or* 5 *or* 6, are the ij average normalized score calculated from the user satisfaction survey.

The steps for calculating the *User Satisfaction Index (USI)* are as follows:

(i) Let U_{ij}, $i = 1,\ldots,6$ *and* $j = 1,\ldots,4$ *or* 5 *or* 6 as the case may be, is the weight of sub-attribute j of attribute i obtained from expert opinion.
(ii) Obtain the average normalized score of factors 1–6 viz. $SSA_1,\ldots,SSA_6$, whose value lies between 0 and 1.
(iii) The User Satisfaction Index is calculated using the following equation.

$$\text{USI} = w_1\text{SSA}_1 + w_2\text{SSA}_2 + w_3\text{SSA}_3 + w_4\text{SSA}_4 + w_5\text{SSA}_5 + w_6\text{SSA}_6 \quad (2)$$

where $w_1,\ldots,w_6$ are the weights of factors 1–6 obtained from expert opinion.

The weights of attributes are calculated using Excel, based on responses from industry experts, using the formula:

$$W_i = \frac{\text{MODE}(W_i)}{\sum_{t=1}^{6} * t}$$

Here, i is an index for factors that varies from 1 to 6 and W_i is normalized weight of the factors. The calculated W_i values are shown in Table 2.

The results for average normalized scores SSA_{ij}, calculated in Excel, are shown in Table 3.

The evaluated USI score, calculated using Eq. (2), comes out as 1.043, indicating positive study highlights.

Table 2 Weight for main factors

W_i	Value
W_1	0.231
W_2	0.347
W_3	0.357
W_4	0.217
W_5	0.245
W_6	0.119

Table 3 Calculation of factors and attributes

		SSAij	Average SSAij
FA1	SSA11	0.674	0.6853
	SSA12	0.688	
	SSA13	0.694	
FA2	SSA21	0.587	0.5726
	SSA22	0.594	
	SSA23	0.537	
FA3	SSA31	0.714	0.739
	SSA32	0.754	
	SSA33	0.749	
FA4	SSA41	0.471	0.4743
	SSA42	0.484	
	SSA43	0.468	
FA5	SSA51	0.741	0.7446
	SSA52	0.758	
	SSA53	0.735	
FA6	SSA61	0.514	0.5622
	SSA62	0.598	
	SSA63	0.574	

5 Conclusion

A nuanced conclusion emerges from the comprehensive analysis of factor weights, USI scores, and their impact on sustainable habitats. The research underscores the transformative implications of sustainable habitat on end-user satisfaction within the construction sector. The study sheds light on the intricate user satisfaction outcomes by systematically evaluating factor weights derived from expert opinions and subsequently assessing factors and their attribute scores. The result indicates that affordable housing is prioritized as a sustainable habitat, which targets economic aspects. The result may also be valid due to data collection from medium-class housing societies. The next priority, with little difference, is given to public service access, which is of utmost importance for end users for their sustainable habitat.

Furthermore, the result shows inclusive support for the environmental aspect of the built environment, along with end user's demand for a safe and secure living environment, which emerge as a critical factor. The accessibility is given a low score as the attributes have been given low emphasis by users, along with the inclusive community spaces. Moreover, considering the contextual differences in regulatory frameworks, cultural norms, and socioeconomic conditions across different geographical settings, conducting region-specific studies would enhance the applicability and relevance of findings.

Appendix

Questionnaire Survey Form

The questionnaire lets end users know their choices when selecting a facility suitable for their sustainable habitat. It comprises 17 items that are used to capture responses from the participants to each question. The respondent must respond on a Likert scale ranging from 1 to 7, where 1 signifies strongly agree and 7 indicates strongly disagree.

1. Does the end user prefer to be distant from a noisy environment?
2. Does the end user prefer to be distant from any hazardous chemical industry?
3. Does the end user prefer a good air quality index?
4. The end user prefers an environment with adequate lighting.
5. The end user prefers a safe and secure acoustics environment.
6. The end user seeks good emergency services provision in the environment.
7. The end user prefers access to water and sanitation facilities to public services.
8. The end user seeks a reliable electricity supply to public services.
9. Does the end user seek accessible health-care facilities in public services?
10. Diverse and inclusive community space is essential to end users.
11. The end user seeks facilities for social interaction.
12. Regional places to be inclusive in community spaces are essential for end users.
13. The end user looks for cost-effective housing options
14. The end user looks for accessible and affordable housing options.
15. The end user prefers accessible, efficient transportation options near the facility.
16. Public transit accessibility and mobility are essential for end users.
17. Does the end user prefer walkable infrastructure in the facility?

1	2	3	4	5	6	7
Strongly agree	Agree	Slightly Agree	Denotes neutral	Slightly disagree	Disagree	Strongly disagree

References

1. The World Business Council for Sustainable Development (WBCSD). (2018). *SDG sector roadmaps*. WBCSD.
2. Brandon, P. S., & Lombardi, P. (2011). Evaluating sustainable development. In *The built environment* (2nd ed.). Blackwell Publishing Ltd.
3. Guinée, J. (2016). life cycle sustainability assessment: What is it and what are its challenges? In R. Clift & A. Druckman (Eds.), *Taking stock of industrial ecology* (pp. 45–68). Switzerland.
4. Parent, J., Cucuzzella, C., & Revéret, J. P. (2010). Impact assessment in SLCA: Sorting the LCIA methods according to their outcomes. *International Journal of Life Cycle Assessment, 15*(2), 164–171.

5. Ding, G. K. C. (2008). Sustainable construction-The role of environmental assessment tools. *Journal of Environmental Management, 86*(3), 451–464.
6. Bragança, L., Mateus, R., & Koukkari, H. (2010). Building sustainability assessment. *Sustainability, 2*(7), 2010–2023.
7. James, P. (2014). *Urban sustainability in theory and practice*. Routledge.
8. Nilsson, M., Chisholm, E., Griggs, D., Howden-Chapman, P., McCollum, D., Messerli, P., et al. (2018). Mapping interactions between the sustainable development goals: Lessons learned and ways forward. *Sustainability Science, 13*(6), 1489–1503.
9. Lafortune, G., Fuller, G., Moreno, J., Schmidt-traub, G., & Kroll, C. (2018). *SDG index and dashboards—detailed methodological paper*. SDSN.
10. Diaz-Sarachaga, J. M., Jato-Espino, D., & Castro-Fresno, D. (2018). Is the Sustainable Development Goals (SDG) index an adequate framework to measure the progress of the 2030 Agenda? *Sustainable Development, 26*(6), 663–671.
11. Gupta, A., & Shukla, A. K. (2024). Optimal approaches in global warming mitigation and adaptation strategies at city scale. *Discover Sustainability, 5*, 272. https://doi.org/10.1007/s43621-024-00497-8
12. Love, P. E. D., Martinez-Peria, M., & Singh, S. (2017). Modeling and forecasting risk and cost for complex construction projects
13. Asif, M., Uddin, N., Khan, A. N., Tareen, I., & Adnan, S. (2013). Sustainable construction practices: A review of current trends and future directions. *Advances in Materials Science and Engineering, 2013*, 1–7.
14. Forte, F., & Russo, Y. (2017). Evaluation of user satisfaction in public residential housing—a case study in the outskirts of Naples, Italy. *IOP Conference Series: Materials Science and Engineering, 245*(5), 0–8. https://doi.org/10.1088/1757-899X/245/5/052063
15. Sharma, S. K., Deeksha, R. P. K., Shukla, S., & Shukla, A. K. (2024). Overview of sustainable development goals. In P. K. Rai, A. Singh, B. Beingachhi, & S. K. Singh (Eds.), *Natural resources management and sustainable livelihoods in the mountainous region. Advances in geographical and environmental sciences*. Springer. https://doi.org/10.1007/978-981-97-2100-9_14
16. Johnsson, F., Karlsson, I., Rootzén, J., Ahlbäck, A., & Gustavsson, M. (2020). The framing of a sustainable development goals assessment in decarbonizing the construction industry—Avoiding "Greenwashing". *Renewable and Sustainable Energy Reviews, 131*. https://doi.org/10.1016/j.rser.2020.110029
17. Wieser, A. A., Scherz, M., Maier, S., Passer, A., & Kreiner, H. (2019). Implementation of Sustainable Development Goals in construction industry—A systemic consideration of synergies and trade-offs. *IOP Conference Series: Earth and Environmental Science, 323*(1). https://doi.org/10.1088/1755-1315/323/1/012177

Optimizing Bentonite Clay Stabilization with Copper Slag for Resilient Infrastructure: Mechanical and Environmental Evaluations

Gaurav Bharti, Bishnu Kant Shukla, and Pushpendra Kumar Sharma

1 Introduction

Bentonite soil, known for its high expansiveness and low bearing capacity, poses considerable issues in engineering construction due to its inclination to undergo substantial volume fluctuations. These characteristics render it inappropriate for foundational applications, resulting in structural instability and possible harm. Ensuring the durability and safety of buildings on difficult soils, such as bentonite soil, requires the implementation of efficient stabilization procedures [1]. An effective approach involves the use of industrial by-products such as steel or copper slag, which provides advantages in terms of both the environment and the economy [2]. Copper slag, a by-product of the copper smelting process, is readily available and commonly regarded as a waste item. The potential of this substance as a soil stabilizer has not been fully investigated, leading to a lack of study in the field of geotechnical engineering. This work seeks to close this gap by conducting experiments to determine the efficacy of copper slag as a stabilizing agent for bentonite soil. Bentonite soil is found in different parts of the world, as shown in Fig. 1. This research is unusual since it focusses on using copper slag to enhance the geotechnical characteristics of bentonite soil, an area that has not been thoroughly researched previously. The experimental program entails blending bentonite soil with different proportions of copper slag and conducting a series of

G. Bharti
Department of Civil Engineering, JSS Academy of Technical Education, Noida, India

B. K. Shukla (✉)
Department of Civil Engineering, JSS Academy of Technical Education, Noida, India

School of Civil Engineering, Lovely Professional University, Phagwara, India

P. K. Sharma
School of Civil Engineering, Lovely Professional University, Phagwara, India

S. Pathak et al. (eds.), *Intelligent Infrastructure and Smart Materials*,
https://doi.org/10.1007/978-3-031-92421-7_2

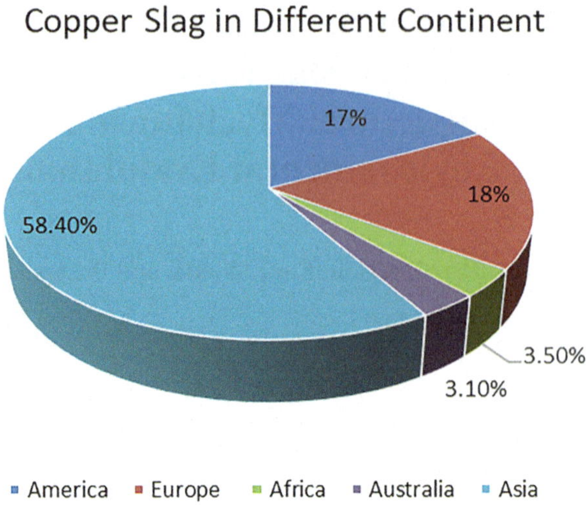

Fig. 1 Availability of copper slag in different part of worlds

laboratory experiments to assess the engineering features of the soil. Tests, including soil stabilization, compaction characteristics, and California Bearing Ratio (CBR), are conducted [3] and the findings are compared to those of untreated bentonite soil. The main goal is to determine the most effective combination of materials that maximizes enhancements in strength and compressibility characteristics. The future potential of this study is substantial, as it not only offers a sustainable option for stabilizing expansive soils but also encourages the use of industrial waste, thereby addressing environmental issues. The discoveries may result in the establishment of fresh protocols and benchmarks for soil stabilization with copper slag, possibly revolutionizing methods in geotechnical engineering and building. The stabilized bentonite soil has a wide range of uses in building projects, such as subgrade [4] and flexible pavements, embankments, and foundations. This method can increase the ability of soils to bear weight, decrease the likelihood of swelling, and promote overall stability, resulting in constructions that are safer and more long lasting [5–7]. Nevertheless, the study clearly recognizes specific constraints. The heterogeneity in the content of copper slag from various origins can impact the uniformity of the stabilization outcomes. Furthermore, conducting extensive investigations over an extended period of time is necessary in order to comprehensively comprehend the longevity and ecological consequences of utilizing copper slag for soil stabilization.

This research has several benefits. It provides a novel application for an industrial by-product, minimizing waste and advocating for sustainable construction methods. This work enhances the engineering characteristics of bentonite soil, thereby providing safer, more stable, and economically efficient building options. In conclusion, the research highlights the capacity of copper slag to significantly transform soil stabilization techniques, thereby opening up possibilities for further progress in the field of geotechnical engineering.

2 Literature Review

Extensive research has been conducted on the stabilization of clayey soils utilizing a variety of materials, such as waste products and synthetic fibres. Akbulut et al. [8] investigated the alteration of clayey soils using used tyre rubber and artificial fibres, illustrating enhanced soil characteristics. In a similar manner, Kavisri et al. [9] examined the impact of utilizing copper slag and ground granulated blast furnace slag (GGBS) to stabilize clayey soils. Their study demonstrated noteworthy enhancements in both soil strength and stability. Bagasse ash, a type of agricultural waste, when blended with synthetic fibres, has demonstrated efficacy in soil stabilization. Arya et al. [10] and Correia et al. [11] documented improvements in the mechanical and geotechnical characteristics of black cotton soil with the use of a combination of bagasse ash and polyester fibre as a stabilizer. In addition, Karthick et al. [12], Joe & Rajesh [13] and Mallikarjuna and Mani [14] specifically examined the use of plastic trash for soil stabilization, emphasizing its effectiveness in enhancing soil characteristics. Extensive research has been conducted on the use of copper slag, which is a by-product of the process of extracting copper, for the purpose of soil stabilization. Qureshi et al. [15] and Bharti et al. [16] have shown that copper slag is efficient in improving the characteristics of expansive soils, making them more appropriate for building. Bharti et al. [17] conducted additional study that demonstrated that a combination of copper and steel slag greatly enhances the mechanical characteristics of black cotton soil. Soil reinforcement has seen extensive utilization of both natural and synthetic fibres. In their study, Yazici and Keskin [18] conducted a thorough examination of soil reinforcing methods that utilize these fibres. Similarly, Hejazi et al. [19] offered a summary of the advantages and uses of both natural and synthetic fibres in soil reinforcement. Galán-Marín et al. [20] examined the stabilization of clay-based composites using natural polymers and fibres, providing an alternate solution to conventional soil stabilization techniques. The efficiency of fibre reinforcement in soil stabilization is further supported by experimental experiments conducted by Eshanthini and Priyadharshini [21]. Gupta and Thomas [22] conducted a study on the use of copper tailings for the purpose of stabilizing soils with sandy and clayey compositions. They discovered enhancements in both ideal moisture content and maximum dry density. Shukla et al. [23] highlighted the advantages of using waste synthetic fibres in pavement building, namely in terms of environmental and economic benefits. Chandrshekhar and Chokshi [24] conducted a review on the use of scrap copper slag in geotechnical applications, emphasizing its ability to improve soil qualities. The research conducted by Qureshi et al. [25], Kumar et al. [26], and Tandel and Patel [27] consistently showed that copper slag has a beneficial effect on soil stabilization, especially for soils that have a tendency to expand. Baraskar and Ahirwar [28] conducted a study precisely examining the CBR of black cotton soil that was stabilized with copper slag. Their research provided vital insights about the efficiency of using copper slag for road building. Subsequent study has built upon these findings. Dhir et al. [29] gathered extensive data on the sustainable use of copper slag in building materials. Yusof [30] investigated the possibility of using tin slag in road pavements, whereas Pavez et al. [31] studied the pozzolanic activity of copper slag, demonstrating its appropriateness for several technical purposes. The Indian Standards (IS) give essential guidelines for assessing soil qualities using standardized testing methodologies. The standards encompass techniques for assessing several properties of the material, such as grain size [26], specific gravity [32], liquid limits (LL), plastic limits (PL) [33], free swell index [34], moisture content and dry density [35], and unconfined compressive strength [36]. Wang [37] and Varadharajan et al. [38] examined the use of slag in civil infrastructure building, highlighting its advantages and practical uses. Mobasher et al. [39] examined the influence of copper slag on the hydration process of blended cementitious mixes, offering significant knowledge on its effects on construction materials.

These comprehensive investigations into soil stabilization methods using waste materials and fibres offer promising solutions for improving soil properties, enhancing construction quality, and promoting sustainable practices in civil engineering.

3 Material

3.1 Bentonite Clay

Bentonite clay is a highly versatile natural material known for its unique properties and numerous applications across various industries. It is primarily composed of montmorillonite, a type of smectite clay mineral, which gives bentonite its characteristic expansive behaviour. The high cation exchange capacity (CEC) and large specific surface area of montmorillonite allow bentonite to absorb and retain significant amounts of water, leading to substantial volume changes. This property makes bentonite exceptionally useful for applications requiring sealing, such as in drilling muds, liners for landfills, and containment barriers for hazardous waste. Additionally, bentonite is used in the production of cat litter, pharmaceuticals, and cosmetics due to its adsorptive capabilities and inert nature. One of the major advantages of bentonite clay is its high plasticity, which allows it to form a flexible, impermeable barrier when hydrated. This characteristic is particularly beneficial in environmental engineering, where bentonite is employed to prevent the leakage of contaminants from landfills and other waste containment sites. Its swelling capacity ensures that any gaps or cracks are effectively sealed, maintaining the integrity of the barrier. Furthermore, bentonite's ability to improve the stability and viscosity of drilling fluids makes it indispensable in the oil and gas industry, where it facilitates the removal of cuttings and stabilizes boreholes. However, the same properties that make bentonite useful in these applications also pose significant challenges when it comes to using it as a foundation material. The high swell-shrink potential of bentonite clay can lead to considerable changes in volume with variations in moisture content. When bentonite absorbs water, it expands, and when it dries out, it contracts. This cyclical expansion and contraction can exert substantial pressure on foundations, leading to differential settlement and structural damage. Buildings and structures erected on bentonite-rich soils are susceptible to cracks, tilting, and even collapse if the soil is not properly managed. The plasticity index (PI) of bentonite clay is another factor contributing to its problematic nature in foundations. The PI is a measure of the range of moisture content over which the soil remains plastic and mouldable. Bentonite has a very high PI, indicating that it can undergo significant deformation without breaking. While this property is advantageous for sealing applications, it means that the soil can shift and deform under load, compromising the stability of foundations built upon it. Moreover, the bearing capacity of bentonite clay is relatively low compared to other soil types. Bearing capacity is the ability of soil to support the loads applied to the ground. The high moisture content and plasticity of bentonite reduce its shear strength, which is the soil's resistance to sliding or collapsing under pressure. As a result, foundations on bentonite clay are prone to excessive settlement, which can lead to structural failure. Engineers often need to implement special foundation designs, such as deep foundations, mat foundations, or the use of soil stabilization techniques, to mitigate these issues. The permeability of bentonite clay, while beneficial for creating impermeable barriers, is a double-edged sword in foundation applications. Its low permeability means that water moves very slowly through the soil, leading to poor drainage and waterlogging. This condition can exacerbate the swelling behaviour of bentonite, as the soil remains saturated for extended periods. The presence of water can also lead to the dissolution of salts and other minerals in the soil, which can further destabilize the foundation. Another disadvantage of using bentonite clay in foundations is its susceptibility to environmental conditions. Changes in temperature, humidity, and groundwater levels can all affect the moisture content and, consequently, the volume of bentonite clay. Seasonal variations, such as wet and dry cycles, can lead to recurrent swelling and shrinkage, causing ongoing stress and strain on foundations. This dynamic behaviour makes it challenging to predict and manage the long-term performance of structures built on bentonite-rich soils. To address these challenges, various soil stabilization techniques are employed to improve the properties of bentonite clay for foundation applications. One common method is the use of chemical stabilizers, such as lime, cement, or fly ash, which react with the clay minerals to reduce plasticity and increase strength. Another approach is mechanical stabilization, which involves mixing bentonite with other materials like sand or gravel to enhance its load-bearing capacity and reduce swell–shrink behaviour. Geosynthetic materials, such as geotextiles and geomembranes, are also used to reinforce bentonite and improve its stability [40, 41]. Despite these stabilization methods, bentonite clay remains a challenging material for foundation construction (Fig. 2). The need for extensive ground preparation and specialized engineering solutions can significantly increase the cost and complexity of building on bentonite-rich soils. Therefore, thorough site investigation and soil testing are essential to determine the suitability of bentonite for any given project. Engineers must carefully evaluate the soil's properties and implement appropriate design and construction techniques to ensure the safety and durability of structures.

Fig. 2 Bentonite clay

Table 1 Chemical composition of bentonite clay

Name	Formula	Percentage
Potassium oxide	K_2O	0.20–1.0
Ferric oxide	Fe_3O_4	2.0–4.0
Sodium oxide	Na_2O	0.50–5.0
Silica	SiO_2	50.0–65.0
Titanium oxide	TiO_2	0.20–0.50
Alumina	Al_2O_3	15.0–25.0
Oxide of magnesium	MgO	3.0–6.0
Calcium oxide	CaO	0.50–2.0

The soil used in this study was brought from Kachchh districts of Gujarat. Table 1 presents the chemical composition of various oxides in the material, detailing their respective formulas and percentage ranges. It includes potassium oxide (K_2O) with a concentration range of 0.20–1.0%, ferric oxide (Fe_3O_4) at 2.0–4.0%, sodium oxide (Na_2O) ranging from 0.50% to 5.0%, and silica (SiO_2) comprising 50.0–65.0%. Titanium oxide (TiO_2) is present in smaller amounts, from 0.20% to 0.50%, while alumina (Al_2O_3) constitutes 15.0–25.0% of the material. Magnesium oxide (MgO) is found in concentrations of 3.0–6.0%, and calcium oxide (CaO) varies from 0.50% to 2.0%. This detailed breakdown provides a comprehensive understanding of the material's chemical makeup, essential for various industrial and research applications.

3.2 Copper Slag

Table 2 (constructed based on information obtained from Ref. [7]) provides an overview of specific physical and chemical properties of a material, likely a type of industrial slag or similar compound. The specific gravity ranges from 3.50 to 3.60, indicating the material's density compared to water. The shape is described as angular with sharp edges, suggesting a coarse texture that could influence its packing and compaction behaviour in construction applications. The material is nonplastic, meaning it does not exhibit plastic deformation when subjected to stress. Chemically, it comprises 55–60% FeO (ferrous oxide), 26–30% SiO_2 (silicon dioxide), 3–6% CaO, and 1–4% MgO. These properties highlight its potential use in various industrial applications, such as a stabilizing agent in soil or as an aggregate in construction; due to its high density, angular shape, and specific chemical composition copper slag is obtained from Quality Spare Centre, Faridabad, India, and being used in the investigation, as shown in Fig. 3.

Table 2 Copper slag specific gravity, shape, and chemical composition

Parameter	Values
Specific gravity	3.50–3.60
Shape	Angular and sharp edges
Plasticity	Nonplastic
FeO	55–60%
SiO2	26–30%
CaO	3–6%
MgO	1–4%

Fig. 3 Copper slag

3.3 Methodology

The methodology for using bentonite clay and copper slag in soil stabilization involves three main stages: materials selection, preparation of specimens, and experimental testing as shown in Fig. 4. Initially, bentonite clay and copper slag are chosen for their respective properties as a highly plastic, expansive clay, and an industrial by-product suitable for stabilization. In the preparation stage, the properties of both materials are determined, and they are mixed in various proportions to create different soil samples. This is followed by experimental testing, which involves conducting baseline tests on the parent bentonite soil to establish control data, including Atterberg limits, compaction tests, CBR, and UCS tests as shown in Table 3. Similar tests are then performed on the combined soil samples to assess the improvements in engineering properties due to the addition of copper slag, aiming to identify the optimal mix for effective soil stabilization.

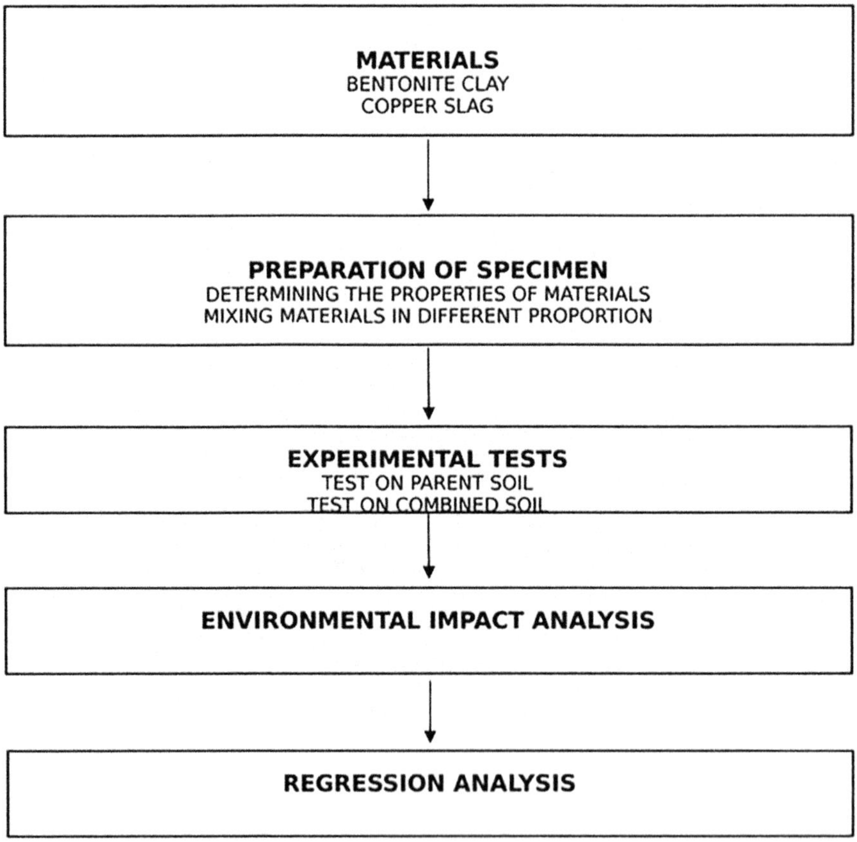

Fig. 4 Methodology flowchart

Table 3 Experiment performed

S. no.	Experiment performed	IS code used
1	Classification	IS: 1498
2	LL test	IS: 2720 (Part 5): 1985
3	PL test	IS: 2720 (Part 5): 1985
4	Specific gravity test (SG)	IS: 2720 (Part 2): 1980
5	Proctor compaction test	IS: 2720 (Part 2): 1980
6	CBR test	IS: 2720 (Part 16): 1987
7	Shrinkage limit test	IS: 2720 Part XL: 1977
8	UCS test	IS 2720 (Part X-1973)

Table 4 Result on parent soil

Experiment (%)	Result
Plasticity Index	51.73
LL test	89.73
PL test (%)	38
Shrinkage limit test (%)	19
SG test	2.35

4 Result and Discussion

The methodology for soil testing in this study involves several key tests to determine the specific properties of the parent soil, adhering to Indian Standards (IS) protocols. The SG test, conducted according to IS: 2720 (Part 3), utilizes density bottles to measure the soil's specific gravity, which is crucial for determining the saturation and unit weight of moist soil. Table 4 shows the SG of the parent soil to be 2.35. The LL test, performed as per IS: 2720 (Part 5)—1985, measures the water content at which fine-grained soil transitions to a liquid state with low shear strength, closing a groove in 25 blows using the Casagrande apparatus. The LL of the parent soil was determined to be 89.73%. The PL test defines the water content at which soil becomes semisolid and brittle, conducted according to IS: 2720 (Part 5)—1985 (Reaffirmed—2006). The PL of the parent soil was found to be 38%. The Plasticity Index Test calculates the plasticity index (PI), which is the numerical difference between the LL and the PL, expressed as a percentage of the soil's dry weight. The PI of the parent soil is 51.73. Lastly, the Shrinkage limit test, performed according to IS 2720—Part XL: 1977, determines the lowest water content at which the soil remains fully saturated, providing critical data for understanding the soil's shrinkage properties. The shrinkage limit of the parent soil was determined.

Figure 5 and Table 5 illustrating the relationship between specific gravity (SG) and the percentage of copper slag added to bentonite clay demonstrate a clear upward trend. Initially, at 0% copper slag, the SG is 2.35, reflecting the density of the parent bentonite soil. As the copper slag percentage increases, the SG steadily rises, reaching around 2.50 at 24% copper slag. This increase indicates that the addition of copper slag enhances the soil's particle density, making it heavier and more compact. Higher specific gravity values suggest a denser soil structure, which is

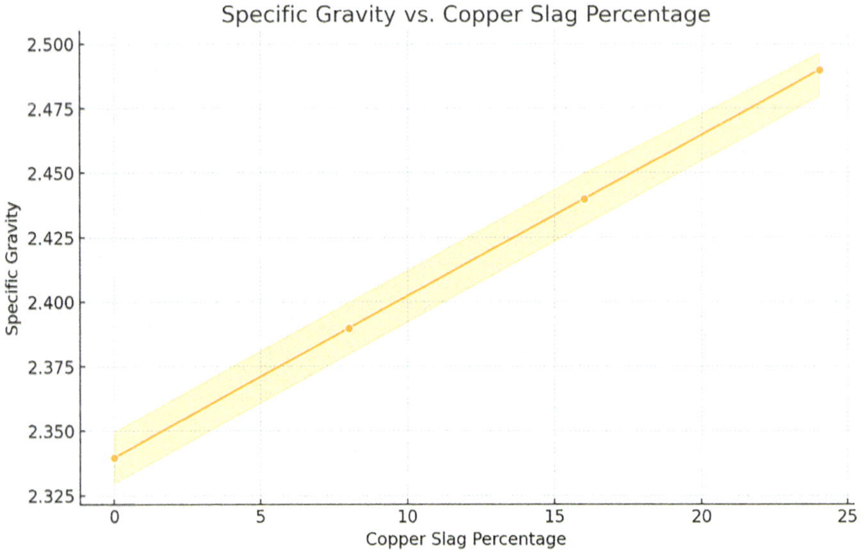

Fig. 5 Specific gravity of treated soil with varying % of copper slag

Table 5 Specific gravity of treated soil with varying % of copper slag

Soil + additive	Specific gravity
Parent soil	2.35
Parent soil + 8% copper slag	2.4
Parent soil + 16% copper slag	2.45
Parent soil + 24% copper slag	2.5

advantageous for construction as it implies greater stability and load-bearing capacity. The consistent rise in SG with increasing copper slag content underscores the effectiveness of copper slag in improving the engineering properties of bentonite clay, making it a more suitable material for building foundations and other structural applications.

Plastic limit (PL): The PL indicates the water content at which soil transitions from a semisolid to a plastic state. Table 6 and Fig. 6 show a decreasing trend in the PL as the copper slag percentage increases. For instance, at 0% copper slag, the PL is around 38%, but it decreases to approximately 32% at 24% copper slag. This reduction implies that with more copper slag, the soil requires less water to become plastic, suggesting improved soil workability and reduced plasticity. Lower plasticity means the soil becomes less sticky and easier to handle, which is advantageous in construction.

Liquid limit (LL): The LL represents the water content at which soil changes from a plastic to a liquid state. Similar to the PL, Table 6 and Fig. 6 show that the LL also exhibits a declining trend with increasing copper slag content. Initially, at 0% copper slag, the LL is around 89.73%, but it drops to about 85.10% at 24% copper slag. This decrease indicates that as more copper slag is added, the soil requires

Table 6 LL and PL of treated soil with varying % of copper slag

Soil + additive	LL (%)	PL (%)
Parent soil	89.73	38
Parent soil + 8% copper slag	88	36
Parent soil + 16% copper slag	86.5	34
Parent soil + 24% copper slag	85	32

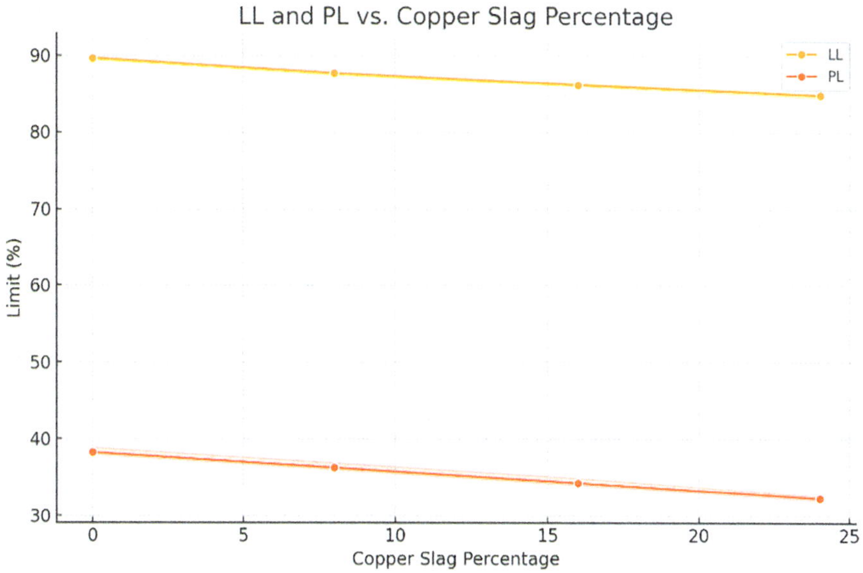

Fig. 6 LL and PL of treated soil with varying % of copper slag

Table 7 CBR value of treated soil with varying % of copper slag

Soil + additive	CBR value (%)
Parent soil	3.03
Parent soil + 8% copper slag	3.91
Parent soil + 16% copper slag	4.39
Parent soil + 24% copper slag	4.94

less water to reach its liquid state, enhancing its stability. Lower LL values suggest that the soil will have a lower tendency to swell and shrink with moisture changes, making it more stable and reliable for construction purposes.

In this study, the CBR values for parent soil and copper slag at different compositions were investigated to evaluate their load-bearing capacity, which is crucial for road construction. The CBR rating, an index of surface hardness, indicates the soil's ability to bear loads. The study aimed to enhance the CBR values through soil stabilization using copper slag as an additive. The results demonstrated that the CBR values increased progressively with higher proportions of copper slag. Notably, a 24% replacement with copper slag yielded the highest CBR values in both soaked and unsoaked conditions as shown in Table 7 and Fig. 7, underscoring its

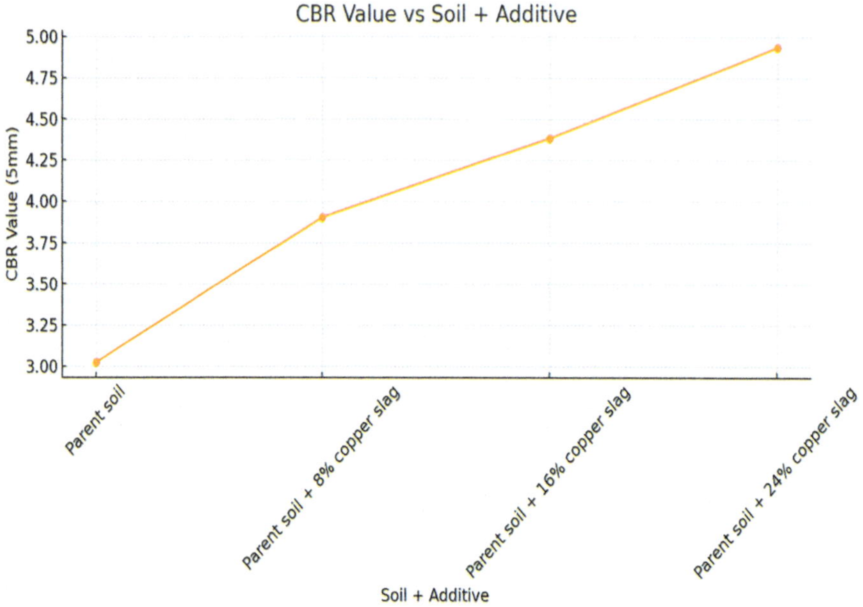

Fig. 7 CBR value of treated soil with varying % of copper slag

effectiveness for subgrade applications in flexible pavements. This improvement in CBR values suggests that incorporating copper slag significantly enhances the physical and engineering properties of bentonite soil, including shear strength and bearing capacity, making it a suitable material for constructing robust road foundations.

Figure 8 illustrates the impact of varying copper slag percentages on the dry density and moisture content relationship of soil samples. The x-axis represents the moisture content (%), while the y-axis indicates the dry density (g/cm^3). Four curves depict soil mixtures with 0%, 8%, 16%, and 24% copper slag. The parent soil (0% copper slag) exhibits the lowest dry density across moisture contents. The addition of 8% copper slag (yellow curve) significantly enhances dry density, achieving the highest peak around 28% moisture content, indicating optimal compaction. With 16% copper slag (grey curve), the dry density remains high, peaking near 30% moisture content. At 24% copper slag (orange curve), although the dry density peak shifts to around 32% moisture content, it is slightly lower than the 8% and 16% mixtures but still superior to the parent soil. Thus, adding copper slag improves soil compaction, with 8% being the most effective proportion.

Table 8 and Fig. 9 indicate the improvement in UCS of soil with the addition of copper slag. The parent soil has a UCS value of 62.3 kPa, which increases to 64.7 kPa with 8% copper slag, 65.8 kPa with 16% copper slag, and reaches a maximum of 66.9 kPa with 24% copper slag. This trend shows a clear enhancement in soil strength with higher copper slag content, suggesting that copper slag is an effective

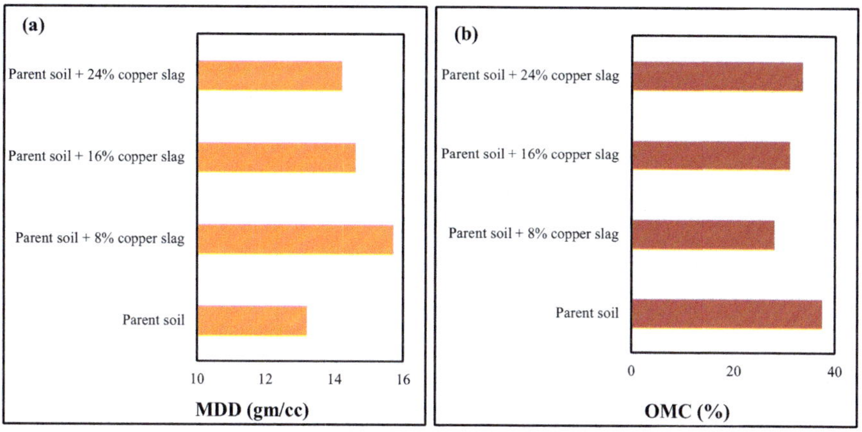

Fig. 8 Compaction test results on treated soil with varying % of copper slag: **(a)** MDD results and **(b)** OMC (%) results

Table 8 UCS value of treated soil with varying % of copper slag

Soil + additive	UCS value (kPa)
Parent soil	62.3
Parent soil + 8% copper slag	64.7
Parent soil + 16% copper slag	65.8
Parent soil + 24% copper slag	66.9

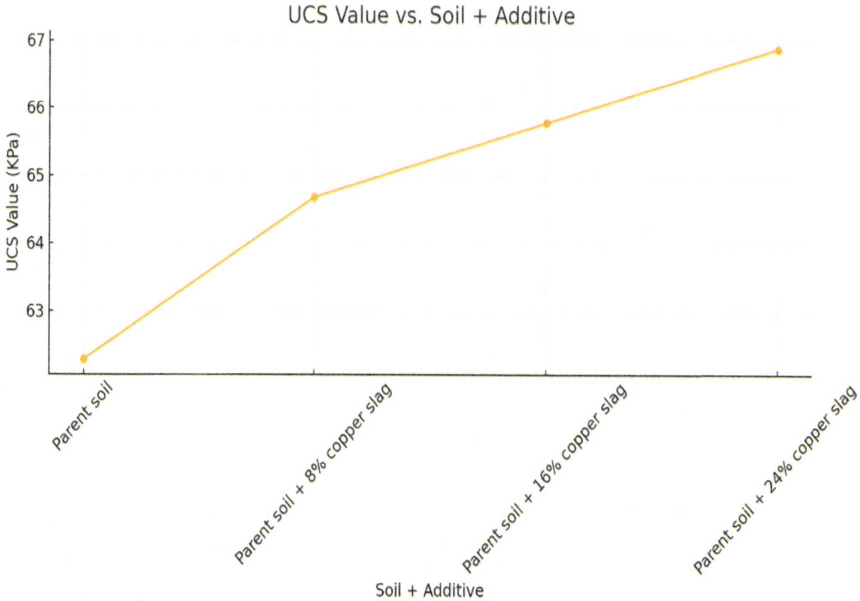

Fig. 9 UCS value of treated soil with varying % of copper slag

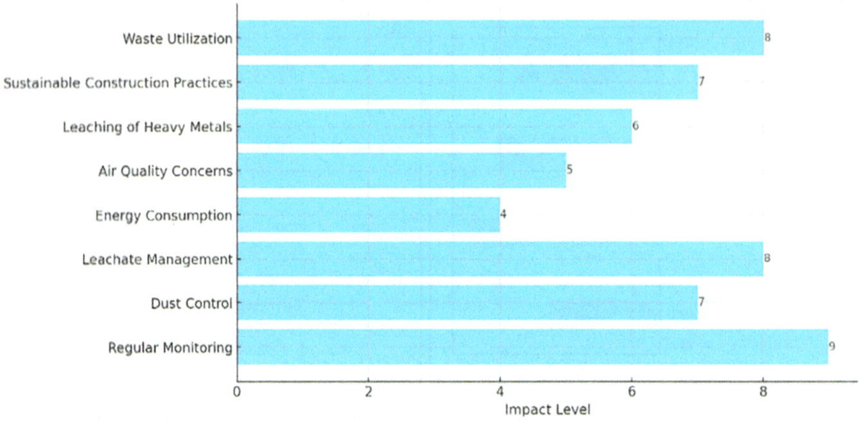

Fig. 10 Environmental impact analysis

stabilizing agent for increasing the load-bearing capacity of soil in construction projects.

The environmental impact analysis shown in Fig. 10 presents a comprehensive assessment of the effects of stabilizing bentonite clay with copper slag. It includes eight key factors: Waste Utilization, Sustainable Construction Practices, Leaching of Heavy Metals, Air Quality Concerns, Energy Consumption, Leachate Management, Dust Control, and Regular Monitoring. Each factor is rated on an impact level scale from 1 to 10, with higher values indicating greater impact. Waste utilization and regular monitoring scored the highest at 9, highlighting their significant positive contributions. Sustainable construction practices and dust control also received high scores, indicating the environmental benefits of these practices. Conversely, leaching of heavy metals and air quality concerns scored lower, suggesting areas where environmental impacts could be more negative and warrant further attention and mitigation strategies. This balanced view underscores the potential environmental benefits and challenges associated with using copper slag for soil stabilization.

The dataset comprises key variables crucial for understanding the stabilization of bentonite clay with copper slag. Copper slag percentage indicates the proportion of copper slag added to the clay. UCS (kPa), or unconfined compressive strength, measures the soil's strength without lateral confinement (Table 9). CBR (%), or California Bearing Ratio, assesses the soil's load-bearing capacity, essential for road construction. LL, signifies the water content at which the soil transitions from a plastic to a liquid state, reflecting its plasticity. PL (%), denotes the water content at which the soil shifts from a semisolid to a plastic state, highlighting its workability. Lastly, SG, or specific gravity, measures the soil particles' density, providing insights into the soil's composition and behaviour. These parameters collectively help evaluate the soil's stability and suitability for construction when mixed with copper slag.

Table 9 Dataset used for linear regression

Trial	Copper slag (%)	UCS (kPa)	CBR (%)	LL (%)	PL (%)	SG
1	0	62.3	3.03	89.73	38	2.35
	8	64.7	3.91	88	36	2.4
	16	65.8	4.39	86.5	34	2.45
	24	66.9	4.94	85	32	2.5
2	0	61.5	2.95	90	39	2.33
	8	64	3.8	87.5	37	2.38
	16	65.5	4.35	86	35	2.43
	24	67	4.9	84.5	33	2.48
3	0	62	3	89.5	38	2.34
	8	64.5	3.85	87.8	36	2.39
	16	66	4.4	86.2	34	2.44
	24	67.5	4.95	85.1	32	2.49

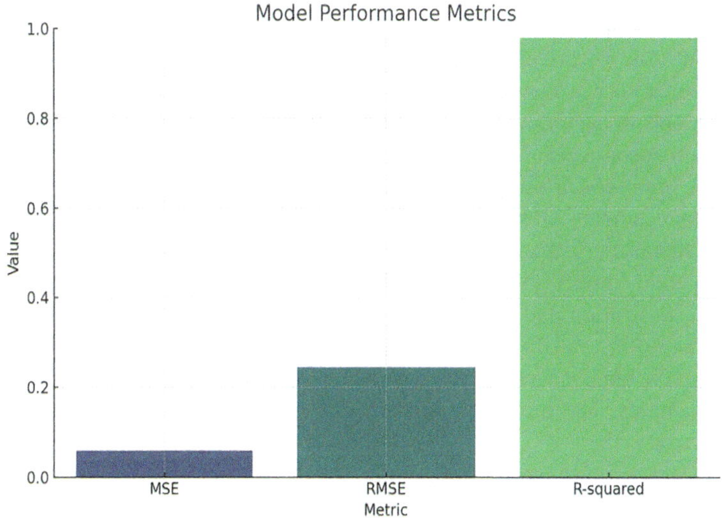

Fig. 11 Result of model metrics

4.1 Predictive Modelling Using Linear Regression

The predictive modelling results for the stabilization of bentonite clay using copper slag demonstrate a high level of accuracy and reliability. The mean squared error (MSE), which measures the average of the squares of the errors, is low at 0.06, indicating minimal discrepancies between the observed and predicted values. The root mean squared error (RMSE), which represents the square root of the average of the squared errors, is also low at 0.245, further affirming the model's precision. The R^2 value, a statistical measure representing the proportion of variance for a dependent variable explained by an independent variable, is exceptionally high at 0.98 as shown in Fig. 11. This suggests that 98% of the variability in the UCS can be explained by the model, highlighting its robustness. Feature importance analysis reveals that SG and CBR have the most significant positive impacts on UCS, with coefficients of 0.30 and 0.24, respectively. Copper slag percentage also contributes positively with a coefficient of 0.12. Conversely, LL and PL negatively impact UCS, with coefficients of −0.15 and −0.10. These insights highlight the crucial role of SG and CBR in soil stabilization, providing a clear pathway to optimize the stabilization process by focusing on these influential factors.

These results demonstrate the effectiveness of copper slag in improving the UCS of bentonite clay, with SG and CBR showing the highest positive impact on UCS. This analysis can help in optimizing the stabilization process by highlighting the most significant factors influencing soil strength.

5 Conclusion

The study on stabilizing bentonite clay with copper slag has provided significant insights into the effectiveness and environmental impact of this method. The addition of copper slag in varying percentages (0%, 8%, 16%, 24%) resulted in notable improvements in unconfined compressive strength (UCS), California Bearing Ratio (CBR), and specific gravity (SG) of the soil. UCS increased from 62.3 to 66.9 kPa, and CBR improved from 3.03% to 4.94% with increasing copper slag content. These enhancements highlight copper slag's potential in improving soil strength and load-bearing capacity, crucial for sustainable construction practices. Environmental impact analysis showed significant positive contributions in areas like waste utilization and sustainable construction practices, with high scores for regular monitoring and dust control. However, concerns regarding leaching of heavy metals and air quality were identified, necessitating careful management and mitigation strategies. The predictive modelling exhibited high accuracy, with a mean squared error (MSE) of 0.06, root mean squared error (RMSE) of 0.245, and an R^2 value of 0.98. Feature importance analysis revealed that specific gravity (SG) and CBR had the most significant positive impacts on UCS, while liquid limit (LL) and plastic limit (PL) had negative impacts. The study confirms that stabilizing bentonite clay using copper slag is a viable and effective approach to enhance soil properties for construction purposes. The improvements in UCS, CBR, and SG, combined with environmental benefits, make copper slag an excellent additive for soil stabilization, contributing to resilient infrastructure. Ongoing monitoring and mitigation strategies are essential to address potential environmental concerns, ensuring a balanced and sustainable approach to soil stabilization.

References

1. Bharti, G., Shukla, B. K., Parashar, B., Kumar, A., & Varadharajan, S. (2023, September). Experimental study on physico-mechanical properties of expansive soil by using wheat straw fiber. In *AIP conference proceedings* (Vol. 2800, No. 1). AIP Publishing.

2. Bharti, G., Shukla, B. K., Kumar, A., & Hurukadli, P. N. (2023, September). Stabilization of black cotton soil using steel slag. In *AIP conference proceedings* (Vol. 2800, No. 1). AIP Publishing.
3. Gowda, S., Kunjar, V., Gupta, A., Kavitha, G., Shukla, B. K., & Sihag, P. (2024). Prediction of the subgrade soil california bearing ratio using machine learning and neuro-fuzzy inference system techniques: A sustainable approach in urban infrastructure development. *Urban Science, 8*(1), 4.
4. Bansal, V., & Shukla, B. K. (2022, May). Application of urea formaldehyde raisin in pavement subgrade stabilization at Hindon Barrage. In *Journal of physics: Conference series* (Vol. 2267, No. 1, p. 012150). IOP Publishing.
5. Shukla, B. K., Bharti, G., Sharma, P. K., Sharma, M., Rawat, S., Maurya, N., Srivastava, R., & Srivastav, Y. (2024). Sustainable construction practices with recycled and waste materials for a circular economy. *Asian Journal of Civil Engineering*, 1–22.
6. Varadharajan, S., Kirthanashri, S. V., Dwivedi, A., Irfan, D., Shukla, B. K., Kumar, A., Bishetti, P., & Bharti, G. (2023, September). Use of fly ash and metakaolin to manufacture sustainable concrete. In *AIP conference proceedings* (Vol. 2800, No. 1). AIP Publishing.
7. Shukla, B. K., Gupta, A., Gowda, S., & Srivastav, Y. (2023). Constructing a greener future: A comprehensive review on the sustainable use of fly ash in the construction industry and beyond. *Materials Today: Proceedings, 93*(3), 257–264.
8. Akbulut, S., Arasan, S., & Kalkan, E. (2007). Modification of clayey soils using scrap tire rubber and synthetic fibers. *Applied Clay Science, 38*(1–2), 23–32.
9. Kavisri, M., Senthilkumar, P., Gurukumar, M. S., & Pushparaj, K. J. (2018). Experimental study on effects of stabilization of clayey soil using copper slag and GGBS. *Rasayan Journal of Chemistry, 11*(1), 111–117.
10. Arya, P., Patel, S. B., Bharti, G., Shukla, B. K., & Hurukadli, P. (2023). Impact of using a blend of bagasse ash and polyester fiber in black cotton soil for improvement of mechanical and geotechnical properties of soil. *Materials Today: Proceedings, 78*, 738–743.
11. Correia, A. A., Oliveira, P. J. V., & Custódio, D. G. (2015). Effect of polypropylene fibres on the compressive and tensile strength of a soft soil, artificially stabilised with binders. *Geotextiles and Geomembranes, 43*(2), 97–106.
12. Karthick, S. Y. S., Vasanthanarayanan, R., Ayswarya, S., & Meenakshi, C. (2019). Soil stabilization using plastics and gypsum. *International Journal of Research in Engineering Science and Management, 2*, 351–360.
13. Joe, M. A., & Rajesh, A. M. (2015). Soil stabilization using industrial waste and lime. *International Journal of Scientific Research Engineering & Technology (IJSRET)., 4*(7).
14. Mallikarjuna, V., & Mani, T. B. (2016). Soil stabilization using plastic waste. *International Journal of Research in Engineering and Technology, 5*(5), 391–394.
15. Qureshi, M. A., Mistry, H. M., & Patel, V. D. (2015). Improvement in soil properties of expansive soil by using copper slag. *International Journal of Advanced Research in Science, Engineering and Technology, 2*(7), 125–130.
16. Bharti, G., Hurukadli, P., Shukla, B. K., Sihag, P., Jagudi, S., & Tripathi, A. (2023). Environmental impact analysis and utilization of copper slag for stabilising black cotton soil. *Materials Today: Proceedings.*
17. Bharti, G., Shukla, B. K., Kumar, A., Nagaraju, M. B., Bansal, V., & Varadharajan, S. (2023September). Impact of copper and steel slag blend in strengthening mechanical properties of black cotton soil. In *AIP conference proceedings* (Vol. 2800, No. 1). AIP Publishing.
18. Yazici, M. F., & Keskin, S. N. (2021). A review on soil reinforcement technology by using natural and synthetic fibers. *Erzincan University Journal of Science & Technology, 14*(2).
19. Hejazi, S. M., Sheikhzadeh, M., Abtahi, S. M., & Zadhoush, A. (2012). A simple review of soil reinforcement by using natural and synthetic fibers. *Construction and Building Materials, 30*, 100–116.
20. Galán-Marín, C., Rivera-Gómez, C., & Petric, J. (2010). Clay-based composite stabilized with natural polymer and fibre. *Construction and Building Materials, 24*(8), 1462–1468.
21. Eshanthini, P., & Priyadharshini, B. Experimental study on soil stabilization using fibres. *Materials Research Proceedings, 19.*

22. Gupta, R. C., & Thomas, B. S. (2013). A study on the optimum moisture content and maximum dry density of sandy and clayey soil stabilized by copper tailings. *RCEE Research in Civil and Environmental Engineering, 1*, 123–138.
23. Shukla, B. K., Bhowmik, A. R., Sharma, P. K., & Abhimanyu. (2020). Use of waste synthetic fiber reinforcement in environmental friendly and economic pavement construction. *Journal of Green Engineering, 10*(1), 62–75.
24. Chandrshekhar, J., & Chokshi, T. A. (2015). A review on utilization of waste material copper slag in geotechnical applications. *International Journal for Innovative Research in Science and Technology, 1*(12), 246–250.
25. Qureshi, M. A., Mistry, H. M., & Patel, V. D. (2015). Improvement in soil properties of expansive soil by using copper slag. *International Journal of Advance Research in Engineering in Science and Technology, 2*(7), 125–130.
26. Kumar, P. R., Kumar, S. P., & Maheshwari, J. (2017). Laboratory study of black cotton soil blended with copper slag and fly ash. *International Journal of Innovative Research in Science, 6*(2), 1960–1967.
27. Tandel, Y. K., & Patel, J. B. (2009). Review of utilisation of copper slag in highway construction. *Australian Geomechanics, 44*(3), 71.
28. Baraskar, T., & Ahirwar, S. K. (2014). Study on California bearing ratio of black cotton soil use waste copper slag. *International Journal of Structural and Civil Engineering Research, 3*(4), 45–56.
29. Dhir, R. K., Brito, J. D., Mangabhai, R. J., & Lye, C. Q. (2017). *Sustainable construction materials: Copper slag*. Woodhead Publishing.
30. Yusof, M. A. W. (2005). *Investigating the potential for incorporating tin slag in road pavements*. Thesis Submitted to the University of Nottingham.
31. Pavez, O., Rojas, F., Palacios, J., & Nazer, A. (2002). Pozzolanic activity of copper slag. In *Proceedings of the VI national conference on clean technologies for the mining industry, Chile*.
32. IS: 2720 (Part III). (1980). *Indian Standard method of test for determination of specific gravity*. Bureau of Indian Standards.
33. IS: 2720 (Part V). (1985). *Indian Standard method of test for determination of liquid & plastic limits*. Bureau of Indian Standards.
34. IS: 2720 (Part XL). (1977). *Indian Standard method of test for determination of free swell index*. Bureau of Indian Standards.
35. IS: 2720 (Part VIII). (1980). *Indian Standard method of test for determination of moisture content & dry density*. Bureau of Indian Standards.
36. IS 2720 (Part X). (1973). *Indian Standard method of test for determination of unconfined compressive strength*. Bureau of Indian Standards.
37. Wang, G. C. (2016). *The utilization of slag in civil infrastructure construction* (pp. 35–61). Woodhead Publishing.
38. Varadharajan, S., Kirthanashri, S. V., Maurya, N., Bishetti, P., Shukla, B. K., & Bharti, G. (2023). Utilization of Fly Ash in concrete: A state-of-the-art review. *Lecture Notes in Civil Engineering, 281*, 189–194.
39. Rehman, M. A., & Jafri, T. (2020). Stabilization of low plastic and high plastic clay using guar gum biopolymer. *Journal of Applied Research on Industrial Engineering, 7*(4), 329–343.
40. Qureshi, M. A., Mistry, H. M., & Patel, V. D. (2015). Improvement in Soil properties of expansive soil by using copper slag. *International Journal of Advance Research in Engineering, Science & Technology (IJAREST), 2*(7), 125–130.
41. IS: 2720 (Part IV). (1975). *Indian Standard method of test for determination of grain size*. Bureau of Indian Standards.

Optimizing Pavement Condition Index Prediction Using PSO-LSTM Hybrid Algorithm: A Data-Driven Approach for Road Maintenance Strategies

Sachin Gowda, Pala Gireesh Kumar, Kesana Naga Suneetha, Gopalapurapu Kavya Sri, Aakash Gupta, and Bishnu Kant Shukla

1 Introduction

India has an extensive road network of about 6.2 million kilometres, the second largest in the world. These roads are crucial for public and goods transportation and are even more important than air and water routes. Despite its magnitude, the sustainability of this vital infrastructure is under threat due to the gradual erosion of structural and functional stability over time. Insufficient maintenance has far-reaching consequences, leading to increased vehicle operating cost (VOC), prolonged travel times, heightened accident rates, and a decline in the overall quality of the road user experience. This triggers the need for extensive rehabilitation and, in extreme cases, calls for complete reconstruction, significantly inflating maintenance costs. To tackle these challenges, a robust maintenance strategy grounded in accurate road condition assessments and unbiased improvement strategies is essential. Efficient road management becomes paramount for ensuring the safety and reliability of transportation systems. Among the critical aspects of road infrastructure, the condition of pavements emerges as a linchpin in maintaining optimal road performance. The Pavement Condition Index (PCI), a benchmark for evaluating road surfaces' holistic well-being and operational efficiency, is pivotal in decision-making

S. Gowda · A. Gupta
CSIR-Central Road Research Institute (CRRI), New Delhi, India
e-mail: sachin08.crri@nic.in; aakash.crri@nic.in

P. G. Kumar · K. N. Suneetha · G. Kavya Sri
Department of Civil Engineering, Shri Vishnu Engineering College for Women (A), Bhimavaram, Andhra Pradesh, India
e-mail: gireeshcivil@svecw.edu.in

B. K. Shukla (✉)
Department of Civil Engineering, JSS Academy of Technical Education, Noida, India

S. Pathak et al. (eds.), *Intelligent Infrastructure and Smart Materials*,
https://doi.org/10.1007/978-3-031-92421-7_3

for maintenance and rehabilitation strategies [1]. This numerical compass, spanning the spectrum from 0 to 100, becomes the arbiter of infrastructure stewardship [2, 3].

Traditionally, assessing pavement conditions relied on labour-intensive visual inspections, which are susceptible to human subjectivity. However, the ongoing technological revolution in the transportation sector has ignited interest in leveraging machine learning techniques to automate PCI prediction. This shift from traditional methods to data-driven approaches is essential for ensuring the accuracy, comprehensiveness, and timeliness of infrastructure assessment. In this evolving landscape of advanced data-driven methodologies, the particle swarm optimized long short-term memory (PSO-LSTM) algorithm emerges as a cutting-edge fusion. Seamlessly integrating optimization techniques with sequential learning, this algorithm represents a transformative leap in predictive modelling, specifically designed for applications demanding temporal dependencies and precision. The PSO-LSTM algorithm combines the strengths of the long short-term memory (LSTM) neural network, known for capturing intricate patterns in sequential data, with particle swarm optimization (PSO) principles. PSO, a nature-inspired optimization algorithm, leverages the collective intelligence of a swarm of particles to navigate the solution space efficiently. This synergy equips the PSO-LSTM model to address complex temporal relationships in data while optimizing model parameters for superior accuracy. Within our research framework, the PSO-LSTM algorithm assumes a central role as we refine the precision of the pavement health assessment. Additionally, we conduct a comparative analysis of the results obtained, contrasting them with outcomes from various existing studies that predict PCI. This approach guarantees precise and consistent assessments and gives transportation agencies the tools to prioritize maintenance strategically. This hybrid algorithm's predictive excellence is validated through quantitative measures, including mean squared error (MSE), root mean squared error (RMSE), mean absolute error (MAE), and coefficient of determination (R^2). These metrics, alongside visualizations of actual vs. predicted PCI values, narrate a tale of accuracy that resonates with the intricacies of road maintenance decision-making.

1.1 Traditional Approach

The flowchart for PCI evaluation is shown in Fig. 1. Calculating the PCI for asphalt pavement involves a systematic approach outlined in the ASTM D6433-23 Standard Practice for Roads and Parking Lots Pavement Condition Index surveys [4].

The standardization of this process is crucial, utilizing the recognized ASTM D6433-23 standard practice as a foundation. The pavement section is divided into smaller sample units to assess the pavement condition. Accurate and reliable data collection methods are emphasized for collecting essential information, including visual inspections, distress surveys, and measurements. The core steps include identifying various distress types, such as cracks, potholes, rutting, surface roughness, and assessing their severity [5]. Distress ratings are assigned based on severity,

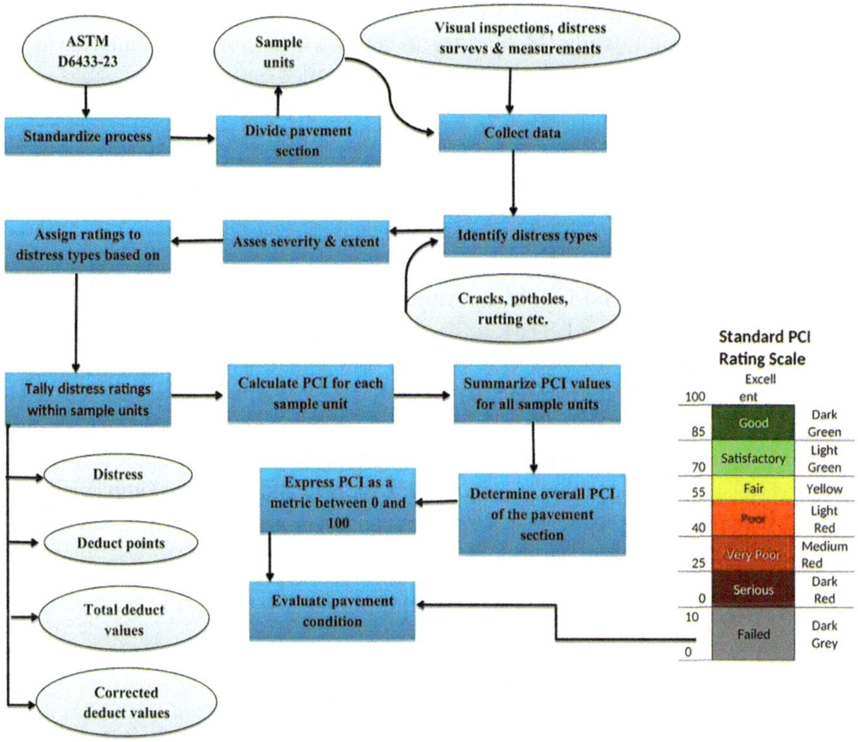

Fig. 1 The critical flow of pavement condition evaluation through PCI

which are tallied for each distress type within the sample units. The PCI for each sample unit is calculated by considering distress tallies and ratings, ultimately reflecting the overall distress quantities and severities. Finally, the summarized PCI values for all sample units provide an overview of the general condition of the asphalt pavement, quantified numerically between 0 and 100.

$$\mathrm{PCI} = e + \frac{s}{\sqrt{N}} \tag{1}$$

$$s = \sqrt{\frac{\Sigma\left(\mathrm{PCI}i - \mathrm{PCI}s\right)^2}{(n-1)}} \tag{2}$$

Equations (1) and (2) encompass critical parameters for assessing pavement condition. 'PCI' stands for the Estimated Pavement Condition Index for a specific section, quantifying quality; 'e' represents the acceptable error range around 65 PCI points for estimating section PCI accurately; 's' signifies the standard deviation of PCI values within the section, measuring pavement condition variability. 'N' reflects the total sample units in the section, impacting statistical significance. Another is the true standard deviation of PCI values within the section. 'PCI_i' is the PCI of each surveyed sample unit, while 'PCI_s' is the mean PCI of all surveyed units; 'n' is the total number of surveyed units [4].

2 Research Objective

The research objective is to develop and evaluate the efficacy of a predictive model for the PCI using the PSO-LSTM algorithm. The study addresses the limitations of traditional models such as LSTM, artificial neural network (ANN), support vector regressor (SVR), and M5 in accurately predicting PCI. The primary focus is on assessing the predictive performance of the PSO-LSTM model based on key evaluation metrics, including mean squared error (MSE), root mean squared error (RMSE), mean absolute error (MAE), and coefficient of determination (R^2). The overarching goal is to provide transportation agencies and infrastructure managers with a robust tool for precise PCI prediction, contributing to effective road maintenance and management strategies.

3 Literature Exploration

Many researchers have directed their efforts toward ensuring efficient road maintenance and transportation infrastructure management, employing diverse methodologies to evaluate pavement performance through various performance indicators. Central to these endeavours is the PCI, a vital tool that offers insights into the state of a pavement. The PCI acts as a quantifiable measure, enabling the assignment of ranks corresponding to the pavement's overall condition.

The valuable and essential recommendations for the sustenance and reconditioning of distressed pavements are put forth by the Markov models and stochastic models [6, 7], where dynamic programming leads to the generation of predictive models. Kimteta et al. [8] investigated bending resistance prediction in rigid fibre-reinforced concrete pavements, finding ANNs superior to RF, RT, and M5P models. For roller-compacted concrete pavements [9], the random forest (RF) model excelled in forecasting material strengths. Ali et al. achieved 80% prediction accuracy [10] for recycled aggregate concrete's elasticity using M5P. With fly ash, self-consolidating concrete flow and structure were well-predicted using M5' and MARS algorithms [11]. Machine learning excels in various applications. Ankita et al. used ML to predict the Marshall stability of carbon fibre-reinforced asphalt concrete, with random forest outperforming other ML algorithms [12]. Given the central objective of this study concerning the prognostication of PCI, it is essential to underscore that PCI is a valuable tool in effectively prioritizing tasks related to maintenance and rehabilitation. This prioritization is facilitated by PCI's capacity to assign rankings corresponding to the pavement's state [12, 13]. Pavement assessment is done with the help of PCI and is frequently used by transportation departments in North America [1]. Numerous researchers have diligently explored diverse facets of pavement-related challenges to advance pavement management systems by implementing AI, ML, and DL algorithms. Among these efforts, predicting PCI has emerged as a frequently undertaken task, showcasing the effectiveness of various computational intelligence techniques. For instance, in a specific study, the degree of pavement deterioration was assessed using PCI, incorporating 14 attributes for predictive analysis, resulting in a commendable % accuracy rate of 70% [1]. Yang et al. also delved into predicting crack indices in asphalt pavements using neural networks [1]. As pavements age, their condition naturally degrades, leading to increased costs for repair and maintenance. With time, pavement quality diminishes, necessitating higher expenditures on restoration and upkeep efforts. A comprehensive analysis of 429 segments from the Pavement Management System (PMS) was undertaken in South Korea. This endeavour involved employing a particle filtering technique rooted in machine learning to predict the Condition Performance Index (CPCI), yielding precise outcomes. This innovative approach offers a reliable means of forecasting future pavement conditions [14]. Additionally, the Pavement Condition Index (PCI) prediction has been conducted through diverse methods. Notably, the employment of ArcGIS 10.2 software and the development of a model via PAVER 6.5.7 software have been leveraged [15, 16]. Several studies have also directed their attention towards investigating the factors contributing to pavement deterioration, mainly stemming from the behaviour of diverse binders employed in pavement construction. Muhammad et al. have conducted research using machine learning methodologies to predict both Marshall stability and Marshall flow for pavements constructed with asphalt concrete [17, 18]. Beyond just the PCI, several other indices valuable for evaluating pavement conditions are calculated using artificial neural networks and various machine learning algorithms [19–21].

Road infrastructure is a nation's essential backbone. However, pavements frequently fall short of the expected lifespan during construction due to traffic, construction practices, design, material quality, and environmental conditions. These variables collectively contribute to pavement deterioration and its subsequent impact on longevity. The accuracy of pavement performance projection hinges on pavement data. Reliable data collection, aided by modern technology, is crucial. Integrating computational intelligence is critical to refining models and data collection methods [5]. In the current research, the fusion of particle swarm optimization with the long short-term memory network is critically used to satisfy the research objective of predicting a specific index used to evaluate the state of asphalt pavements. Yu Sun et al. introduced a research endeavour in which the PSO–SVM model exhibited remarkable outcomes in prognosticating the fatigue longevity of an adapted asphalt road surface. This aspect is pivotal as it empowers road engineers to secure the operational functionality and robustness of the pavement [22]. A study has introduced an innovative hybrid prediction technique to estimate shear wave velocity, a critical factor in reservoir characterization and fluid identification. This approach harnesses the power of an LSTM network optimized by the PSO algorithm. With data from three wells, Jun Wang et al. proposed that this study demonstrates the model's robustness and highlights its superior accuracy compared to conventional MLR models [23]. An LSTM network model enhanced by PSO optimization has also exhibited enhanced performance in projecting flood dynamics, showcasing improved outcomes, and contributing to mitigating future flood risks.

The PSO-LSTM algorithm optimized using NSE demonstrated superior performance across Bangladesh's river network, offering valuable utility in forecasting floods for a 15-day horizon [24]. This optimized network model is also used in predicting the reliability of power transformers based on their ageing process and predicting wind power's interval to ensure that the wind turbine works efficiently [25, 26]. A novel hybrid model, namely the PSO LSTM-AE model, is presented in energy load forecasting. This model incorporates inventive strategies, including a nonlinear assignment method and a sine function, to optimize the PSO algorithm, which fine-tunes the parameters of the LSTM model. The improved PSO-optimized LSTM model showcases a remarkable 9% and 13% reduction in prediction errors compared to traditional LSTM and ARIMA models, emphasizing its reliability and practical insights [27]. For precise PM2.5 concentration prediction, this study proposes an innovative hybrid system involving the spectral enhancement of the wavelet transform (SWT) method, quantum particle swarm optimization (QPSO) algorithm, and LSTM model [28, 29]. In the context of cyber security, detecting attacks is crucial, and this research introduces a hybrid optimization algorithm that synergizes PSO with Grey Wolf optimization (GWO), enhancing security mechanisms against potential threats [30]. Moreover, within the Internet of things (IoT) landscape, devices possess unique IP addresses facilitating external communication with entities like smart home users [30]. These combined findings enrich the literature by providing novel perspectives on energy forecasting, air quality prediction, and IoT security enhancement.

The proposed methodologies exhibit diverse applications. They showcase promise in predicting stock market trends [31] and contribute to precise atmospheric wind speed forecasts in near space, aiding scientific research and engineering applications [32]. Additionally, the hybrid PSO-LSTM-FCN algorithm enhances the lithologic identification of complex reservoirs [33]. In the realm of text classification, the novel spotted hyena optimizer (SHO)-long short-term memory (SHO-LSTM) model offers a fresh approach to multilabel text classification (MLTC) [34]. In the context of COVID-19 data prediction, an improved ARIMA-LSTM hybrid model adapts to the virus's mutation [35]. Addressing railway freight volume and power load forecasting, the PSO-LSTM method proves effective [36, 37]. Collectively, these contributions enhance various domains and enrich the existing literature. Aiming to enhance wastewater treatment efficiency, this study delves into predicting COD levels within treatment plants. Simultaneously, it advances financial risk prediction through an FA-PSO-LSTM deep learning model, incorporating several benchmark models for comprehensive comparative analysis across diverse evaluation indices [38, 39]. The LSTM network model has consistently exhibited strong and precise outcomes, especially when optimized with PSO across diverse domains. Traditional machine learning methods like MLR models and decision trees often fall short in similar tasks. This study leverages the potential of the LSTM network and optimal PSO algorithm to predict PCI values. The approach yields enhanced accuracy by exploring the nonlinear connections between input distress features and the desired PCI outcomes. The research underscores the robustness of this proposed approach, as supported by the performance indicators under examination, which will be comprehensively elaborated upon.

4 Proposed Methodology

The methodology employed in this chapter for predicting the PCI unfolds as a systematic journey through key stages. Beginning with the meticulous procurement and refinement of road infrastructure data, the approach delves into exploratory analysis to uncover valuable insights. Model selection is approached thoughtfully, considering the intricacies of PCI prediction. The core of this context lies in an integrated model PSO and LSTM architecture, aiming for optimal predictive precision. The model is fabricated and fine-tuned, anticipating PCI by decoding distress parameters. Hyperparameter optimization with PSO further refines the model, ensuring peak efficiency. Calibration and validation round off the methodology, verifying the model's accuracy against real-world conditions. The entire methodology is summarized in Fig. 2.

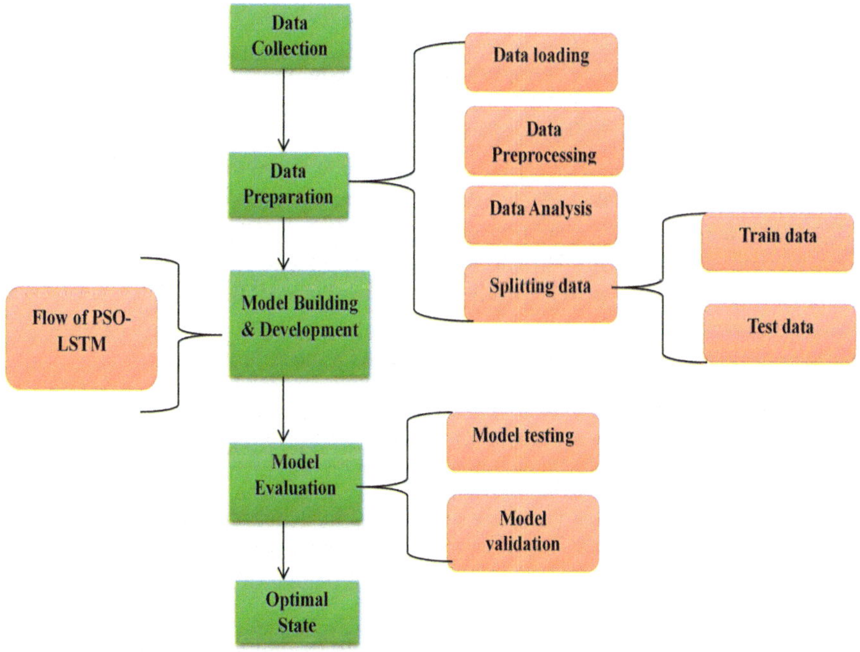

Fig. 2 The flow of proposed methodology

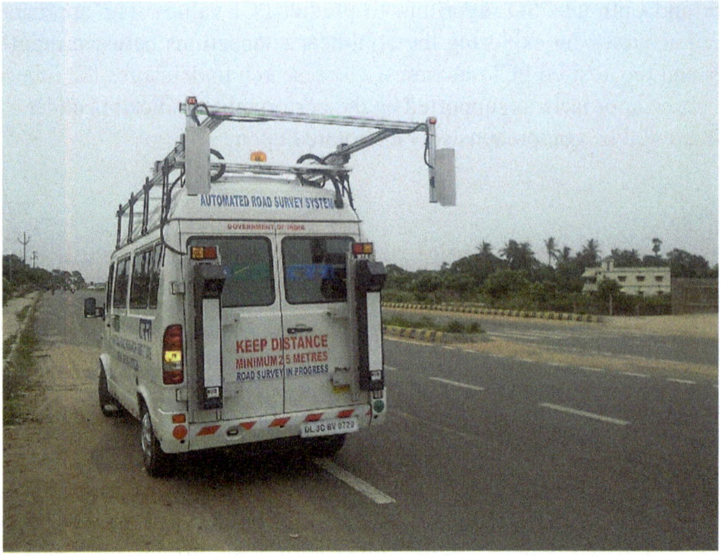

Fig. 3 The Hawkeye 2000 professional network survey vehicle (NSV)

4.1 Data Procurement and Analysis

4.1.1 Data Collection

The data collection process involved gathering comprehensive information from state highways on Indian roads (Andhra Pradesh State), covering 9478 km. The Hawkeye 2000 professional network survey vehicle (NSV) equipped with an integrated Hawkeye 2000 data acquisition system was utilized to acquire the necessary data (Fig. 3). This system facilitated the measurement of distress parameters such as alligator cracking, block cracking, edge cracking, patching, potholes, ravishing, rusting, and shoving (Fig. 4). Additionally, road inventory details, including the area, were collected to provide contextual information. The careful collection of these variables ensures a robust dataset for accurate PCI prediction, ultimately contributing to effective road maintenance and management strategies.

4.1.2 Basic Study of Data

Mean and standard deviation are essential statistical measures that offer valuable insights into the dataset. The mean represents the average value of a particular attribute, while the standard deviation quantifies the degree of variability around that mean. For example, take the attribute alligator cracking, which has an average (mean) of 61.38 and a substantial spread (standard deviation) of 241.04. This indicates a wide range in the severity of alligator cracking on the roads. Similarly, the

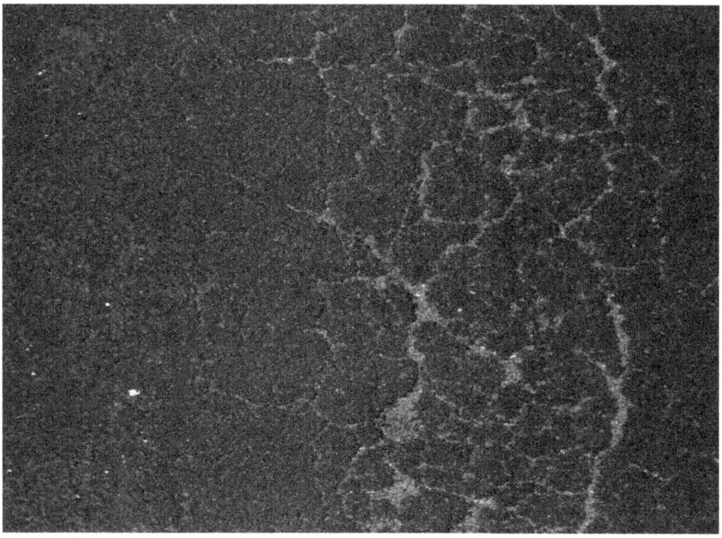

Fig. 4 Pavement surface distress images captured by NSV

Table 1 Descriptive statistics of data

Attribute	Mean	Standard deviation	Minimum	Maximum
Alligator cracking	61.38	241.04	0	3581.5
Area	5483.31	1746.1	3080.8	21227.2
Block cracking	0.027	0.44	0	15
Edge cracking	67.45	179.95	0	2616.2
Patching	108.84	228.2	0	2456.2
Potholes	0.54	2.43	0	45
Ravelling	35.99	80.23	0	1158.2
Rutting	2.287	4.45	0	45
Shoving	0.033	0.61	0	23
PCI	89.64	13.2	19.8	100.3

attribute area has an average of 5483.31 and a moderate variation (standard deviation) of 1746.1, signifying differences in the size of road sections. These statistics provide a clear picture of both typical values and the extent of variability within the dataset. This knowledge is instrumental in understanding road conditions and making informed decisions regarding maintenance and repair efforts. These measures provide valuable insights into the distribution characteristics, facilitating the identification of areas of concern and prioritizing maintenance efforts (Table 1). Examining the minimum and maximum values reveals the range of values observed for each attribute. For instance, alligator cracking ranges from a minimum value of 0 to a maximum value of 3581.5, indicating the presence of both sections with no crack and areas with extensive damage. Similarly, the Area attribute varies from a minimum of 3080.8 to a maximum of 21,227.2, representing the diversity in the sizes of road sections. These minimum and maximum values help establish the lower and upper bounds of the data, providing crucial information about the severity, extent, and scale of the road conditions.

4.2 Data Preprocessing

Data preprocessing is crucial in preparing raw data for analysis, especially in determining the PCI in a transportation infrastructure study. The process involves several key steps, each aimed at ensuring that the dataset is clean, well-structured, and suitable for the application of various machine learning algorithms. Data preprocessing, also called data processing, is an essential stage in data preparation for determining PCI. This involves cleaning and transforming raw data to create a dataset suitable for diverse machine learning algorithms. The prepared dataset includes attributes related to pavement distress types, severity levels, and target variables, facilitating comprehensive analysis.

4.2.1 Steps in Data Preprocessing

Data cleaning: Data cleaning is a critical aspect of preprocessing where incomplete, incorrect, irrelevant, duplicated, or improperly framed data is addressed. The goal is to enhance data quality by removing noise and ensuring accuracy. Data cleaning involves removing, modifying, or deleting data to standardize the dataset for accurate analysis without compromising important information.

Normalization: Normalization is employed to handle incomplete or uncertain data. The Min–Max normalization method is utilized to scale feature values within a specified range (e.g. 0–1). This process involves transforming each feature value, ensuring that minimum values become 0 and maximum values become 1. The normalization equation (Eq. 3) is used to calculate normalized values, filling in missing data and improving data quality.

$$X_{\text{norm}} = \frac{x_i - x_{\min}}{x_{\max} - x_{\min}} \quad (3)$$

where X_i is the data point, $X_{\min}$ is the minimum value of the data point, and $X_{\max}$ is the maximum value of the data point or the batch instances.

Discretization: Discretization is performed on continuous functions expressed in terms of variables, converting them into discrete counterparts. This process involves modifying variables to categorize granularity, particularly when dealing with multiple discrete variables. The objective is to reduce the complexity of the model by considering fewer levels of variables during the modelling process.

Data transformation: Data transformation is a pivotal stage in the data preprocessing pipeline, playing a crucial role in shaping raw data into a format conducive to effective mining and analysis. This multifaceted process involves several key techniques aimed at refining the dataset for enhanced compatibility with machine learning algorithms, thereby bolstering the robustness of subsequent analyses. One fundamental aspect of data transformation is normalization, where attribute data undergoes scaling to conform to a predetermined range, such as −1.0 to 1.0 or 0 to 1.0. This standardization ensures that all variables contribute proportionally to the analysis, preventing the dominance of certain features due to disparate scales. Smoothing, another facet of data transformation, entails the removal of noise from the dataset. This noise, often stemming from irregularities or outliers in the raw data, can obscure patterns and compromise accuracy. Smoothing techniques are employed to mitigate these disturbances, resulting in a more refined dataset conducive to accurate analysis. Aggregation represents a strategic operation in data transformation, involving the application of summary or aggregation functions to the dataset. This step simplifies the dataset's structure by condensing information, facilitating a more streamlined analysis. Aggregation is particularly valuable when dealing with large datasets or when seeking overarching trends and patterns. Generalization, the final transformation technique, entails replacing low-level or raw data with higher-level concepts using concept hierarchies. This process aids in simplifying complex structures and reduces the dataset's intricacy, making it more amenable to analysis. By abstracting detailed information into broader categories, generalization enhances the dataset's interpretability and contributes to the development of more effective models.

4.2.2 Data Exploration

Data exploration is a pivotal step involving thoroughly examining and analysing the loaded dataset to comprehend its inherent patterns, relationships, and characteristics. This process is indispensable for uncovering the underlying structure of the data, identifying patterns, detecting outliers, and preparing the dataset for subsequent modelling efforts. Exploratory data analysis (EDA) is crucial for gaining valuable insights and making informed decisions before applying complex analytical techniques. In machine learning, EDA is typically conducted using established libraries like matplotlib and seaborn. These tools facilitate the generation of scatter plots, pair plots, and heatmaps, providing visual representations of the data's relationships and correlations. Scatter plots and pair plots offer insights into the distribution of variables, while heatmaps reveal the strength and nature of relationships between variables (Figs. 5 and 6). Such visualizations are instrumental in feature selection and understanding how different variables impact the target variable.

This research's core is a comprehensive dataset containing attributes essential to pavement assessment. This dataset is structured into discrete entities: input features and the target variable, forming the foundation for meticulous analysis and model development. The initial stage demarcates the transition from raw data to insightful

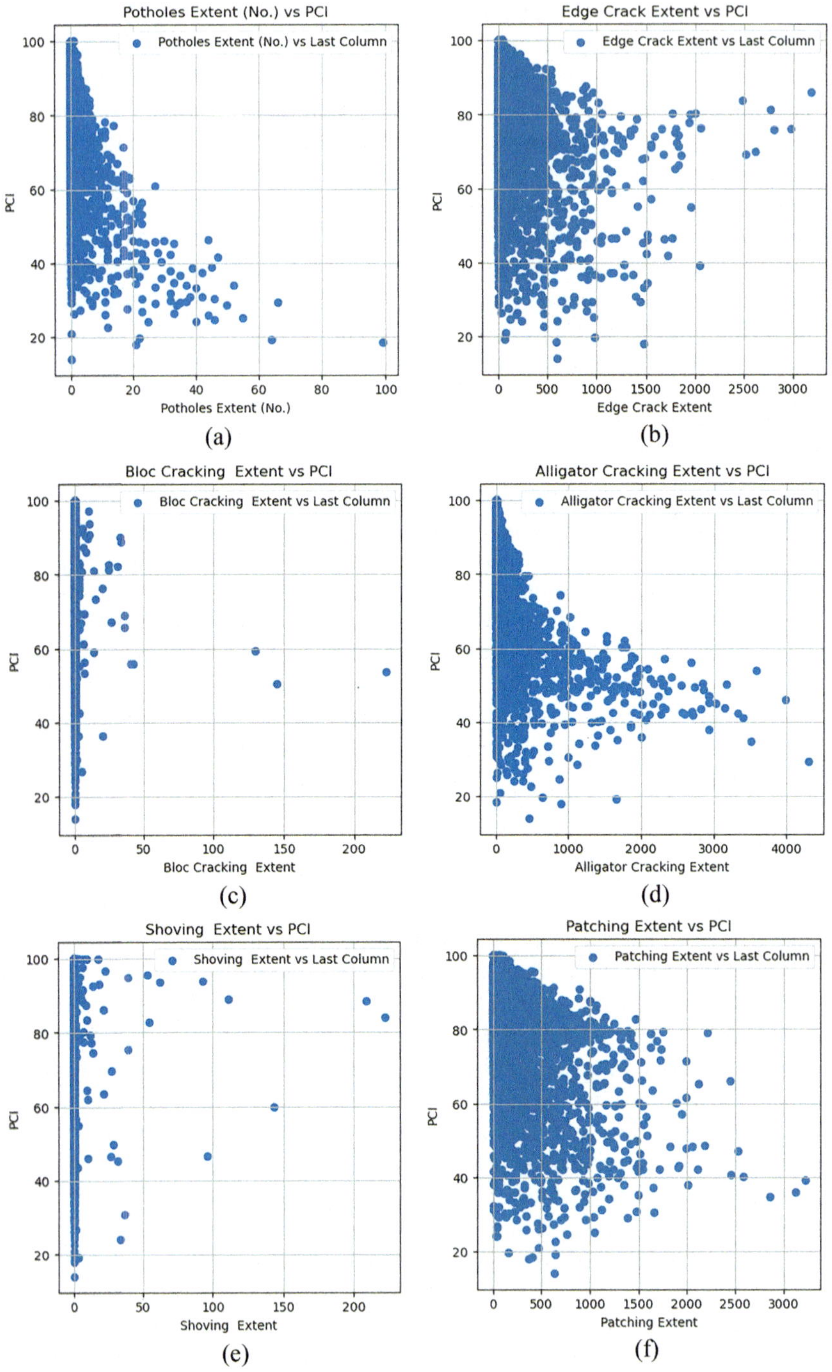

Fig. 5 (**a–h**) Relationship between input features (distresses) and target variable (PCI)

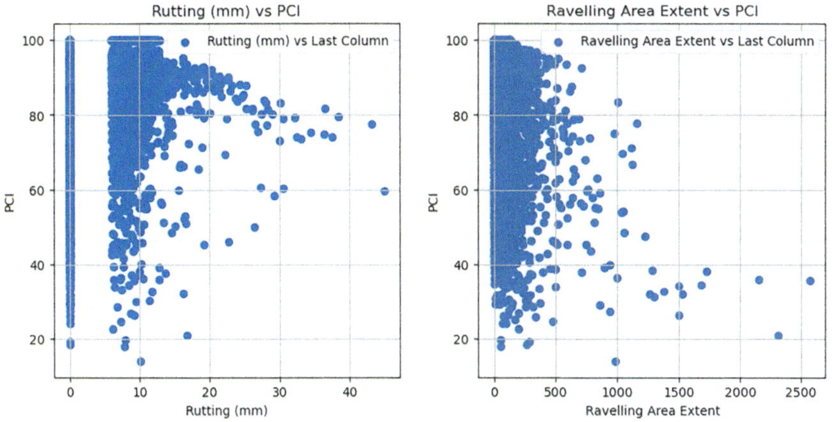

(g) (h)

Fig. 3.5 (continued)

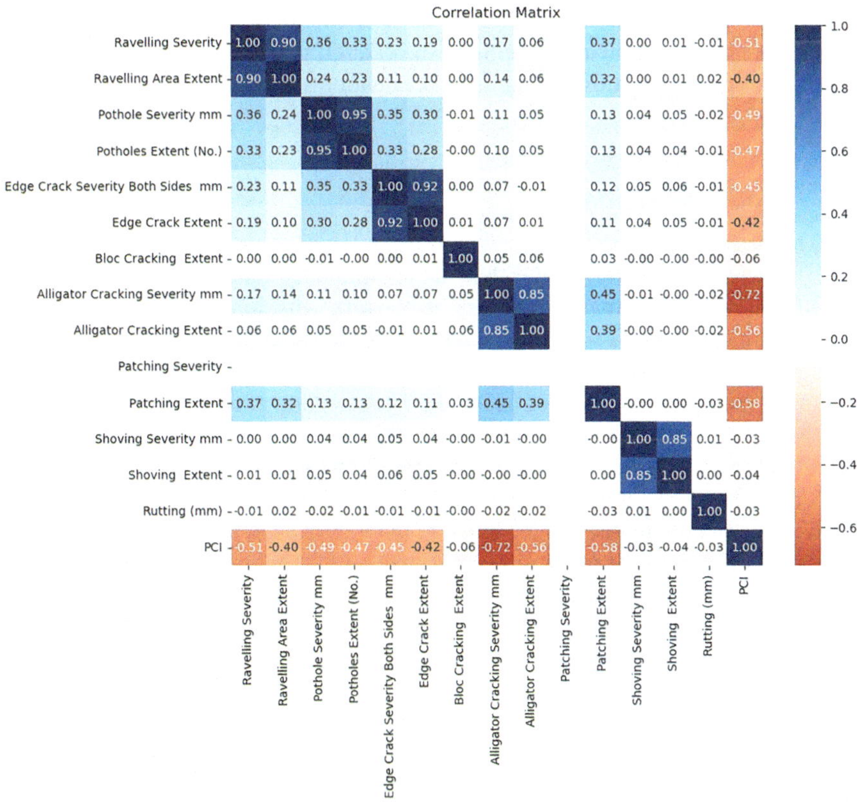

Fig. 6 Relation between input features through a heatmap

knowledge. A meticulous data division process is implemented to create robust predictive models. The dataset is strategically partitioned into distinct training and testing subsets, enhancing the model's predictive capabilities and allowing for realistic assessments. Simultaneously, a standardization process ensures that input features harmonize across various parameters, facilitating coherent and efficient model learning. The analytical journey depicted in Figs. 5(a–h) visually illustrates the interplay between input features, encompassing various distress parameters and the target variable—PCI. This graphical representation serves as a roadmap for understanding the relationships within the dataset, guiding subsequent steps in the analysis and model development process.

Crucially, the potency of the PSO-LSTM network resides in the intricate reshaping of data, aligning it harmoniously with the LSTM's inherent structure. Through meticulous transformation, the training data is transmuted into a three-dimensional representation. This strategic conversion—moulding the data into the dimensions of 'batch size', 'time steps', and 'numb features'—empowers the LSTM to navigate and unveil the temporal intricacies and dependencies inherently ingrained within the dataset. Additionally, Figure 6 offers a visual testament to the intricate relationships that underlie the diverse input features, serving as a visual compass to navigate the multidimensional terrain of data interdependencies. Through this multifaceted exploration and refining of the dataset, we lay the groundwork for the subsequent model creation, training, and evaluation stages.

4.3 Model Selection

The PSO-LSTM network algorithm, a revolutionary amalgamation of swarm-based optimization and recurrent neural network with long short-term memory, presents an avant-garde solution for predicting the PCI. This pioneering approach harnesses the synergy of optimization and sequential learning, transcending conventional methods for enhanced accuracy. The selection of an appropriate model is paramount in successfully predicting the PCI through the PSO-LSTM network algorithm. Model building is a pivotal stage in the methodology, where the intricate fusion of PSO with LSTM methodology is harmoniously orchestrated to derive accurate predictions. The choice of the PSO-LSTM model architecture significantly influences the precision and efficacy of PCI predictions, as it determines how well the model captures the underlying patterns and temporal dependencies within the data. This section delves into the construction of the PSO-LSTM network, encompassing the intricacies of architecture design, hyperparameter tuning, and model validation. By meticulously selecting and configuring the PSO-LSTM model, the methodology ensures that it aligns with the complex nature of pavement distress data, thus paving the way for more accurate and insightful PCI predictions. Utilizing the prowess of both PSO and LSTM neural networks, the integrated PSO-LSTM model presents a sophisticated methodology for precisely forecasting the PCI through analysis of historical data. The comprehensive steps below depict the complete process, from data collection to result evaluation and visualization.

This model seamlessly integrates PSO's hyperparameter optimization capabilities with LSTM's ability to capture temporal patterns, resulting in an enhanced predictive tool for pavement condition assessment. The PSO-LSTM model employs a twofold process:

4.3.1 Hyperparameter Optimization with PSO

PSO is an iterative optimization strategy that draws inspiration from the communal dynamics exhibited by bird flocks or fish schools [24]. In the context of the PSO-LSTM model, PSO plays a crucial role in fine-tuning the LSTM's hyperparameters, such as the number of LSTM units, activation functions, and learning rates. This optimization process navigates the hyperparameter space, seeking configurations that maximize the model's anticipatory correctness. A clearly defined sequence of steps characterizes the PSO progression. The initialization phase marks the commencement, where a populace of particles is instantiated, each representing a unique hyperparameter configuration for the model. These particles embark on a journey through the hyperparameter space, driven by the interplay of historical knowledge and collective wisdom. This movement, known as particle movement, is orchestrated by adjusting their positions based on their best performance and the pinnacle performance achieved by the entire swarm (Eqs. 4 and 5).

$$vi+1=\left(w\times vi\right)\left(c1\times r1\times\left(\text{pbest}-xi\right)+\left(c2\times r2\times\left(\text{gbest}-xi\right)\right.\right. \tag{4}$$

$$xi+1=xi+vi+1 \tag{5}$$

In the intricate dance of optimization, the inertia factor (w) emerges as the conductor, orchestrating the distribution of weights between a particle's personal best (pbest) and the global best (gbest). Pictures $c1$ and c2 emerge as the virtuoso composers, fine-tuning the harmonies of flight steps. Random numbers, $r1$ and $r2$, play the role of serendipitous muses, infusing unpredictability. Enter vi and xi, embodiments of velocity and position, leading each particle dimension in a dynamic ballet. Meanwhile, pbest and gbest are the prima donnas, embodying local and global optima for each dimension, respectively [39, 40]. The schematic framework of PSO-LSTM Network in predicting PCI is as shown in Fig. 7.

At every juncture of this odyssey, the heart of the PSO-LSTM model comes into play: the evaluation stage. Each particle contributes to the grand tapestry of exploration by orchestrating the training and evaluation of the LSTM model [40]. These models, tailored according to the specific hyperparameters embodied by each particle, engage in predictive exercises. Serving as the evaluator of their endeavours, frequently embodied by the MSE, Eq. (6), this performance measure passes a verdict on the precision of their prognostications relative to the PCI values. Evolving through this process, particles continually reassess their trajectories, engaging in an

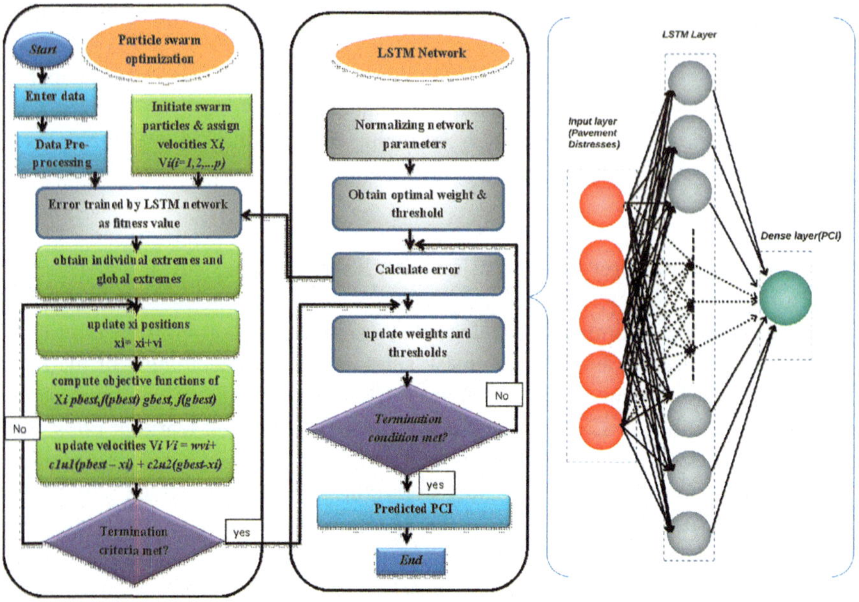

Fig. 7 The schematic framework of PSO-LSTM network in the prediction of PCI

eternal dance of adaptation. If a particle achieves a new personal best in its predictions, it meticulously updates its historical best position to remember its moment of triumph. Beyond individual victories, the collective striving for excellence leads to updating the swarm's best position—the culmination of the most remarkable predictions across all particles. This communal striving amplifies the swarm's ability to exploit fruitful regions of the hyperparameter space. Yet, even as particles progress, the pursuit is not boundless. The ultimate objective is convergence, the juncture where particles find the best possible hyperparameter configuration or fulfil a predetermined number of iterations. This convergence may hinge upon reaching a desired level of performance or witnessing the particles harmoniously cluster around a promising domain of the hyperparameter landscape. Intersection marks the termination of PSO's dynamic choreography, having accomplished its role in sculpting the optimal hyperparameters that will guide the LSTM model's predictive prowess in the journey to forecast PCI values accurately.

4.3.2 Temporal Pattern Learning with LSTM Network

The strength of the LSTM network lies in its proficiency in apprehending sequential patterns and interdependencies within datasets. Within the PSO-LSTM model framework, the central role of learning temporal relationships from historical PCI data is assumed by the LSTM layer. In the PSO-LSTM model's structure, each layer uniquely harnesses the predictive capabilities presented by amalgamating the particle swarm optimization and long short-term memory paradigms.

Input layer: At a significant part of the process, the input layer assumes a crucial function by assimilating historical data to facilitate predictive analysis. The layer accommodates the reshaped historical data, configured to align with the dimensions. This intricate design allows the model to process sequences of data effectively, preserving the temporal characteristics inherent in the PCI dataset.

LSTM layer: Positioned at the heart of the model's architecture, the LSTM layer, with its 50 units, embodies the essence of sequence analysis. Armed with the ability to capture intricate temporal patterns, this layer deciphers the dynamic relationships within the PCI data. Including the rectified linear unit (ReLU) activation function introduces vital nonlinearity, enhancing the model's capacity to recognize complex dependencies and variations in the data. This layer's pivotal role in unveiling hidden patterns is at the core of the model's power to forecast PCI values.

Dense layer: Following the LSTM layer's perceptive analysis, the dense layer appears as a bridge to the final predictions. The layer's singular output unit consolidates the insights extracted from the previous layers, channelling them into a cohesive representation for prediction. By blending learned temporal nuances and optimized hyperparameters, this layer distils the complex patterns into a single prediction for the PCI value.

As the model assembles its architectural components, the choice of optimizer and loss function is pivotal in guiding its training. Here, the Adam optimizer takes the reins, a choice celebrated for its efficiency in navigating high-dimensional spaces during training. Stepping into prominence, the MSE loss function emerges as the guiding compass that quantifies the disparity between the projected and factual PCI values. This metric guides the model's efforts in refining its predictions to align with ground truth, ultimately steering the model toward precision. Collectively, these layers orchestrate a symphony of intelligence, where the PSO-optimized hyperparameters infuse potency into the LSTM's intricate analysis. The harmonious interplay of these architectural components culminates in a model that not only learns the temporal dynamics of the PCI dataset but also optimizes itself through PSO to provide accurate and reliable predictions of pavement condition. The PSO-LSTM model seamlessly integrates the optimized hyper parameters from PSO with the LSTM's sequential pattern learning. This fusion results in a predictive tool that can accurately forecast PCI values while accounting for global and local optima in the hyperparameter space. The PSO-LSTM model signifies a synergy between optimization and deep learning, attaining optimal hyper parameters while comprehensively capturing the dynamic nature of pavement condition data. This methodology exemplifies a cutting-edge approach to infrastructure management, promoting informed decisions based on precise PCI predictions.

4.4 Model Calibration and Validation

The core of the PSO-LSTM network's functionality unfolds during its training phase, where the intricate interplay between data, epochs, batch size, and parameter optimization takes centre stage. The training data, input features, and target PCI values are the foundation upon which the model refines its internal parameters. The training process operates iteratively over a specified number of epochs, each encompassing multiple batches of data controlled by the chosen batch size. Through the guidance of Particle Swarm Optimization and the learning capabilities of LSTM, the model's parameters dynamically evolve to enhance its predictive accuracy. Having matured through the rigors of training, the PSO-LSTM model confidently advances into its prediction phase, where its acquired insights metamorphose into tangible forecasts of PCI values for the testing dataset.

This projection establishes the platform for an exhaustive model assessment that highlights its performance. An extensive array of precisely calculated performance metrics highlights this assessment. The MSE, Eq. (6), is central among these pivotal metrics. MSE quantifies the average squared deviation between the model's predicted and actual PCI values. Its counterpart, the RMSE, Eq. (7), reveals the mean magnitude of prediction discrepancies, reflecting the model's overall precision. Additionally, the mean absolute error (MAE), Eq. (8), a metric resilient to outliers, measures the mean absolute differences between projected and factual PCI values. Beyond these error metrics, the emergence of the R^2, Eq. (9) metric adds depth. R^2 exposes the model's ability to clarify variance in PCI values through its forecasts. R^2 showcases the proportion of variance the model explains, offering evidence of its predictive capability (Table 2). This ensemble of meticulously computed metrics coalesces, presenting an all-encompassing picture of the PSO-LSTM model's aptitude to distil data into meaningful forecasts.

The culmination of rigorous training, combined with the fusion of particle swarm optimization and LSTM-driven temporal analysis, culminates in a model proficient

Table 2 Performance metrics and their results of the PSO-LSTM model

Sl. no.	Metric	Formula	Result
1	MSE	$\frac{1}{N}\sum_{k=1}^{N}\left(yk-\hat{y}k\right)^2$ (6)	6.86
2	RMSE	$\sqrt{\frac{\sum_{k=1}^{N}\left(y_k-\hat{y}_k\right)^2}{N}}$ (7)	2.62
3	MAE	$\frac{\sum_{k=1}^{N}\left\|y_k-\hat{y}_k\right\|}{N}$ (8)	1.40
4	R^2	$1-\frac{\sum_{k=1}^{N}\left(Pk-Ok\right)^2}{\sum_{k=1}^{N}\left(Ok-Om\right)^2}$ (9)	0.97

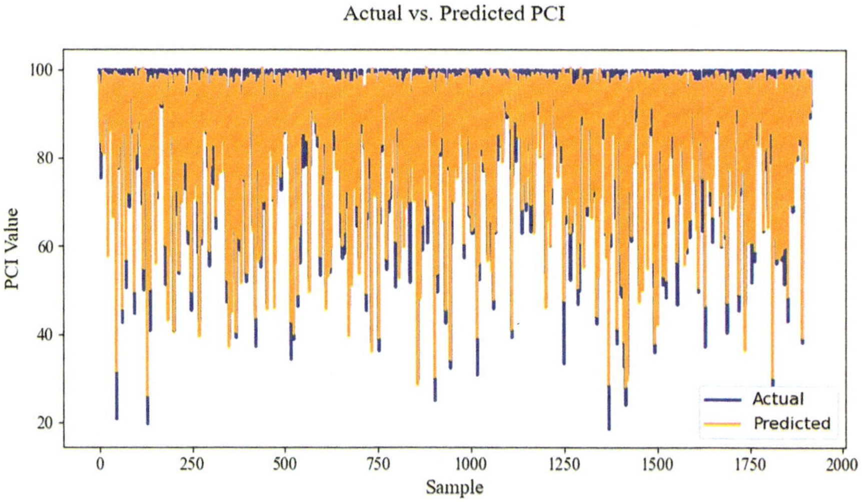

Fig. 8 Comparison of actual vs. predicted PCI for each sample fragment

in grasping intricate patterns and converting them into accurate forecasts of PCI values. Consequently, this model significantly contributes to informed decision-making in infrastructure management. Figure 8 illustrates the alignment between the PSO-LSTM model's predicted PCI values and the actual PCI values for each sample fragment. This visualization provides a quick overview of the model's predictive accuracy, bridging numerical metrics with real-world outcomes. It showcases the model's ability to interpret distress patterns and accurately forecast PCI, underscoring its effectiveness in pavement condition prediction.

5 Results and Discussions

The performance of different models in predicting the PCI was rigorously evaluated using key metrics, namely MSE, RMSE, MAE and coefficient of determination (R^2) that are demonstrated in Table 3.

Despite LSTM being a widely utilized model, its higher errors reveal challenges in capturing intricate patterns and dependencies within the PCI data. The strong correlation indicated by an R^2 value of 0.94 falls short of the PSO-LSTM model's performance. While ANN outperforms LSTM, it still exhibits elevated errors, and its strong correlation does not surpass the accuracy and precision achieved by PSO-LSTM. SVR shows higher errors and a lower R^2 value than LSTM and ANN, signalling suboptimal performance in handling the complexity of PCI prediction. M5 performs well, but PSO-LSTM surpasses it in terms of lower errors and a higher R^2 value, showcasing PSO-LSTM's predictive excellence. PSO-LSTM, with the lowest MSE, RMSE, and MAE, stands out for its superior accuracy. Its impressive R^2

Table 3 Comparative performance metrics of various algorithms

Model	MSE	RMSE	MAE	R-squared
LSTM	11.94	3.45	12.84	0.94
ANN	10.77	3.28	2.21	0.94
SVR	23.22	4.82	2.68	0.88
M5	7.99	2.83	1.58	0.96
PSO-LSTM	6.86	2.62	1.40	0.97

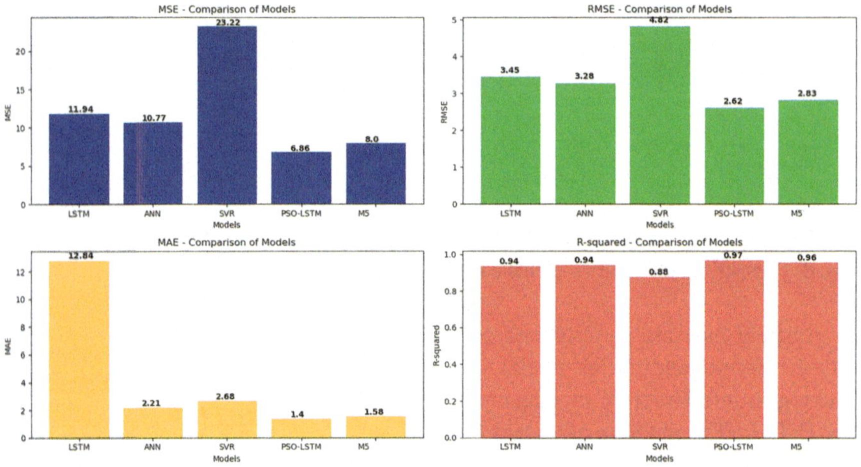

Fig. 9 Performance metrics of LSTM, ANN, SVR, PSO-LSTM, and M5 models

value of 0.97 underscores its efficacy in capturing complex temporal relationships, positioning it as the most effective model in PCI prediction. To know the relation among different models discussed earlier, their performance metrics are plotted as shown in Fig. 9.

The PSO-LSTM model demonstrates clear superiority across all metrics, boasting the lowest MSE, RMSE, and MAE. Its R^2 value nearing 1 indicates a strong correlation between predicted and actual PCI values shown in Fig. 10. This underscores the efficacy of combining particle swarm optimization and long short-term memory networks in enhancing predictive accuracy. Traditional models like LSTM, ANN, and SVR exhibit higher errors and lower R^2 values. This suggests their limitations in capturing the complex patterns and temporal dependencies inherent in PCI prediction tasks. PSO-LSTM's outperformance highlights the inadequacy of these traditional approaches for this specific application. The study emphasizes the holistic importance of road infrastructure, particularly pavements, in socioeconomic progress.

Traditional methods of PCI determination are highlighted as manual and error-prone, paving the way for the introduction of advanced computational intelligence. The PSO-LSTM algorithm is positioned as a transformative leap in predictive modelling. It adeptly combines the sequential learning ability of LSTM with the optimization capabilities of PSO, enabling accurate prediction of PCI, which can be

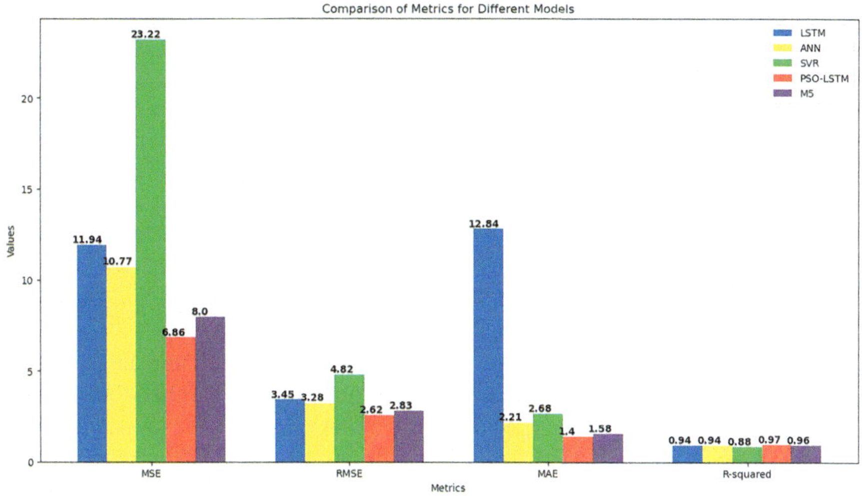

Fig. 10 Grouped bar chart of predictive models used for Pavement Condition Index (PCI) prediction

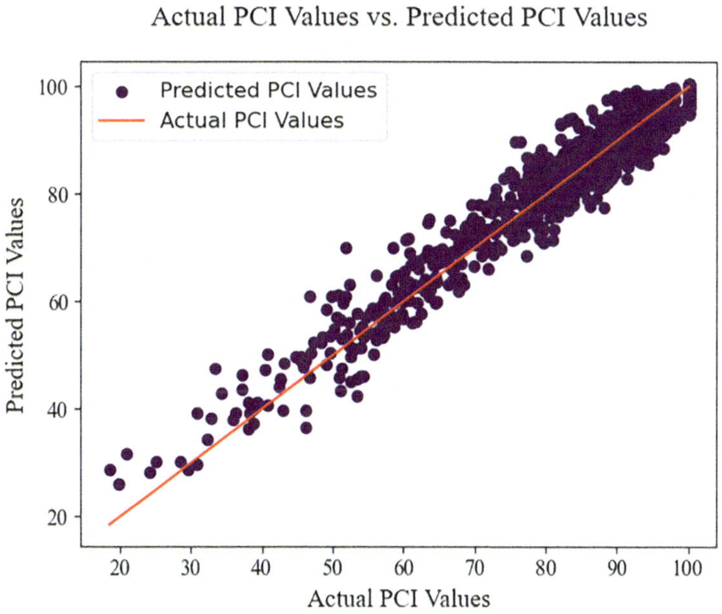

Fig. 11 Scatter plot of actual and predicted PCI

observed in Fig. 11. The algorithm's excellence is underscored by its ability to navigate complex temporal relationships in data. The context of India's vast road network, the second largest globally, adds significance to the research. The deteriorating condition of roads is identified as a critical issue, emphasizing the need for accurate

PCI prediction. PSO-LSTM's potential to contribute to effective road maintenance aligns with the urgent demand for sustainable infrastructure management. As the research concludes, the PSO-LSTM algorithm is a pivotal tool for the future. It promises accurate PCI predictions and sets the stage for smarter, safer, and more resilient roadways. The integration of technology is highlighted as indispensable for elevating the efficiency of road infrastructure management.

6 Conclusion

In the pursuit of accurate Pavement Condition Index (PCI) prediction, this study has unequivocally demonstrated the superior performance of the PSO-LSTM (Particle Swarm Optimization for Long Short-Term Memory) algorithm over traditional models like LSTM, ANN, and SVR. The metrics evaluation, encompassing mean squared error (MSE), root mean squared error (RMSE), mean absolute error (MAE), and coefficient of determination (R^2), consistently positions the PSO-LSTM model as the most effective in capturing complex temporal relationships inherent in PCI data. While the PSO-LSTM algorithm exhibits remarkable predictive excellence, it is essential to acknowledge potential drawbacks and identify avenues for future research. One limitation lies in the algorithm's sensitivity to hyperparameter tuning, which can impact its performance across diverse datasets. Future studies should investigate optimizing these parameters to enhance the model's robustness and applicability to varied road conditions. Additionally, despite the PSO-LSTM model's success, its interpretability may pose challenges, especially in explaining complex predictions to non-technical stakeholders. Addressing this limitation could involve incorporating explainable AI techniques or developing user-friendly interfaces to make the model insights more accessible. The future scope of research lies in refining the PSO-LSTM algorithm, exploring ensemble techniques, or combining it with other emerging technologies such as reinforcement learning. These advancements could contribute to more accurate predictions and a nuanced understanding of pavement conditions.

Acknowledgments We sincerely thank CSIR—Central Road Research Institute (CRRI)—in New Delhi for their invaluable assistance in our internship. Additionally, our heartfelt gratitude to the Department of Civil Engineering and the administrative board of SVECW (A) for their invaluable backing in exploring and implementing soft computing methodologies to advance the PCI prediction model. Their unwavering commitment significantly contributed to this study's successful culmination, and we deeply acknowledge their contributions.

References

1. Ali, A., Heneash, U., Hussein, A., & Eskebi, M. (2022). Predicting pavement condition index using fuzzy logic technique. *Infrastructures, 7*(7), 91. https://doi.org/10.3390/infrastructures7070091

2. Batrakova, A. G., Troyanovsky, V. V., Batrakov, D. O., Pilicheva, M. O., & Skrypnyk, N. S. (2020). Prediction of the road pavement condition index using stochastic models. *Roads and Bridges—Drogi i Mosty*, (3), 225–242. https://doi.org/10.7409/rabdim.020.015
3. Shukla, B. K., Bhowmik, A. R., Sharma, P., & Abhimanyu, K. (2020). Use of waste synthetic fiber reinforcement in environmental friendly and economic pavement construction. *Journal of Green Engineering, 10*(1), 62–75.
4. E17 Committee. (n.d.). *Practice for roads and parking lots pavement condition index surveys.*
5. Lee, J. H., Jung, D. H., Lee, M. S., & Jeon, S. I. (2022). A feasibility study for the prediction of concrete pavement condition index (CPCI) based on machine learning. *Applied Sciences, 12*(17), 8731. https://doi.org/10.3390/app12178731
6. Ashrafian, A., Taheri Amiri, M. J., Masoumi, P., Asadi-shiadeh, M., Yaghoubi-chenari, M., Mosavi, A., & Nabipour, N. (2020). Classification-based regression models for prediction of the mechanical properties of roller-compacted concrete pavement. *Applied Sciences, 10*(11), 3707. https://doi.org/10.3390/app10113707
7. Chen, X., & Long, Z. (2023). E-Commerce enterprises financial risk prediction based on FA-PSO-LSTM neural network deep learning model. *Sustainability, 15*(7), 5882. https://doi.org/10.3390/su15075882
8. Althoey, F., Akhter, M. N., Nagra, Z. S., Awan, H. H., Alanazi, F., Khan, M. A., Javed, M. F., Eldin, S. M., & Özkılıç, Y. O. (2023). Prediction models for Marshall mix parameters using bio-inspired genetic programming and deep machine learning approaches: A comparative study. *Case Studies in Construction Materials, 18*, e01774. https://doi.org/10.1016/j.cscm.2022.e01774
9. Behnood, A., Olek, J., & Glinicki, M. A. (2015). Predicting modulus elasticity of recycled aggregate concrete using M5′ model tree algorithm. *Construction and Building Materials, 94*, 137–147. https://doi.org/10.1016/j.conbuildmat.2015.06.055
10. Boyapati, B., & Kumar, R. P. (2015). Prioritisation of pavement maintenance based on pavement condition index. *Indian Journal of Science and Technology, 8*(14). https://doi.org/10.17485/ijst/2015/v8i14/64320
11. Butt, A. A. (1994). Application of Markov process to pavement management systems at network level. In *Conference proceedings 1* (Vol. 2).
12. Bharti, G., Shukla, B. K., Kumar, A., Nagaraju, M. B., Bansal, V., & Varadharajan, S. (2023, September). Effect of implementation of waste shredded plastic on construction of flexible pavement. In *AIP conference proceedings* (Vol. 2800, No. 1). AIP Publishing.
13. Du, M., Chen, Y., Liu, Y., & Yin, H. (2022). A novel hybrid method to predict PM2.5 concentration based on the SWT-QPSO-LSTM hybrid model edited by S. Du. *Computational Intelligence and Neuroscience, 2022*(1), 7207477. https://doi.org/10.1155/2022/7207477
14. Huang, W., Sun, Y., Zhou, Y., Cheng, X., Xiang, Z., Shao, C., Bo, H., & Xie, K. (2023). A hybrid ensemble learning approach for the aging-dependent reliability prediction of power transformers. *Energy Reports, 9*, 219–230. https://doi.org/10.1016/j.egyr.2023.04.046
15. He, Y., Li, W., Dong, Z., Zhang, T., Shi, Q., Wang, L., Lei, W., Qian, S., Wang, Z., Liu, Z., & Lei, G. (2023). Lithologic identification of complex reservoir based on PSO-LSTM-FCN algorithm. *Energies, 16*(5), 2135. https://doi.org/10.3390/en16052135
16. Hosseini, S. A., Alhasan, A., & Smadi, O. (2020). Use of deep learning to study modeling deterioration of pavements a case study in Iowa. *Infrastructures, 5*(11), 95. https://doi.org/10.3390/infrastructures5110095
17. Gul, M. A., Islam, M. K., Awan, H. H., Sohail, M., Al Fuhaid, A. F., Arifuzzaman, M., & Qureshi, H. J. (2022). Prediction of marshall stability and marshall flow of asphalt pavements using supervised machine learning algorithms. *Symmetry, 14*(11), 2324. https://doi.org/10.3390/sym14112324
18. Kaveh, A. (2017). "M5" and mars based prediction models for properties of self-compacting concrete containing fly ash. *Periodica Polytechnica Civil Engineering, 62*(2), 281–294. https://doi.org/10.3311/PPci.10799
19. Jin, Y.-C., Cao, Q., Wang, K.-N., Zhou, Y., Cao, Y.-P., & Wang, X.-Y. (2023). Prediction of COVID-19 data using improved ARIMA-LSTM hybrid forecast models. *IEEE Access, 11*, 67956–67967. https://doi.org/10.1109/ACCESS.2023.3291999

20. Kimteta, A., Thakur, M. S., Sihag, P., Upadhya, A., & Sharma, N. (2022). Prediction of flexural strength of FRC pavements by soft computing techniques. *Archives of Materials Science and Engineering, 117*(1), 13–24. https://doi.org/10.5604/01.3001.0016.1393
21. Moalla, H., Elloumi, W., & Alimi, A. M. (2017). H-PSO-LSTM: Hybrid LSTM trained by PSO for online handwriter identification. In D. Liu, S. Xie, Y. Li, D. Zhao, & E.-S. M. El-Alfy (Eds.), *Neural information processing* (Lecture notes in computer science) (Vol. 10637, pp. 41–50). Springer International Publishing.
22. Mubaraki, M. (2014). Identification of pavement distress types and pavement condition evaluation based on network level inspection for Jazan city road network. *The Journal of Engineering Research [TJER], 10*(2), 44. https://doi.org/10.24200/tjer.vol11iss1pp44-54
23. Osman, S. A., Almoshaogeh, M., Jamal, A., Alharbi, F., Al Mojil, A., & Dalhat, M. A. (2022). Intelligent assessment of pavement condition indices using artificial neural networks. *Sustainability, 15*(1), 561. https://doi.org/10.3390/su15010561
24. Pranolo, A., Mao, Y., Wibawa, A. P., Utama, A. B. P., & Dwiyanto, F. A. (2022). Robust LSTM with tuned-PSO and bifold-attention mechanism for analyzing multivariate time-series. *IEEE Access, 10*, 78423–78434. https://doi.org/10.1109/ACCESS.2022.3193643
25. Pinatt, J. M., Chicati, M. L., Ildefonso, J. S., & Filetti, C. R. G. D.'a. (2020). Evaluation of pavement condition index by different methods: Case study of Maringá, Brazil. *Transportation Research Interdisciplinary Perspectives, 4*, 100100. https://doi.org/10.1016/j.trip.2020.100100
26. Qiu, Y.-Y., Zhang, Q., & Lei, M. (2020). Forecasting the railway freight volume in China based on combined PSO-LSTM model. *Journal of Physics: Conference Series, 1651*(1), 012029. https://doi.org/10.1088/1742-6596/1651/1/012029
27. Sansine, V., Ortega, P., Hissel, D., & Hopuare, M. (2022). Solar irradiance probabilistic forecasting using machine learning, metaheuristic models and numerical weather predictions. *Sustainability, 14*(22), 15260. https://doi.org/10.3390/su142215260
28. Saoud, A., & Recioui, A. (2023). Load energy forecasting based on a hybrid PSO LSTM-AE model. *Algerian Journal of Environmental Science and Technology, 9*(1).
29. Yu, Z., Sun, H., & Zhang, B. (2022). Research on power load forecasting based on PSO-LSTM. In *2022 4th International Conference on Artificial Intelligence and Advanced Manufacturing (AIAM)* (pp. 167–170). IEEE.
30. Shah, H. A., Nehdi, M. L., Khan, M. I., Akmal, U., Alabduljabbar, H., Mohamed, A., & Sheraz, M. (2022). Predicting compressive and splitting tensile strengths of silica fume concrete using M5P model tree algorithm. *Materials, 15*(15), 5436. https://doi.org/10.3390/ma15155436
31. Shao, B., Li, M., Zhao, Y., & Bian, G. (2019). Nickel price forecast based on the LSTM neural network optimized by the improved PSO algorithm. *Mathematical Problems in Engineering, 2019*, 1–15. https://doi.org/10.1155/2019/1934796
32. Shtayat, A., Moridpour, S., Best, B., & Abuhassan, M. (2023). Using supervised machine learning algorithms in pavement degradation monitoring. *International Journal of Transportation Science and Technology, 12*(2), 628–639. https://doi.org/10.1016/j.ijtst.2022.10.001
33. Sun, Y., He, D., & Li, J. (2021). Research on the fatigue life prediction for a new modified asphalt mixture of a support vector machine based on particle swarm optimization. *Applied Sciences, 11*(24), 11867. https://doi.org/10.3390/app112411867
34. Upadhya, A., Thakur, M. S., Sihag, P., Kumar, R., Kumar, S., Afeeza, A., Afzal, A., & Ahamed Saleel, C. (2023). Modelling and prediction of binder content using latest intelligent machine learning algorithms in carbon fiber reinforced Asphalt concrete. *Alexandria Engineering Journal, 65*, 131–149. https://doi.org/10.1016/j.aej.2022.09.055
35. Wang, X., Wang, Y., Yuan, P., Wang, L., & Cheng, D. (2021). An adaptive daily runoff forecast model using VMD-LSTM-PSO hybrid approach. *Hydrological Sciences Journal, 66*(9), 1488–1502. https://doi.org/10.1080/02626667.2021.1937631
36. Xu, S., Li, W., Zhu, Y., & Xu, A. (2022). A novel hybrid model for six main pollutant concentrations forecasting based on improved LSTM neural networks. *Scientific Reports, 12*(1), 14434. https://doi.org/10.1038/s41598-022-17754-3

37. Yang, S., Yang, H., Li, N., & Ding, Z. (2023). Short-term prediction of 80–88 km wind speed in near space based on VMD–PSO–LSTM. *Atmosphere, 14*(2), 315. https://doi.org/10.3390/atmos14020315
38. Zhang, G., Hamzehkolaei, N. S., Rashnoozadeh, H., Band, S. S., & Mosavi, A. (2022). Reliability assessment of compressive and splitting tensile strength prediction of Roller compacted concrete pavement: Introducing MARS-GOA-MCS. *International Journal of Pavement Engineering, 23*(14), 5030–5047. https://doi.org/10.1080/10298436.2021.1990920
39. Zhang, J.-R., Zhang, J., Lok, T.-M., & Lyu, M. R. (2007). A hybrid particle swarm optimization–back-propagation algorithm for feedforward neural network training. *Applied Mathematics and Computation, 185*(2), 1026–1037. https://doi.org/10.1016/j.amc.2006.07.025
40. Yang, X., Maihemuti, B., Simayi, Z., Saydi, M., & Na, L. (2022). Prediction of glacially derived runoff in the Muzati River watershed based on the PSO-LSTM model. *Water, 14*(13), 2018. https://doi.org/10.3390/w14132018

Innovative Light-Emitting Concrete: Enhancing Energy Efficiency and Aesthetic Appeal in Sustainable Infrastructure

Gaurav Bharti, Parikshit Hurukadli, B. Bharath, Jyothi Chandrakantha Nagavi, Pushpendra Kumar Sharma, and Bishnu Kant Shukla

1 Introduction

One of the most significant developments in the field of construction materials is the introduction of translucent concrete, which is sometimes referred to as light-transmitting concrete. Through the incorporation of optical components, typically optical fibers, this concrete-based material possesses light-transmissive capabilities. These features enable light to be transmitted from one side of the concrete to the other side of the material. It is possible to obtain the distinctive light-conducting property of translucent concrete by inserting fibers that run the full length of the structure made of concrete. The light is directed through the material by these fibers, which results in the formation of different light patterns on the surface that is opposite to the source of the light and the casting of shadows in the form of silhouettes. Because of its exceptional quality, transparent concrete is a tempting choice for modern architecture, notably for use in facades and interior wall cladding, as well as for use in a variety of creative items. As a consequence of the production process, which entails blending fine-grain concrete with a tiny proportion of light-conducting components, the end product is a material that not only preserves the structural integrity of traditional concrete but also offers aesthetic and practical benefits through the transmission of light.

G. Bharti · P. Hurukadli · B. Bharath · J. C. Nagavi
Department of Civil Engineering, JSS Academy of Technical Education, Noida, India

P. K. Sharma
School of Civil Engineering, Lovely Professional University, Phagwara, India

B. K. Shukla (✉)
Department of Civil Engineering, JSS Academy of Technical Education, Noida, India

School of Civil Engineering, Lovely Professional University, Phagwara, India

S. Pathak et al. (eds.), *Intelligent Infrastructure and Smart Materials*,
https://doi.org/10.1007/978-3-031-92421-7_4

The concepts of nano-optics, which investigate the behavior of light on a small scale, are the foundation upon which the creation of transparent concrete is based. The optical fibers that make up transparent concrete are uniformly dispersed throughout the matrix of the concrete. Optical fibers generally account for around 4% of the total weight of the material. The numerical aperture of these fibers is rather high, which enables them to effectively absorb and transmit light. Significant light transmission is achieved despite the inherent light losses that are caused by fiber bends and surface roughness. This is made possible by the perfect placement of optical fibers within the concrete, which guarantees that light is transported across the material. Concrete is transformed from a material that is solely used for structural purposes into one that may also contribute to the illumination of interior spaces when optical fibers are included into the material. Because of its dual use, transparent concrete has become a popular choice in the construction of energy-efficient buildings. This is because it has the ability to considerably reduce the reliance on artificial lighting, which in turn results in a reduction in both energy consumption and emissions of greenhouse gases.

For the purpose of achieving sustainable infrastructure development, energy conservation is becoming an increasingly important problem. The consumption of energy by buildings accounts for around 34% of the overall energy consumption in the globe, with artificial lighting alone accounting for approximately 19% of the total power provided. There has been an increase in the amount of greenhouse gas emissions that are linked to the generation of electricity as a result of the rising demand for electric illumination, which has been driven by population expansion, urbanization, and the development of high-rise structures. Efforts have been made to reduce these negative effects on the environment, which include the creation of lighting systems that are more energy-efficient, the improvement of lighting designs, and the development of advanced lighting control technology. The utilization of natural light to brighten the interiors of buildings is made possible through the utilization of translucent concrete, which provides an innovative approach to further reduce energy usage as shown in Fig. 1. When transparent concrete is utilized as a structural facade or architectural wall, it enables natural sunshine to penetrate deep into buildings. This reduces the need for artificial lighting, which in turn contributes to a reduction in energy consumption and carbon footprints. This quality establishes transparent concrete as a crucial material in the development of green building techniques, thus lending assistance to the worldwide push toward construction methods that are more sustainable and kind to the environment.

Fig. 1 Glimpse of translucent concrete

2 Translucent Concrete: A Sustainable Solution

Energy conservation is an essential and growing concern for the development of sustainable infrastructure. The building sector represents about 34% of the global energy demand. Artificial lighting alone consumes approximately 19% of the total electricity delivered worldwide. This demand is continuously increasing due to the rising population, urbanization, and construction of high-rise buildings. The US Department of Energy reports that lighting accounts for 15% of global electricity consumption and 5% of worldwide greenhouse gas emissions. To address this, extensive efforts have been made to reduce energy consumption in lighting. These include the creation of energy-efficient lighting devices, enhancement of lighting design, and the use of advanced lighting control systems. These strategies aim to reduce the environmental impact of lighting while satisfying the escalating energy needs of modern infrastructure. It enhances architectural aesthetics with unique light-transmitting properties, suitable for both functional and artistic applications. The material offers privacy while maintaining transparency, making it ideal for interior partitions and building facades. Additionally, it supports energy efficiency by harnessing daylight effectively and meets structural durability standards typical of concrete. Despite its benefits, translucent concrete can be costly due to specialized materials and manufacturing processes. Installation complexity is another consideration, requiring careful handling of optical fibers and specialized expertise. Maintenance needs may be specific to preserve optical and structural integrity over time. Design options may also be somewhat limited compared to other materials, and variations in light diffusion could affect uniformity of illumination across surfaces. These factors should be weighed against the benefits when considering its use in construction and design projects.

3 Literature Review

The concept of translucent concrete has emerged as a revolutionary development in the field of construction materials, combining the structural benefits of traditional concrete with the aesthetic and functional advantages of light transmission. This literature review synthesizes research on the development, properties, and applications of translucent concrete, highlighting key studies and findings from various scholars and practitioners.

Early studies by Paul and Dutta (2013) laid the foundation for understanding the fundamental properties and potential applications of translucent concrete in architecture. Their research highlighted the material's aesthetic benefits, particularly in enhancing natural light usage within buildings [1]. This initial work was expanded upon by Ahuja et al. (2014), who utilized computational modeling to simulate the behavior of translucent concrete panels, providing insights into their structural performance under various conditions. These studies collectively underscored the importance of integrating optical fibers within the concrete matrix to achieve the desired translucency while ensuring the material's durability and strength [2]. Further experimental investigations into translucent concrete were conducted by Pradheepa and Krishnamoorthi (2015), who explored the material's light transmission capabilities through controlled laboratory experiments [3]. Their findings confirmed that the inclusion of optical fibers significantly enhances light transmittance, making the concrete suitable for various architectural applications. Additionally, Rahimi (2013) investigated the use of fiber-optic sensors in ferro concrete structures, demonstrating the potential for real-time monitoring of structural health [4]. This approach was seen as a valuable addition to the material's functionality, offering both aesthetic and practical benefits. Kashiyani et al. (2013) further contributed to this field by studying transparent concrete as a novel architectural material, emphasizing its potential to revolutionize construction practices through enhanced light utilization and energy efficiency. The integration of optical fibers into the concrete matrix is a critical aspect that influences the performance of translucent concrete [5]. Shukla et al. (2014) conducted experimental work focusing on light-transmitting concrete, highlighting the effectiveness of various types of optical fibers in achieving optimal light transmission [6]. Their research provided a comparative analysis of different fiber materials and their impact on the overall performance of the concrete. Similarly, Ansari (2007) explored the practical implementation of optical fiber sensors in civil structural health monitoring, reinforcing the dual functionality of translucent concrete as both a building material and a medium for structural health monitoring [7]. The incorporation of such sensors can significantly enhance the safety and longevity of structures by enabling continuous monitoring and early detection of potential issues. The study by Salih et al. (2014) examined the effects of plastic optical fibers on the properties of translucent concrete. Their research indicated that the type and arrangement of optical fibers play a crucial role in determining the mechanical properties and light transmittance of the concrete [8]. He et al. (2011) expanded on this by developing smart transparent concrete products and assessing their performance. They focused on the integration of advanced materials and technologies to create a more efficient and functional product [9]. The research by Zhou et al. (2009) on plastic optical fiber-based smart transparent concrete further reinforced these findings, showcasing the potential for innovative applications in intelligent building systems [10].

The practical applications of translucent concrete extend beyond its aesthetic appeal to include significant functional advantages in modern construction. Momin et al. (2014) investigated the light transmittance properties of concrete using optical fibers and glass rods, highlighting the versatility of the material in various architectural contexts. Their work demonstrated that translucent concrete could be effectively utilized in facades, interior walls, and other structural elements to enhance natural lighting and reduce energy consumption [11]. The combination of aesthetic and functional benefits makes translucent concrete an attractive option for sustainable building practices. The ongoing research in this field suggests a promising future for the development and application of translucent concrete. Future studies are expected to focus on optimizing the material's performance through advanced modeling techniques and experimental validation. Innovations in optical fiber technology and the integration of smart sensing capabilities are likely to drive further advancements, making translucent concrete a viable option for a wide range of construction applications. The collective body of work reviewed here provides a solid foundation for future research, highlighting the material's potential to transform the construction industry by offering a sustainable, aesthetically pleasing, and functionally versatile building material. The areas of architecture and civil engineering have shown a large amount of interest in the revolutionary material known as translucent concrete. This material combines the strength of concrete with the light-transmitting qualities of optical fibers. Drawing from a wide variety of sources, this literature review investigates the creation of transparent concrete panels, as well as their applications and computer modeling.

Ahuja et al. (2015) present a thorough computational model for transparent concrete panels [2]. Their work, which was published in the Journal of Architectural Engineering, provides insights into the structural behavior and light transmission capabilities of these panels. As a result, it serves as a fundamental reference for later research. The accuracy with which the model predicts the performance of transparent concrete under a variety of loading circumstances highlights the potential for practical applications of the model. The fundamentals of optical network design and execution are discussed in Alwayn (2004). These ideas are essential for comprehending the function that optical fibers play in translucent concrete [12]. The principles that are described in this book by Cisco Press are applicable to the design and optimization of light-transmitting materials, despite the fact that the primary focus of the book is on telecommunications. In order to guarantee effective light transmission and low signal loss, the incorporation of optical fibers into concrete is largely dependent on the principles of optical network architecture. The media coverage, such as that provided by Bates (2010), which highlights the aesthetic and practical benefits of transparent cement, demonstrates that the general public is fascinated by translucent concrete [13]. The material has the potential to revolutionize interior lighting and architectural design, as shown in this story from the Daily Mail, which highlights real-world applications of the material. This kind of coverage helps bridge the gap between academic research and public understanding, which in turn encourages a broader acceptance and implementation of novel construction materials.

According to Chaves (2008), nonimaging optics is a field that has a substantial influence on the development of transparent concrete. This article gives an introduction to the area [14]. The field of nonimaging optics is concerned with the effective transfer of light without the formation of an image. This is a key aspect of optimizing the light-transmitting qualities of optical fibers that are implanted in concrete. This book, which was published by CRC Press, provides both theoretical and practical insights that are directly applicable to the design of transparent concrete panels. Chen et al. (2012) investigate the solar heat gain coefficient of semitransparent photovoltaic modules. Their findings provide useful insights that may be utilized in the designing of translucent concrete that is efficient in terms of energy consumption [15]. Their study, which was published in the journal *Energy and Buildings*, includes the use of indoor calorimetric hot box and solar simulator techniques for the purpose of measuring solar heat gain. These approaches may be modified to evaluate the thermal performance of transparent concrete. This study demonstrates the dual functioning of transparent concrete as a material that is both structurally sound and efficient in terms of energy consumption. In 2004, the Commissioner of Building Control published rules on thermal energy transfer values for buildings [16]. These criteria are vital for evaluating the performance of transparent concrete since they offer standards for thermal energy transfer. In order to guarantee that transparent concrete satisfies legal requirements, the Building and Construction Authority has published these rules, which provide benchmarks for thermal insulation and energy efficiency. Increasing the practicality of transparent concrete as a sustainable construction material may be accomplished by the incorporation of these recommendations into the design and testing of the material. In 2002, Edwards and Torcellini conducted a literature evaluation on the effects of natural light on building occupants [17]. This study was released by the National Renewable Energy Laboratory. The results of their study highlight the positive effects that natural light has on both health and productivity, highlighting the significance of materials such as transparent concrete that increase natural illumination. Through the findings of this study, a persuasive argument is presented in favor of introducing transparent concrete into building designs in order to enhance the well-being of occupants and improve energy efficiency. Galesemann and Castilone (2002) investigate the mechanical reliability of optical fibers in bending, which is an essential component in determining the longevity of transparent concrete [18]. Because sustaining fiber integrity under stress is essential for the long-term performance of transparent concrete panels, their white paper for Corning illustrates the problems that must be overcome as well as the solutions that may be implemented. It is helpful to have an understanding of the mechanical behavior of optical fibers in order to optimize their incorporation into concrete. This helps to ensure that the structure is stable and that light transmission is dependable. Goff (2002) provides a thorough reference book on fiber optics, which is essential for the creation of transparent concrete since it provides core information [19]. The concepts, technology, and applications of fiber optics are discussed in this book, which was published by Focal Press. These aspects are directly important to the design and optimization of light-transmitting concrete. The book written by Goff is an indispensable resource for engineers and architects who work with transparent concrete.

Prentice Hall of India is the publisher of the textbook that Gupta (2005) presented, which is about optical fiber communication and the uses of this technology [20]. This study dives into the technical features of fiber optics and provides in-depth descriptions of how they may be integrated into a variety of materials, including concrete. It is a helpful reference for both scholars and practitioners since the textbook written by Gupta gives the theoretical underpinnings that are important for developing light-transmitting concrete structures that are efficient. Rajak et al. (2024) studied the creation and performance of smart transparent concrete. Specifically, they focus on the performance of the concrete. Their research investigates the possibility of combining optical fibers with intelligent materials in order to produce tangible buildings that are both responsive and adaptable [21]. The findings of this study indicate the possibility for merging transparency with intelligent technology in order to improve the utility and adaptability of building materials. Hipstomp (2010) provides a report on the unique usage of translucent concrete at the World Expo, highlighting the many functional possibilities of this material as well as its overall aesthetic appeal [22]. The adaptability of translucent concrete in a variety of architectural forms is demonstrated in this article from Core77, which also demonstrates the potential of this material to revolutionize contemporary building. This kind of media coverage is extremely important because it helps to popularize novel materials and encourages their use in building projects that are considered to be mainstream. In their book, which was released by Elsevier Academic Press in 2009, Hui and O'Sullivan compiled a collection of approaches for measuring fiber-optic connections. When it comes to evaluating the quality and performance of optical fibers that are implanted in concrete, these procedures are very necessary [23]. In order to guarantee that the optical fibers work as intended, accurate measurement and analysis are required. This ensures that the optical fibers provide constant light transmission and improve the overall performance of transparent concrete. A variety of industries, including building, can benefit from the utilization of optical fibers, and this site provides both practical insights and technical standards for their implementation [24]. Enhancing the integration of optical fibers into transparent concrete may be accomplished through a better understanding of industry standards and best practices, which in turn ensures the material's longevity and performance. Kahsay et al. (2019) performed a numerical analysis of the convective heat transfer coefficient for building facades. The results of this research are published in the Journal of Building Physics [25]. The significance of thermal performance in construction materials is brought to light by their results, which may be directly applied to transparent concrete. It is possible for transparent concrete to significantly contribute to the development of more energy-efficient building designs by optimizing its thermal qualities.

In an article that was taken from Concrete Decor, Klemenc (2011) examines the inventive usage of twenty-first-century concrete, including types that are translucent [26]. This work highlights the innovative applications of transparent concrete in contemporary architecture, demonstrating the material's capacity to reimagine the aesthetics and functionality of buildings. In order to better explain the practical benefits and potential of employing transparent concrete in contemporary architecture, it is helpful to highlight instances from outside the actual world. In his book that was released by Prentice-Hall in 1982, Lacy offers a historical viewpoint on the field of fiber optics. For the purpose of contextualizing its present uses in materials such as transparent concrete, it is helpful to have an understanding of the history of fiber optics [27]. The research conducted by Lacy provides instructive and insightful information on the technical developments that have made it possible to create novel building materials. A study that was published in the journal *Building* Research & Information in 1998 by Littlefair investigates the prediction of lighting energy usage under daylight-linked lighting controls [28].

Shreyas (2018) conducted an experimental study on translucent concrete, focusing on the reinforcement of optical fibers at various percentages. The materials used included 53-grade cement, sand (2.36 mm sieve passing), and optical fiber cables (200 μm diameter) [29]. The mix ratio for the concrete was 1:2 with a water-cement ratio of 0.45. The results showed that the compressive strength of light-transmitting concrete ranged between 20 and 23 N/mm^2, meeting the requirements for M20 grade concrete. The integration of optical fibers not only enhances the aesthetic value of buildings but also offers an eco-friendly way to utilize natural sunlight, reducing energy consumption. Omkar Kadamb (2014) emphasized the importance of energy efficiency and environmental sustainability in modern construction. His study highlighted the significant energy consumption in buildings, particularly in densely populated countries like India, where lighting accounts for a substantial portion of total electrical energy use [30]. Translucent concrete, incorporating optical fibers, offers a solution to reduce energy consumption by utilizing natural sunlight. The optical fibers embedded in the concrete form a matrix that allows light to pass through, maintaining the material's structural integrity. This technology can significantly reduce reliance on artificial lighting, thereby saving electricity. Additionally, fiber-optic sensors embedded in the concrete can monitor structural conditions, providing valuable data for maintenance and durability. Awasthi et al. (2020) reviewed the concept of Light Transmitting Concrete (LTC) as a smart and green building material. The review summarized research on LTC's strength, aesthetic appearance, and light-transmitting properties [31]. The materials used in transparent concrete production include optical fibers, Portland cement, fine aggregates, and coarse aggregates. The study detailed the manufacturing process, which involves casting concrete cubes with varying percentages of optical fibers. The compressive strength of these cubes was tested, and the light-transmitting properties were measured using sensors and light-detecting resistors. The results indicated a significant potential for reducing electricity consumption in buildings by using LTC. Wahane et al. (2022) explored the properties and applications of light-transmitting concrete in architecture and interior design. Studies by Soumyajit Paul and Avik Dutta found that the amount of light transmission in transparent concrete was directly proportional to the quantity of optical fiber used. Monika et al. experimented with different mix ratios and found that the compressive strength of light-transmitting concrete met the requirements for M20 grade [32]. Shah et al. investigated the effects of adding rice husk ash and steel fibers to translucent concrete, concluding that the optical fibers did not significantly affect compressive strength. Joao Manuel et al. explored the use of lightweight aggregates, such as expanded granulated cork, in producing translucent lightweight concrete. This research also described the process of developing transparent concrete by incorporating optical fibers into the concrete structure. The fibers are placed within a specially designed mold, extending through both ends of the concrete block [33]. This arrangement allows light to pass through without significant loss, enhancing the aesthetic appeal of the material. The study noted that incorporating materials like perlite into the mortar mixture could reduce the weight of the concrete blocks, although there was a slight reduction in strength. Nonetheless, the resulting strength was considered acceptable for certain applications.

Translucent concrete, an innovative material integrating optical fibers or resins into the concrete matrix, allows light to pass through while maintaining structural integrity. Research in this area has explored various aspects, including the mechanical properties, environmental impact, and potential applications of translucent concrete in sustainable construction practices. Studies have demonstrated that incorporating recycled and waste materials into the concrete mix can enhance sustainability and contribute to a circular economy [34]. Experimental investigations have shown that using absorbent materials as internal curing agents can improve the mechanical properties and durability of cement mortar, which is relevant for the development of translucent concrete [35]. Additionally, designing energy-efficient and environmentally friendly residential buildings for tropical climates can benefit from the integration of translucent concrete, providing natural lighting and reducing energy consumption [36]. The behavior of fiber-reinforced prestressed concrete and the inclusion of steel fibers in concrete beams have been critically reviewed, highlighting the potential of such materials to enhance the performance and aesthetic appeal of translucent concrete structures [37, 38, 39]. Further reviews have discussed the influence of replacing industrial wastes with fine aggregate in concrete, emphasizing the environmental benefits and structural performance improvements [40]. Studies on the behavior of fiber-reinforced polymer laminates in strengthening prestressed concrete beams have also contributed to understanding the applicability of these techniques in enhancing translucent concrete [39, 41]. These comprehensive investigations underline the significance of translucent concrete in modern construction, offering a blend of functionality and sustainability.

The findings of his research highlight the potential for materials that increase natural illumination, such as transparent concrete, to save energy. Reduced reliance on artificial illumination and increased energy efficiency are both attainable outcomes that may be achieved via the use of transparent concrete into building designs. The literature that was studied highlights the multifarious potential of transparent concrete, including its aesthetic and functional benefits, as well as its role in improving energy efficiency and the well-being of occupants. An important step forward in the development of building materials is represented by the incorporation of optical fibers into concrete matrices, as described in the computational models and the actual implementations. Nevertheless, there are still a number of issues that require more investigation. These include the optimization of the mechanical dependability of optical fibers, the enhancement of the thermal characteristics of transparent concrete, and the investigation of the incorporation of intelligent technologies. The focus of future research should be on filling up these gaps, making use of recent developments in fiber optics and material science to create transparent concrete products that are even more effective and adaptable. It is possible for academics and practitioners to open up new avenues of possibility in environmentally responsible and aesthetically beautiful building design if they continue to investigate and improve upon this cutting-edge material.

4 Experimental Methodology

The initial step was preparing a mold in which the concrete can be poured and the fibers could be infused, and thus the required cube of the optical fiber diffused concrete is obtained as shown in Fig. 2.

4.1 The Preparation of Mold

The prototype requires a wooden mold. Within this mold, precise markings are made to match the dimensions of the cube, allowing for the use of perforated plates. The diameter and number of holes in these plates are calculated based on the percentage of fiber intended for use.

The mold is a container that you pour a liquid or substance into. The liquid then becomes solid (sets) in the same shape as the container is. A wooden frame is worked as a mold at the central workshop of JSSATEN under the guidance of the workshop attendants and two thermocol sheets attached at the base.

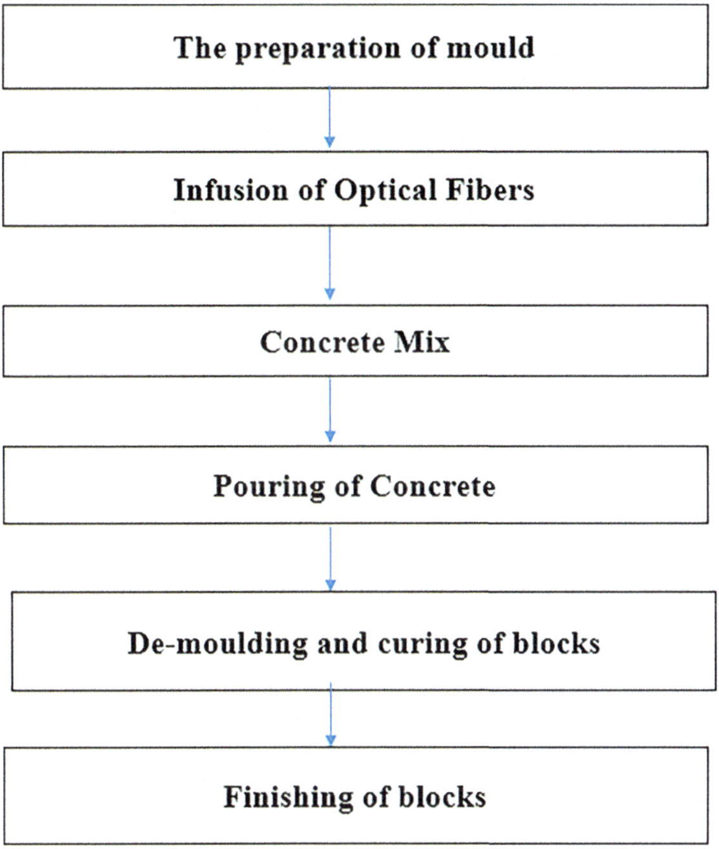

Fig. 2 Flowchart showing methodology

The mold is constructed to handle excess concrete, which helps to reduce material waste when disposing of unused concrete at the end of the day's pour. The mold dimensions are 22 cm in length, 18 cm in width, and 5 cm in depth, as depicted in Fig. 3. Each side of the mold is fastened using nails to ensure stability.

4.2 *Infusion of Optical Fibers*

Optical fibers are cut into 7 cm length each and with the help of adhesive are inserted into the thermocol sheet and erected to obtain uniformly spaced fiber length and the sheet is then placed into the mold. Some length of optical fiber, approximately 1 cm, is above the mold so that the free end is outside of the block. Cutting and infusion of optical fibers are illustrated in Figs. 4, 5, 6 and 7. These fibers are slender, cylindrical strands made of glass or other transparent dielectric materials, comparable in

Fig. 3 Mold for concrete

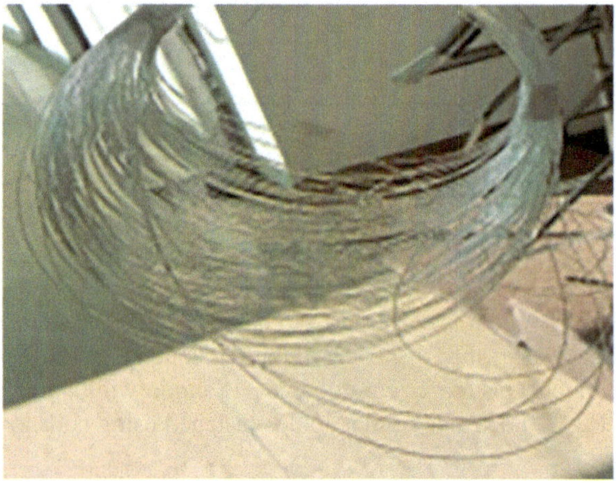

Fig. 4 Full-length optical fibers cut

thickness to a human hair. The primary goal of incorporating optical fibers is to harness sunlight as a light source, thereby reducing the power consumption needed for illumination. Additionally, optical fibers are utilized to monitor the stress within structures and enhance the aesthetic appeal of buildings when used in concrete for architectural purposes.

Fig. 5 7 cm pieces of optical fibers

Fig. 6 Embedded optical fibers

Fig. 7 Optical fiber and mold

Table 1 Concrete mix design component

Cement	500 g
Sand	1500 g
Water	230 mL

4.3 Concrete Mix: Specifications

Table 1 presents the quantities of materials used in a typical mix design for concrete, specifically for the production of a small batch or a test sample. The materials listed are cement, sand, and water, which are the primary components in concrete mix design, making the standard 1:3 ratio mortar.

4.4 Pouring of Concrete

The concrete mix hence prepared is poured into the mold with the help of a shovel, and the concrete is compacted with vibration as shown in Fig. 8.

Fig. 8 Pouring of concrete

- The formwork or the surface set to receive the fresh concrete must be thoroughly cleaned, adequately prepared, and well watered to ensure that the concrete is deposited as close as possible to its final position.
- When depositing concrete, it is essential to place large quantities at once. Failing to do so may cause the concrete to flow along the formwork, leading to a nonuniform composition. To maintain uniformity, small portions of concrete should be poured and spread evenly.
- Concrete should be dropped vertically from an appropriate height. When laying concrete vertically, it is crucial to use a stiff mix to prevent the concrete from bleeding through cracks in the formwork. Bleeding refers to the diffusion or leakage of concrete through the formwork.

The concrete mix was placed vertically from a reasonable height using stiff mix, in order to avoid bleeding of concrete.

Concrete should be deposited in horizontal layers approximately 150 mm high. For mass concrete applications, these layers can be between 400 and 500 mm in height. To prevent the formation of laitance, which is the accumulation of excess water in the upper layers, shallow layers with a stiff mix should be used, or dry batches of concrete should be added to absorb the excess water.

- Whenever possible, concrete should be placed in a single thickness. For deep sections, it is essential to place the concrete in successive horizontal layers, ensuring proper bonding between the fibers and the concrete. This method ensures that the concrete achieves a strong bond by placing it in successive horizontal layers of single thickness, promoting proper bonding between the layers.

4.5 Demolding and Curing of Blocks

After 24 h of casting, the demolding of the block was done. Demolding should be done under proper supervision and carefully by using appropriate tools. After that, the proper curing of block was done by the water bath method. The water that is used for curing should be away from alkali and chlorine. The curing process involves positioning the blocks in such a way that water can be poured into recesses on one side of each block, allowing the water to reach an overflowing point. This method ensures that the blocks remain in a consistently moist condition throughout the curing period.

4.6 Finishing of Blocks

The cutting of extra fiber was done by using a hot knife and another sharp object. After 28 days of curing, the finishing of block is done by using sandpaper as shown in Fig. 9.

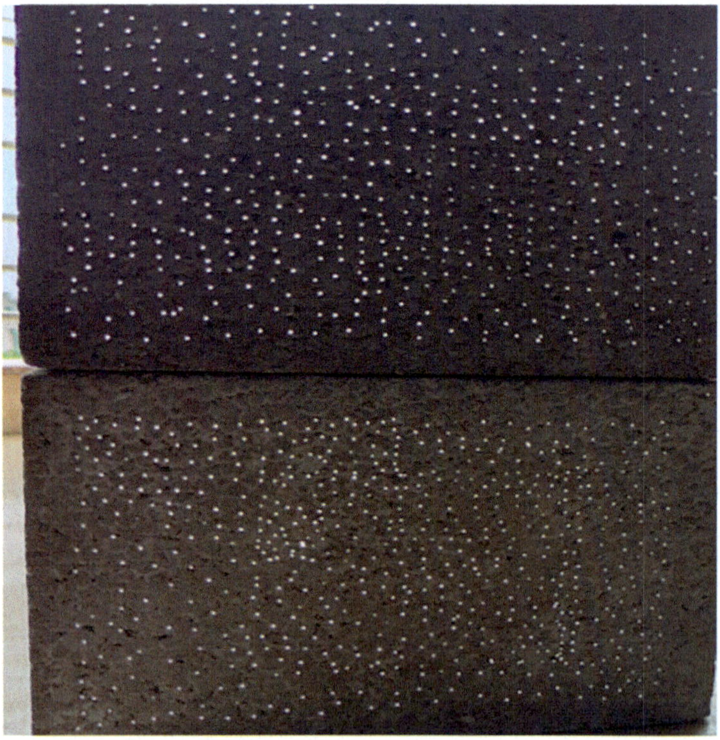

Fig. 9 Translucent concrete after finishing

5 Tests Conducted

To see the difference between traditional concrete and translucent concrete, a compression test was conducted and it was seen that the effect of including the optical fiber in compression strength was nearly the same.

For the 7-day compressive quality, ordinary concrete tests show values of 14.0, 16.0, and 15.5 MPa (Table 2), which are comparable to the 14.5 and 15.4 N/mm^2 obtained for translucent concrete with 4% and 5% optical fiber mixture (Table 3), individually. This demonstrates that the consideration of optical strands in translucent concrete does not altogether compromise the early quality advancement compared to typical concrete as shown in Fig. 10. Be that as it may, at 28 days, the compressive quality of ordinary concrete tests ranges from 23.0 to 25.0 MPa. In

Table 2 Compressive strength values for normal concrete

Serial no.	7 Days (N/mm^2)	28 Days (N/mm^2)
1	14	23
2	16	24.4
3	15.5	25

Table 3 Compressive strength values for translucent concrete

% of optical fiber infused (%)	7 Days (N/mm^2)	28 Days (N/mm^2)
4%	14.5	24.4
5%	15.4	23.9

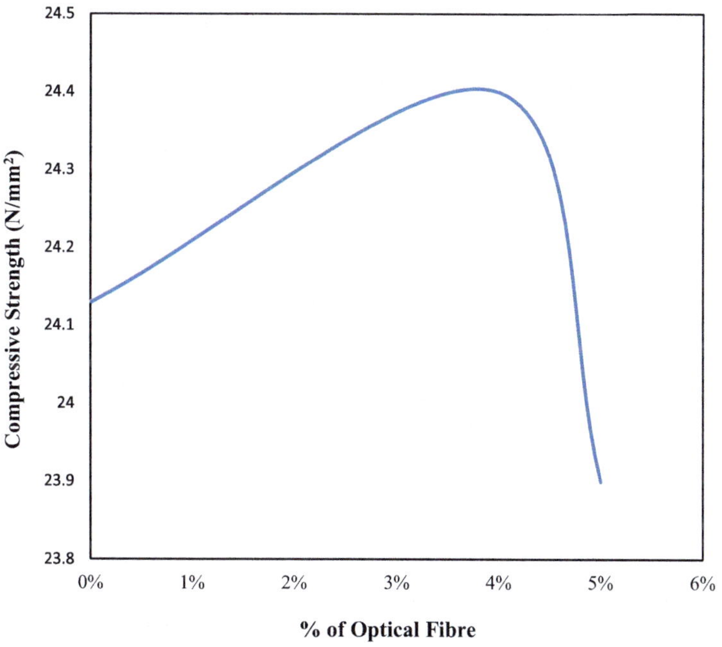

Fig. 10 Compressive strength trend for translucent concrete

contrast, translucent concrete with 4% and 5% optical fiber implantation shows marginally lower qualities of 24.4 and 23.9 N/mm^2, individually. In spite of the fact that the lessening is not considerable, it proposes that the implantation of optical filaments may marginally influence the long-term compressive quality of the concrete.

In outline, translucent concrete with optical fiber mixture illustrates comparative early-age quality as typical concrete, with a slight decrease in compressive quality at 28 days. This demonstrates that whereas optical strands improve the tasteful and utilitarian properties of concrete by empowering light transmission, they have a negligible effect on the material's generally compressive quality.

6 Conclusion

Light-emitting concrete (translucent concrete) represents a significant advancement in sustainable construction, offering superior light-guiding capabilities and aesthetic appeal. The study identifies the optimal volume ratio of optical fibers as crucial for maximizing light transmission while preserving compressive strength. TC, despite the inclusion of optical fibers, maintains strength characteristics comparable to conventional concrete, demonstrating its practical applicability. Initial costs, primarily due to plastic fibers, can be mitigated by the material's long-term energy savings and aesthetic benefits. Key findings include:

- Achieving approximately 4% light transmissibility with optimal fiber content.
- Enhanced light transmissibility with reduced structural thickness.
- Comparable compressive strength to traditional concrete, with minor reductions.
- Significant contributions to the aesthetic and energy efficiency of buildings, supporting green construction initiatives.
- Potential as a window replacement, combining decorative and structural functions.
- Health benefits from increased natural light exposure, addressing vitamin D deficiencies.

Translucent concrete's integration into diverse construction applications aims to optimize its properties, promoting the development of sustainable and smart infrastructure. Future research should focus on further enhancing TC's performance and exploring its full potential in various building contexts.

References

1. Paul, S., & Dutta, A. (2013). Translucent concrete. *International Journal of Scientific and Research Publications, 3*(10), 1–10.
2. Ahuja, A., Mosalam, K. M., & Zohdi, T. I. (2015). Computational modeling of translucent concrete panels. *Journal of Architectural Engineering, 21*(2), B4014008.

3. Pradheepa, R., & Krishnamoorthi, S. (2015). An experimental study on translucent concrete. *International Journal for Scientific Research & Development, 3*(03), 174–177.
4. Rahimi, S. (2013). Investigating the use of fibre optic sensors in ferro concrete structures. *Technical Journal of Engineering and Applied Sciences*, 2796–2798.
5. Kashiyani, B. K., Raina, V., Pitroda, J., & Shah, B. K. (2013). A study on transparent concrete: A novel architectural material to explore construction sector. *International Journal of Engineering and Innovative Technology, 2*(8), 83–87.
6. Shukla, A., Poriya, T., & Zala, J. (2014). An experimental work on light transmitting concrete. *International Journal of Advance Engineering and Research Development (IJAERD), 1*(5).
7. Ansari, F. (2007). Practical implementation of optical fiber sensors in civil structural health monitoring. *Journal of Intelligent Material Systems and Structures, 18*(8), 879–889.
8. Salih, S. A., Joni, H. H., & Mohamed, S. A. (2014). Effect of Plastic Optical Fibre on Some Properties of Translucent Concrete. *Engineering & Technology Journal, 32*(A12), 2846–2861.
9. He, J., Zhou, Z., Ou, J., & Huang, M. (2011, July). Study on smart transparent concrete product and its performances. In *Proceedings of the 6th international workshop on advanced smart materials and smart structures technology*. Asian-Pacific Network of Centers for Research in Smart Structure Technology (ANCRISST), Harbin Institute of Technology.
10. Zhou, Z., Ou, G., Hang, Y., Chen, G., & Ou, J. (2009, April). Research and development of plastic optical fiber based smart transparent concrete. In *Smart sensor phenomena, technology, networks, and systems 2009* (Vol. 7293, pp. 107–112). SPIE.
11. Momin, A., Kadiranaikar, R., Jagirdar, V., & Inamdar, A. (2014). Study on light transmittance of concrete using optical fibers and glass rods. In *Proceedings: International conference on advances in engineering & technology–2014*.
12. Alwayn, V. (2004). *Optical network design and implementation, ciscopress.com*. United State of America.
13. Bates, D. (2010). *Mortar this than meets the eye: The 'transparent' cement that lets daylight flood into a room*. Retrieved Nov, 24, 2012.
14. Chaves, J. (2008). *Introduction to nonimaging optics*. CRC Press.
15. Chen, F., Wittkopf, S. K., Ng, P. K., & Du, H. (2012). Solar heat gain coefficient measurement of semi-transparent photovoltaic modules with indoor calorimetric hot box and solar simulator. *Energy and Buildings, 53*, 74–84.
16. Commissioner of Building Control. (2004). *Guidelines on thermal energy transfer values for buildings*. Building and Construction Authority.
17. Edwards, L., & Torcellini, P. (2002). Literature review of the effects of natural light on building occupants.
18. Glaesemann, G. S., & Castilone, R. J. (2002). *The mechanical reliability of Corning® optical fiber in bending*. Γ Corning White Paper WP3690. Γ.
19. Goff, D. R. (Ed.). (2002). *Fiber optic reference guide* (3rd ed.). Focal Press.
20. Gupta, S. C. (2018). *Textbook on optical fiber communication and its applications*. PHI Learning Pvt.Ltd.
21. Rajak, M. K., Kumar, R., Biswas, R., & Singh, S. H. (2024). Light transmitting concrete: A systematic review of research trends. *Recent Advances in Material, Manufacturing, and Machine Learning*, 219–226.
22. Hipstomp. (2010). *Transparent concrete and cement at World Expo*. Retrieved Dec 25, 2011, from http://www.core77.com/blog/materials/transparent_concrete_and_cement_at_world_expo_16562.asp
23. Hui, R., & O'Sullivan, M. (2022). *Fiber-optic measurement techniques*. Academic Press.
24. Min, R., Hu, X., Pereira, L., Soares, M. S., Silva, L. C., Wang, G., et al. (2022). Polymer optical fiber for monitoring human physiological and body function: A comprehensive review on mechanisms, materials, and applications. *Optics & Laser Technology, 147*, 107626.
25. Kahsay, M. T., Bitsuamlak, G., & Tariku, F. (2019). Numerical analysis of convective heat transfer coefficient for building facades. *Journal of Building Physics, 42*(6), 727–749.
26. Klemenc, S. E. (2011). *Wave of the future: 21st century concrete on display*. Retrieved Nov, 17, 2011.

27. Lacy, E. (1982). *Fiber optics Prentice-Hall* (p. 70). Englewood Cliffs.
28. Littlefair, P. J. (1998). Predicting lighting energy use under daylight linked lighting controls. *Building Research & Information, 26*(4), 208–222.
29. Shreyas, K. (2018). Transparent concrete. *International Journal of Engineering Technology Science and Research (IJETSR), 5*(3), 220–232.
30. Kadam, O. (2017). Translucent concrete. *IOSR Journal of Mechanical and Civil Engineering, 14*(03), 18–31.
31. Aswathi, R., Athira, B., Devika, M., & Arun, T. R. (2020). Transparent concrete as a green material for building. *International Journal of Scientific & Engineering Research, 11*(3), 144–147.
32. Wahane, A., Sahu, A. K., Jain, D., & Rao, R. S. (2022). Experimental study on translucent concrete. *International Journal for Research in Applied Science & Engineering Technology (IJRASET), 10*(1), 789–792.
33. Shah, D., Patel, H., & Vaghasia, N. (2018). A study on translucent concrete. *International Journal of Scientific Research, 7*(5), 49–51.
34. Shukla, B. K., Bharti, G., Sharma, P. K., Sharma, M., Rawat, S., Maurya, N., & Srivastav, Y. (2024). Sustainable construction practices with recycled and waste materials for a circular economy. *Asian Journal of Civil Engineering*, 1–22.
35. Hurukadli, P., Kumar, A., Shukla, B. K., & Bharti, G. (2023, September). Experimental study on the influence of available absorbent as internal curing agent on cement mortar. In *AIP conference proceedings* (Vol. 2800, No. 1). AIP Publishing.
36. Shukla, B. K., Parashar, B., Bharti, G., Sharma, P. K., & Kumar, A. (2023, September). A sustainable design of energy efficient and environment friendly residential building for tropical climate. In *AIP conference proceedings* (Vol. 2800, No. 1). AIP Publishing.
37. Hurukadli, P., Kumar, A., Shukla, B. K., & Bharti, G. (2023, September). A critical review on behavior of fiber reinforced pre stressed concrete. In *AIP conference proceedings* (Vol. 2800, No. 1). AIP Publishing.
38. Hurukadli, P., Bharti, G., Shukla, B. K., Garg, P., & Tripathi, A. (2023). Deflection of reinforced concrete beams: An analytical study on the impact of steel fibers. *Materials Today: Proceedings*.
39. Parikshit, N. H., Garg, P., Bharti, G., Tripathi, A., & Shukla, B. K. (2023 February). Analytical study on the effect of steel fiber in the deflection of pre-stressed concrete beams. In *IOP conference series: Earth and environmental science* (Vol. 1110, No. 1, p. 012027). IOP Publishing.
40. Varadharajan, S., Kirthanashri, S. V., Samim, M., Shukla, B. K., & Bharti, G. (2021, November). A review on influence of replacing industrial wastes with fine aggregate in concrete. In *Indian geotechnical and geoenvironmental engineering conference* (pp. 183–188). Springer Nature Singapore.
41. Hurukadli, P., Bharti, G., & Shukla, B. K. (2023). Behavior of fiber reinforced polymer laminates strengthening prestressed concrete beams. *Materials Today: Proceedings, 78*, 731–737.

Cost–Benefit of Different Insulation Alternatives for Indian Residential Buildings in the Warm and Humid Regions

Kishor S. Kulkarni, Garima Singh, and Ankita Paul

1 Introduction

Energy consumption from primary sources of energy at the world level for building construction and operation is attributed to 36% at 149 KJ, of which operational energy accounts for 85% and energy for construction takes up to 15%. In this, residential energy consumption adds to 22% for operational needs like lighting, cooling, heating, etc. The energy consumption rate in the building sector is gradually increasing due to urbanisation, industrial growth, and rapid population growth. This increased energy consumption has led to issues like increased greenhouse gas (GHG) emissions, fast depletion of fossil fuels, etc. The carbon emissions related to building operation and construction increased to 37% at the global level in 2020, of which 45% is from the residential sector directly or indirectly. Global initiatives to increase sustainability and efficiency are needed for mitigation [1].

India, the second-most populous nation in the world, is also the third-largest consumer of energy and emitter of CO_2. Its emphasis on urbanisation, which is projected to reach 40% by 2030, drives up demand for housing, infrastructure, and rural–urban migration [2]. India has pledged to double its urban building space by 2040 in response to the demands of urbanisation, with 70% of construction occurring in urban areas. Building construction is one of the major energy consumers in the country, accounting for 33% of total energy demand. Just residential buildings make up 24% of this total. The residential sector's heating and cooling costs are expected to increase three to five times over the next 10 years due to changing

K. S. Kulkarni (✉) · G. Singh
CSIR-Central Building Research Institute, CBRI Colony, Roorkee, Uttarakhand, India
e-mail: kishorsk@cbri.res.in

A. Paul
Vellore Institute of Technology, Katpadi, Vellore, Tamil Nadu, India

S. Pathak et al. (eds.), *Intelligent Infrastructure and Smart Materials*,
https://doi.org/10.1007/978-3-031-92421-7_5

conditions. By 2040, this might lead to about 244 million metric tonnes of CO_2 emissions from new construction, of which the residential sector would account for almost half [3]. Thus, it is seen that emissions from energy consumption are increasing every decade leading to environmental degradation. Increased consumption patterns, population growth, and the need to reduce GHG emissions highlight the importance of energy conservation and efficiency [4].

The sustainable development goals for India discuss energy efficiency and energy optimisation as important areas of research, with a focus on carbon reduction, improving building performance, and the use of renewable resources with decreased demand for fossil fuels as part of policies and acts like the ECBC, NAPCC, etc. [5, 6]. The integration of strategies for energy efficiency at the conceptual stage of passive design strategies and innovative construction material technologies in residential buildings like shading devices, natural ventilation systems, thermal insulation, building form and orientation, smart window systems, and building energy management systems resulted in a reduction of energy consumption by 50% and also decreased levels of carbon emissions [7–13]. It can be seen that the studies conducted previously have no major relevance to energy-saving building codes and bylaws, and economic constraints in terms of individual income, material cost, electricity cost, embodied energy at construction and operational phases, etc. have not been majorly considered for evaluating the energy consumption for human comfort [14–16].

Through different research studies, it was observed that external wall insulation materials (5.08%), which is followed by roof insulation (2.59%) and floor insulation (2.24%) when it comes to methods for ensuring thermal comfort and energy efficiency, achieve the best average energy consumption reduction. When combined, these insulation techniques perform better than natural ventilation (2.43%), green roofs (0.39%), and shading (2.86%) in passive strategies [17]. Thus, regarding energy-saving codes and previous studies, a major focus can be placed on insulation materials as a strategy for energy reduction in residential buildings. The focus should be placed on determining an appropriate insulation material with its thermal properties and establishing an ideal thickness for energy efficiency in buildings [18]. Based on their place of origin, substance composition, and easy access, insulation materials can be divided into three categories: advanced, cutting-edge, and conventional. Thermal parameters, including material density, specific heat capacity, non-steady-state thermal diffusivity, and steady-state thermal conductivity, are used to evaluate them. Because conventional materials are readily applicable, affordable, and have access to trained labour for handling, they are frequently utilised. The best thermal performance among widely used materials is found in polyisocyanurate (0.022 W/mK), polyurethane (0.025 W/mK), recycled glass fibre (0.030 W/mK), XPS (0.034 W/mK), and stone wool (0.040 W/mK). These materials are excellent insulation materials because they have a high volumetric heat capacity and a low thermal conductivity [19, 20]. Reducing operational energy demand in the residential sector can be achieved with great benefits by applying insulation materials at the proper thickness relative to the wall materials. The location of insulation material is another important factor influencing performance. Insulation is most effective when placed on the outside of walls in buildings with air conditioning; however, for systems that run on and off during the day, insulation placed on the inside of walls has a greater potential to reduce heat [21, 22]. While it is associated with an additional construction cost, external wall insulation can successfully lower the operational load for heating and cooling. For this reason, figuring out the ideal insulation thickness requires doing an economic evaluation or cost-benefit analysis. The goal of such an analysis should be to minimise long-term energy consumption costs while also lowering the project's initial cost [23]. The primary goal of this research is to identify which building envelope system is best adapted to India's warm and humid climate. Furthermore, the study compares the cost-effectiveness of different building envelope configurations while accounting for cost analysis and thermal comfort.

2 Methodology

The study aims to examine how building typology and climate-specific building envelope systems can be integrated within the context of India's diverse climates namely warm-humid climate, with an emphasis on using BIM modelling to assess the systems' appropriateness, affordability, and influence on thermal comfort. The study was conducted in three stages. Firstly, a preliminary literature review was conducted to understand the parameters that influence building design for energy efficiency. Secondly, a simulation study was conducted using a 3D Revit software model and energy modelling in Design Builder. Thirdly, to determine the ideal thickness of insulation materials and energy efficiency, a comparative analysis using the techno-economic method was conducted.

2.1 Study Area

The research was conducted in Kolkata, which is the metropolitan and capital city of the state of West Bengal, India, situated along the coast of the Hooghly River (22°34′03″N 88°22′12″E). The climate is considered warm-humid as per the Indian climate classification system [24]. The annual recorded low temperature reaches 19.1 °C and the highest temperature goes to 35.6 °C, while the monthly mean temperature ranges from 15 °C to 30 °C. Relative humidity throughout the year ranges from 38.8% to 99.2% with annual rainfall recorded as 1600 mm [25]. The population density as per the Kolkata Municipal Authority website is 24,000/km^2, which is one of the highest densities in the world and the seventh largest city in the country. The nation's metro centres of productivity and economic opportunity are its metropolitan areas. As such, they are ideal settings in which policymakers can introduce novel ideas and spearhead development projects [26]. Thus, the housing areas in such environments help in understanding the requirements of the present and future population.

2.2 The Building Model

A typical single-storey residential building was identified, which was in operation in Kolkata, West Bengal, with a plot area of 256 m^2 and a plinth area of 145 m^2. The plot has an entrance oriented towards the south, with openings in the east–west direction. Figure 1 presents the sun path of the model during the two different times: one in the forenoon and the other in the afternoon. The building consists of two bedroom spaces, two bathroom units, one dining room, and one hall with kitchen, store, balcony, and utility area. The building construction is loading bearing with a traditional hip roof due to heavy rainfall in a warm-humid climate and an additional foyer area in the entrance, as shown in Fig. 2.

The external walls are made of burnt brick with cement plaster of 277 mm thickness; the internal partition walls are also done with burnt brick with light cement plaster of 254 mm thickness. The casement windows with fixed clear glass are being used in the building along with wooden doors. The flooring is of ceramic tiles and the roofing is done with RCC, with an external finish of gypsum plaster, as shown in Table 1.

2.3 Insulated Case for Building

Insulation materials play a major role in the reduction of heat gain in a building envelope. For the city of Kolkata, various materials were identified based on a market study and data on materials used in similar climates. The materials used in the investigation are extruded polystyrene (XPS) and polyurethane foam (PUF). The wall section and roof slab were modelled using XPS and PUF materials of varying thickness, ranging from 20 to 80 mm for eight different model configurations with insulation materials applied to external walls, internal partition walls, and the roof

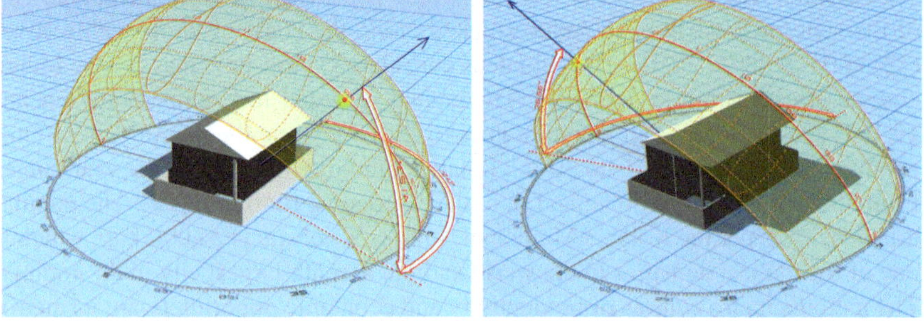

a) Sunpath at 9 am b) Sunpath at 3 pm

Fig. 1 The front view of the model prototype, oriented towards south for Kolkata city

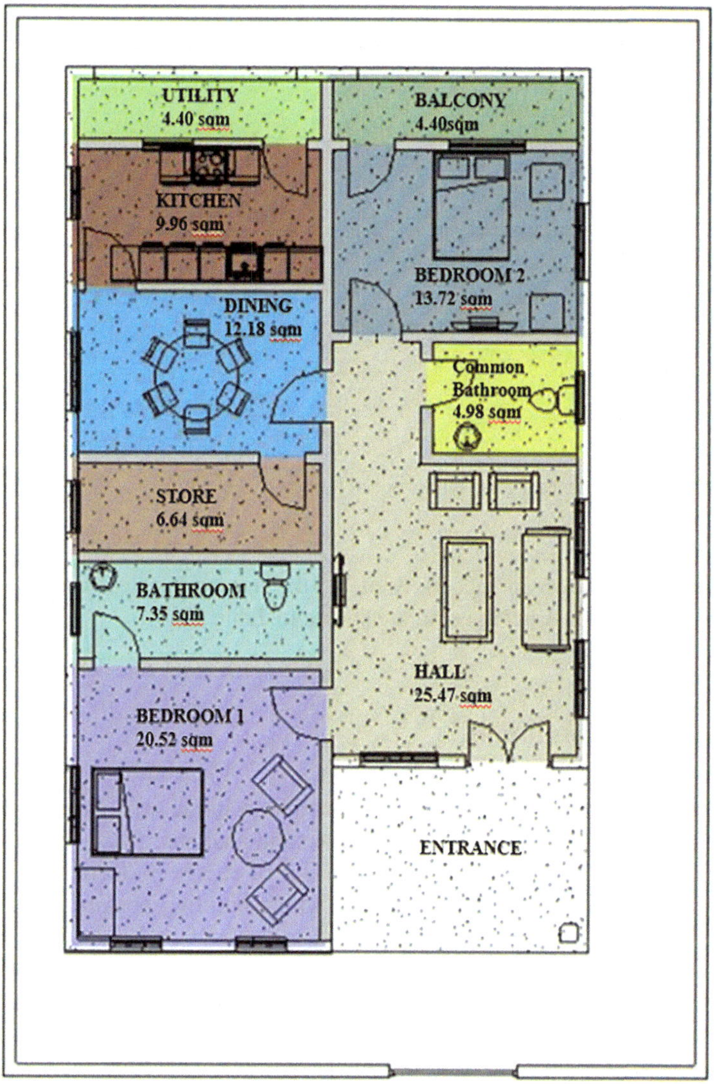

Fig. 2 Base-case single-storey residential building—floor plan

Table 1 Building envelope details for the base case

Envelope	Material layers (outer to inner)	Thickness (in mm)	U-value (W/m^2 K)
External wall	Cement plaster	15	1.78
	Burnt clay bricks	250	
	Cement plaster	12	
Internal wall	Cement plaster	12	1.28
	Burnt clay bricks	230	
	Cement plaster	12	
Roof	Ceiling tiles	20	1.52
	Cement plaster	12	
	Reinforced cement concrete slab	125	
	Gypsum plaster	13	
Flooring	Ceramic tiles	12	3.53
	Cement plaster	15	
	RCC slab	100	
Window	Generic clear glass with wooden frame	6	5.73

of the building (shown in Fig. 3). Some of the thermo-physical attributes of the materials used to construct the wall are shown in Table 2.

2.4 *Energy Simulation*

The base study model is a traditional construction in the urban area of Kolkata, where façade detailing impacts performance and analysis. Thus, for the simulation study, we require a user-friendly interface of energy performance software. The Bureau of Energy Efficiency for the whole building performance analysis suggests software like Energy Plus, Design Builder, eQUEST, IDA-ICE, IES-VE, Open Studio, etc. [27]. The Design Builder software was used for the study, as it is an advanced user interface of Energy Plus, the software works with the thermal heat balance method ASHRAE approved for each wall surface. The software has fewer limitations compared to other programs regarding the geometry of basements and central spaces. Energy Plus is a cutting-edge building performance simulation programme that serves as the foundation for Design Builder. It helps model the geometry, materials, and division of the building into discrete areas with different HVAC, lighting, and equipment. In addition, it can estimate energy use, carbon emissions, solar gains, and passive building performance, among other things. It can also simulate dynamic thermal models hourly [28–30].

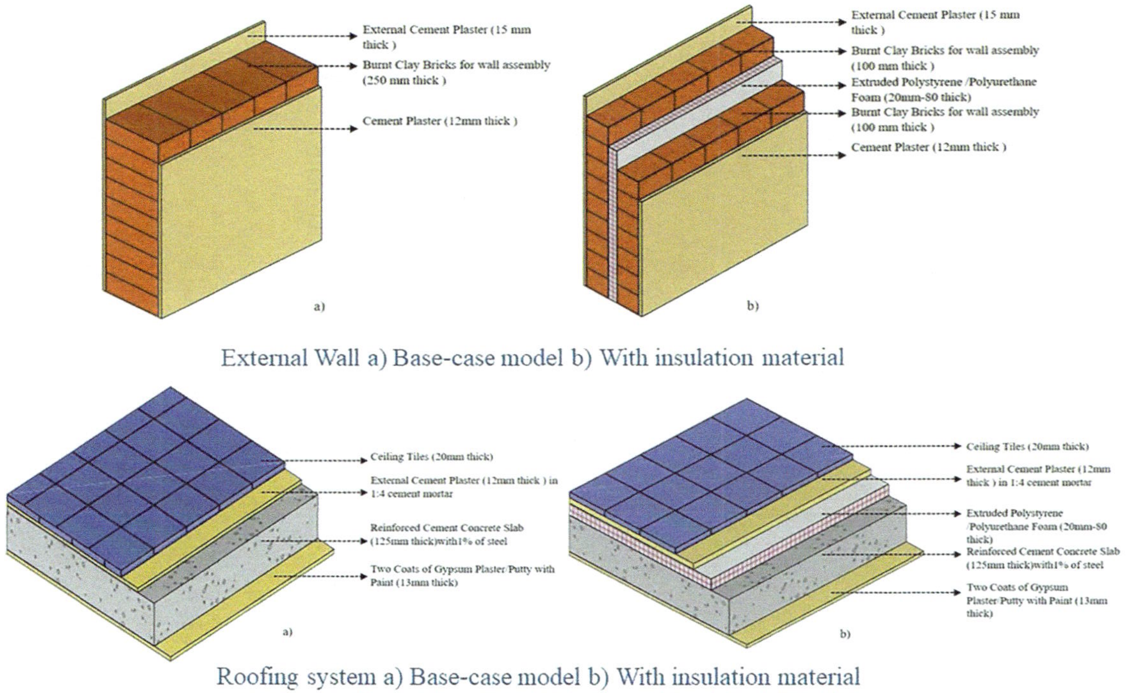

Fig. 3 The building sectional details for base-case and insulation materials. External wall: (**a**) base-case model and (**b**) with insulation material. Roofing system: (**a**) base-case model and (**b**) with insulation material

Table 2 Thermo-physical properties of Insulation Materials

Insulation material	Density ρ (kg/m^3)	Thermal conductivity k (W/m K)	Specific heat capacity c (kJ/kg K)
XPS	35	0.034	1.34
PUF	30	0.028	1.00

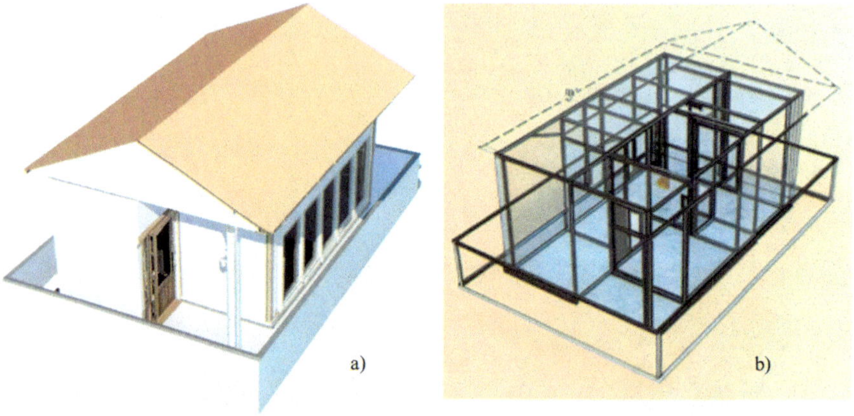

Fig. 4 3D model of building with Revit architecture 2021: (**a**) 3D model with site details and (**b**) 3D wireframe showing the casing of the internal structure

2.5 The Simulation Process

The base-case 3D model was created using Revit Architecture 2021, with input parameters including plot area, plinth area, number of floors, dimensions of the rooms, number of occupants, and other required data all being collected from the site and then proceeded to the modelling of this sample building in REVIT software. The base-case building materials were applied while creating the 3D model, as shown in Fig. 4.

A 3D model of the single-storey residential building was created and imported into Design Builder in gbXML format. Additional data were then incorporated for energy simulation. In Design Builder, the site location was provided so that the weather data can be automatically applied to the model. The Constant Volume DX system was considered as a cooling system for bedrooms and dining areas, with an indoor temperature of 25 °C; the rest of the house was in a natural ventilation system with openable windows.

To improve the accuracy of the simulation software results, any complex detailing was reduced for the base case, and the ground floor level was set at ±0.00. The influence of the environment was not considered in this simulation. To understand the thermal performance of the insulation materials, a comparative analysis was performed with eight different models using conventional and insulation materials as envelopes of buildings. The data used for the simulation are presented in Table 3. In this study, the evaluation is divided into three main categories: (a) validation of

Table 3 Design builder simulation input parameters

Characteristics	Details	Remarks
Plinth area	174 m^2	
Gross wall area	142 m^2	
Window-to-wall ratio	40%	Orientation-south facing structure
No. of occupants	5	24-h occupancy schedule is considered for 7 days a week
Lighting power density	4.09 W/m^2	
Application power density	3.81 W/m^2	
HVAC type	Constant volume DX	Cooling set point 25 °C

the simulation model, (b) building energy performance, and (c) techno-economic analysis.

3 Results and Discussions

3.1 Simulation Model Validation

The simulation model can be validated in various ways including analytical, theoretical, and actual technique methods. The actual technique method gives the best precise results, as it compares actual values with simulated data [31]. The percentage of percentage error method was used to access the simulated base-case model to understand how close the simulated model results are to the actual values at the site. The correct values of energy consumption in terms of electricity bills were collected from the owner of the building for the whole year. The average annual energy consumption was found to be 264 kWh. The simulated model gave an average annual energy consumption of around 284 kWh. The simulated base-case model's monthly energy consumption using conventional materials is shown in Fig. 5, along with the site's actual billed energy consumption for the residential building. The percentage of error in the values was determined using the following equation:

$$\%\text{Error} = \frac{|T - M|}{|T|} \times 100\% \quad (1)$$

where T = simulated value; M = experimental value [32].

The result of the base case shows that maximum energy consumption takes place during the month of April–June, which is the peak time to maintain a comfortable condition in the residential building. The percentage of error is 7.5%, which leads to more than 90% acceptable data. It is also important to note that the pattern of energy consumption closely resembles the energy consumption for the residential building that was billed. Thus, the simulation model gives a near-correct value for evaluation.

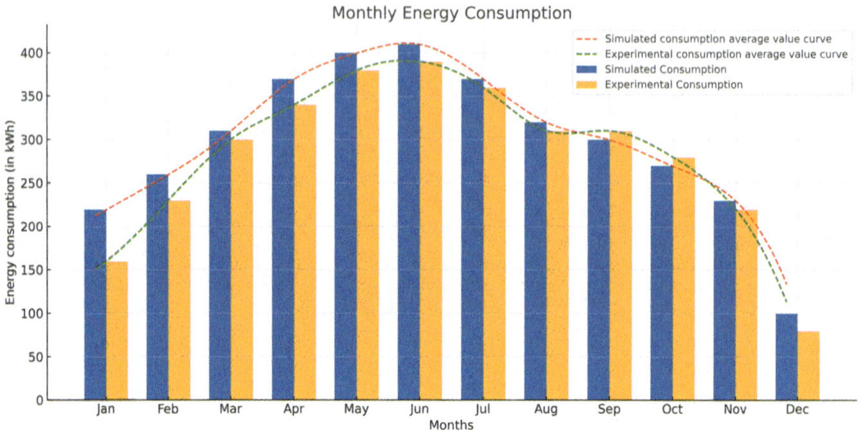

Fig. 5 Monthly energy consumption

To reduce energy consumption during the peak months of April–June, passive design strategies such as insulation to envelope and enhanced shading need to be incorporated or the active load like lighting and cooling needs to be decreased through the incorporation of energy-efficient rated electric lights and air-conditioning systems [33].

3.2 Building Energy Performance

Table 4 shows the energy performance of XPS and PUF based on the material thickness, along with their respective *U*-values for wall composition. The base case with conventional building material has 90 kWh/m^2 energy consumption. The results show that with the increased thickness of insulation material, the energy consumption decreases, suggesting that the thickness of 80 mm gives the best results for both XPS and PUF materials. It can also be seen that XPS has better energy performance than PUF for the same thickness. Thus, an ideal solution will be an 80 mm XPS insulation material application in a wall composite, which gives energy savings of 10.75%.

Determining the optimum thickness along with the optimum thickness cost is an important factor. The cost incurred includes the purchasing cost and installation cost of the insulation materials. Thus, it is necessary to perform an economic analysis to determine the optimum thickness of wall insulation to minimise energy and insulation costs [34].

Table 4 Energy consumed and percentage saved as per different thicknesses of insulation materials

		XPS		PUF	
Case structure	Insulation material thickness (in mm)	Energy consumed in kwh/m^2	Energy conserved in %	Energy consumed in kwh/m^2	Energy conserved in %
Base case	**0**	90	0	90	0
Case 1	**20**	86.22	4.2	86.49	3.9
Case 2	**40**	83.52	7.2	83.97	6.7
Case 3	**60**	81.27	9.7	81.72	9.2
Case 4	**80**	80.32	10.75	80.73	10.3

3.3 *Techno-Economic Analysis*

The optimum insulation thickness depends on the cost benefits of each insulation material over the lifetime depending on thermal conductivity and cost of insulation material. To reduce the GHG emissions and enhance the cost-effectiveness, the drawbacks and benefits of the materials used have to be analysed systematically. For this, a simple payback period (in years) can be calculated to propose material thickness alternatives for economic feasibility [35, 36]. The payback period is the amount of time needed for all of the savings accrued to cover a product's initial investment. The factors taken up here are the cost of insulation and the energy expenditure as per the simulated data obtained from each combination. The simple payback period [37] was calculated as follows:

$$Qce = \frac{Qbc - Qsc}{Qbc} \times 100$$

$$Cce = \left(Qbc - Qsc\right) * Et$$

$$PP = Cim / Cce \tag{2}$$

where Qce = percentage of conserved energy; Qbc = energy consumption for base-case building; Qsc = energy consumption of each insulated building envelope for different thicknesses; Cce = cost of conserved energy; Et = electricity tariff; Cim = construction cost of materials; PP = payback period for insulated materials.

The XPS and PUF unit costs in ₹/m^2 were collected from the local market, as stated in Table 5. The electricity tariff for the location was ₹7.31 per unit as per CESC Limited (Calcutta Electric Supply Corporation), Kolkata. The lifetime of the materials is assumed to be 30 years, and the cost of energy consumption for a total 30 years in the different simulated cases is calculated by the following equation:

$$\text{Total Energy Consumption} / \text{m}^2 = Qsc * Et * 30 \tag{3}$$

Table 5 The cost per unit of insulation materials

Case structure	Thickness of insulated material in mm	XPS			PUF		
		Energy cost in ₹/m²	Insulation cost in ₹/m² *(Cim)*	Total cost in ₹/m²	Energy cost in ₹/m²	Insulation cost in ₹/m² *(Cim)*	Total cost in ₹/m²
Base case	**0**	19,737	0	19,737	19,737	0	19,737
Case 1	**20**	18,909	210	19,118	18,967	180	19,147
Case 2	**40**	18,316	420	18,736	18,415	390	18,805
Case 3	**60**	17,823	620	18,443	17,921	590	18,511
Case 4	**80**	17,616	870	18,485	17,704	830	18,534

Table 6 Summary of the techno-economic analysis

Case structure	Thickness of insulated material in mm	XPS				PUF			
		Qsc in kwh/m²	*Qce* in %	*Cce* in ₹/m²	*PP* in years	*Qsc* in kwh/m²	*Qce* in %	*Cce* in ₹/m²	*PP* in years
Base case	0	90	0	0	0	90	0	0	0
Case 1	20	86.22	4.20	27.63	7.60	86.49	3.90	25.66	7.00
Case 2	40	83.52	7.20	47.37	8.90	83.97	6.70	44.08	8.80
Case 3	60	81.27	9.70	63.82	9.71	81.72	9.20	60.53	9.74
Case 4	80	80.32	10.75	70.72	12.3	80.73	10.30	67.76	12.2

The difference in energy consumption multiplied by the plinth area and the rate of electricity consumption per unit is used to compute the cost of conserved energy that a consumer saves by using insulation material [38]. Consequently, how much money saved annually by applying insulation during the preconstruction stage can be determined. The total energy consumption, percentage of energy conserved, cost of conserved energy, and payback period are shown in Table 6. In terms of cost savings for both insulation materials, a thickness of 60 mm is found to be the most promising; at this thickness level, XPS shows lower costs than PUF. This optimisation successfully strikes a balance between overall project costs and insulation cost savings. The payback period is the shortest for a 20 mm insulation thickness, with PUF having a smaller payback period in comparison to XPS, as shown in Fig. 6. The above results show that the application of insulation as a passive strategy in the preconstruction phase can help reduce the energy load on mineral resources.

The cost saving is a maximum of 6.56% when XPS is used as an insulation material at 60 mm thickness, while PUF gives 6.21% cost savings at the same thickness. We can determine that as per the climatic condition and location of the building, which is a warm and humid climate zone of India, XPS can be used at a thickness of 60 mm in walls and roofs to make a building energy-efficient and cost-effective.

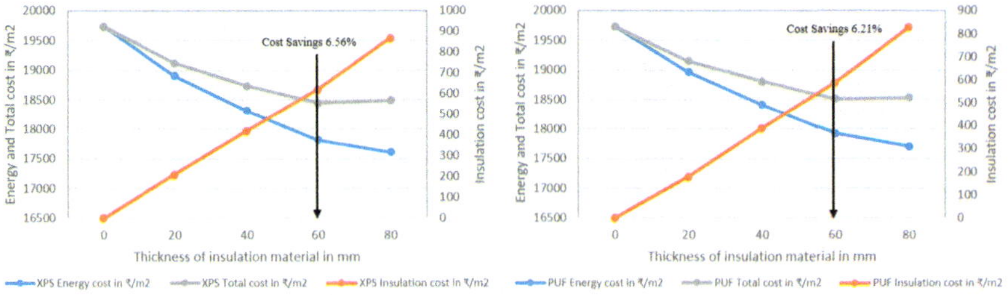

a) Effect of XPS insulation thickness on cost and energy

b) Effect of PUF insulation thickness on cost and energy

Fig. 6 The cost-saving potential of XPS and PUF insulation material

4 Conclusion

The study specifies the optimum insulation material and thickness to be used in the construction of a building envelope to achieve the goal of energy conservation and utilises a strong methodological process of energy simulation and validation from actual energy consumption. The importance of insulating materials in Kolkata's warm, humid climate is emphasised. The results show that in air-conditioned buildings, a decrease in energy consumption can be achieved with the application of insulation materials. The cost of insulation rises linearly with material thickness; for an ideal solution thickness, the total cost will be the lowest, regardless of the insulation material used.

The study showed that thicker insulation material with 60 mm of XPS specifically leads to appreciable energy savings. The techno-economic analysis demonstrated how cost-effective insulation materials could save a significant amount of money while preserving energy efficiency. This study has importance in practical implications to provide insightful information to the stakeholders, policymakers, and builders. India can achieve its goals for sustainability and energy efficiency by choosing the appropriate insulation materials and thickness. The practical implications of this study could result in more energy-efficient residential construction, lessening the negative effects on the environment for India's quickly expanding metropolitan areas. Subsequent investigations may examine the combined effects of merging intelligent building technologies, resident conduct, and insulating materials to augment energy efficiency and economy in warm and humid residential structures.

References

1. Ian Hamilton, D. H. K. (2021). *Global status report for buildings and construction: Towards a zero-emission, efficient and resilient buildings and construction sector* (p. 2021). United Nations Environment Programme.
2. Aayog, N. (2021). *Urban transformation sector report*. Development Monitoring and Evaluation Office (DMEO).

3. Tim Gould, S. S., & Zeniewski, P. (2021). *India energy outlook*. International Energy Agency.
4. Tushar, Q., Bhuiyan, M., Sandanayake, M., & Zhang, G. (2019). Optimizing the energy consumption in a residential building at different climate zones: Towards sustainable decision making. *Journal of Cleaner Production, 233*, 634–649. https://doi.org/10.1016/j.jclepro.2019.06.093
5. Yu, S., Tan, Q., Evans, M., Kyle, P., Vu, L., & Patel, P. L. (2017). Improving building energy efficiency in India: State-level analysis of building energy efficiency policies. *Energy Policy, 110*, 331–341. https://doi.org/10.1016/j.enpol.2017.07.013
6. Ministry of Environment, Forest & Climate Change. (2019). *India cooling action plan*. Ministry of Environment, Forest & Climate Change.
7. Elnagar, E., & Köhler, B. (2020). Reduction of the energy demand with passive approaches in multifamily nearly zero-energy buildings under different climate conditions. *Frontiers in Energy Research, 8*, 1–8. https://doi.org/10.3389/fenrg.2020.545272
8. Mostafavi, F., Tahsildoost, M., & Zomorodian, Z. S. (2021). Energy efficiency and carbon emission in high-rise buildings: A review (2005–2020). *Building and Environment, 206*, 108329. https://doi.org/10.1016/j.buildenv.2021.108329
9. Mushtaha, E., et al. (2021). The impact of passive design strategies on cooling loads of buildings in temperate climate. *Case Studies in Thermal Engineering, 28*, 101588. https://doi.org/10.1016/j.csite.2021.101588
10. Wang, X., Chen, D., & Ren, Z. (2010). Assessment of climate change impact on residential building heating and cooling energy requirement in Australia. *Building and Environment, 45*(7), 1663–1682. https://doi.org/10.1016/j.buildenv.2010.01.022
11. Feng, W., Zou, L., Gao, G., Wu, G., Shen, J., & Li, W. (2016). Gasochromic smart window: Optical and thermal properties, energy simulation and feasibility analysis. *Solar Energy Materials & Solar Cells, 144*, 316–323. https://doi.org/10.1016/j.solmat.2015.09.029
12. Papadopoulos, A. M., & Giama, E. (2007). Environmental performance evaluation of thermal insulation materials and its impact on the building. *Building and Environment, 42*(5), 2178–2187. https://doi.org/10.1016/j.buildenv.2006.04.012
13. Al Naqbi, A., Alyieliely, S. S., Talib, M. A., Nasir, Q., Bettayeb, M., & Ghenai, C. (2021). Energy reduction in building energy management systems using the Internet of Things: Systematic literature review. In *2021 Int Symp Networks Comput Commun ISNCC 2021* (pp. 1–7). https://doi.org/10.1109/ISNCC52172.2021.9615641
14. Zabalza Bribián, I., Valero Capilla, A., & Aranda Usón, A. (2011). Life cycle assessment of building materials: Comparative analysis of energy and environmental impacts and evaluation of the eco-efficiency improvement potential. *Building and Environment, 46*(5), 1133–1140. https://doi.org/10.1016/j.buildenv.2010.12.002
15. Shoubi, M. V., Shoubi, M. V., Bagchi, A., & Barough, A. S. (2015). Reducing the operational energy demand in buildings using building information modeling tools and sustainability approaches. *Ain Shams Engineering Journal, 6*(1), 41–55. https://doi.org/10.1016/j.asej.2014.09.006
16. Venkatarama Reddy, B. V. (2009). Sustainable materials for low carbon buildings. *International Journal of Low-Carbon Technologies, 4*(3), 175–181. https://doi.org/10.1093/ijlct/ctp025
17. Khalid, H., Thaheem, M. J., Malik, M. S. A., Musarat, M. A., & Alaloul, W. S. (2021). Reducing cooling load and lifecycle cost for residential buildings: A case of Lahore, Pakistan. *Int J Life Cycle Assess, 26*(12), 2355–2374. https://doi.org/10.1007/s11367-021-02000-1
18. Yu, J., Yang, C., Tian, L., & Liao, D. (2009). A study on optimum insulation thicknesses of external walls in hot summer and cold winter zone of China. *Applied Energy, 86*(11), 2520–2529. https://doi.org/10.1016/j.apenergy.2009.03.010
19. Schiavoni, S., D'Alessandro, F., Bianchi, F., & Asdrubali, F. (2016). Insulation materials for the building sector: A review and comparative analysis. *Renewable and Sustainable Energy Reviews, 62*, 988–1011. https://doi.org/10.1016/j.rser.2016.05.045
20. Kumar, D., Alam, M., Zou, P. X. W., Sanjayan, J. G., & Memon, R. A. (2020). Comparative analysis of building insulation material properties and performance. *Renewable and Sustainable Energy Reviews, 131*, 110038. https://doi.org/10.1016/j.rser.2020.110038

21. Al-Sanea, S. A., Zedan, M. F., & Al-Hussain, S. N. (2012). Effect of thermal mass on performance of insulated building walls and the concept of energy savings potential. *Applied Energy, 89*(1), 430–442. https://doi.org/10.1016/j.apenergy.2011.08.009
22. Aditya, L., et al. (2017). A review on insulation materials for energy conservation in buildings. *Renewable and Sustainable Energy Reviews, 73*, 1352–1365. https://doi.org/10.1016/j.rser.2017.02.034
23. Daouas, N., Hassen, Z., & Ben Aissia, H. (2010). Analytical periodic solution for the study of thermal performance and optimum insulation thickness of building walls in Tunisia. *Applied Thermal Engineering, 30*(4), 319–326. https://doi.org/10.1016/j.applthermaleng.2009.09.009
24. G. of I. Ministry of Power, Energy Conservation Building Code for Residential Buildings, 1. 2018. Retrieved from: https://www.beeindia.gov.in/sites/default/files/ECBC_BOOK_Web.pdf.
25. Dru Crawley and Linda Lawrie (2014). *Climate*. OneBuilding.Org.
26. I. Haque, S. Mehta, & A. Kumar (2019). Attribution: Ismail Haque, Simi Mehta and Arjun Kumar, *Towards sustainable and inclusive cities: The case of Kolkata*. ORF special report No Towar. Sustain. Incl. Cities Case Kolkata, 83. Retrieved from: https://www.orfonline.org/series/urbanisation-and-its-discontents/
27. Ministry of Power (MoP), Government of India, Energy Conservation Building Code 2017 (with Amendments upto 2020). (2020).
28. DesignBuilder, DesignBuilder Software Ltd.—Simulation. (2021). Retrieved from https://designbuilder.co.uk/simulation.
29. Conference, F. N., Madison, I., & August, W. (2012). *Comparison of two different simulation programs while calibrating the same building* (pp. 128–135). University of Southern California.
30. Weytjens, L., Attia, S., Verbeeck, G., & De Herde, A. (2011). The 'architect-friendliness' of six building performance simulation tools: A comparative study. *International Journal of Sustainable Building Technology and Urban Development, 2*(3), 237–244. https://doi.org/10.5390/SUSB.2011.2.3.237
31. Fathalian, A., & Kargarsharifabad, H. (2018). Actual validation of energy simulation and investigation of energy management strategies (Case Study: An office building in Semnan, Iran). *Case Studies in Thermal Engineering, 12*, 510–516. https://doi.org/10.1016/j.csite.2018.06.007
32. Geeen, K., & Tashman, L. (2009). Percentage error: What denominator? *Foresight: The International Journal of Applied Forecasting, 12*, 36–40.
33. Bureau of Energy Efficiency (BEE). (2022). *Star labelled appliances*. Ministry of Power, Government of India.
34. Daouas, N. (2011). A study on optimum insulation thickness in walls and energy savings in Tunisian buildings based on analytical calculation of cooling and heating transmission loads. *Applied Energy, 88*(1), 156–164. https://doi.org/10.1016/j.apenergy.2010.07.030
35. Kessler, W. (2017). Comparing energy payback and simple payback period for solar photovoltaic systems. *E3S Web of Conferences, 22*, 0–5. https://doi.org/10.1051/e3sconf/20172200080
36. Abdelrady, A., Abdelhafez, M. H. H., & Ragab, A. (2021). Use of insulation based on nanomaterials to improve energy efficiency of residential buildings in a hot desert climate. *Sustain, 13*(9). https://doi.org/10.3390/su13095266
37. Ragab, A., et al. (2022). An assessment of external wall retrofitting strategies using GRC materials in hot desert regions. *Buildings, 12*(8). https://doi.org/10.3390/buildings12081194
38. Bolattürk, A. (2006). Determination of optimum insulation thickness for building walls with respect to various fuels and climate zones in Turkey. *Applied Thermal Engineering, 26*(11–12), 1301–1309. https://doi.org/10.1016/j.applthermaleng.2005.10.019

Sustainable Water Management in the Digital Age: Intelligent Solutions for a Precious Resource

Balbir Nagal and A. K. Prabhakar

1 Introduction

Water plays a crucial role in supporting life and many types of human activities. Owing to threats such as overextraction of groundwater, expansion of the agricultural sector, rapid urbanization, industrial growth, and impacts of climate change, sustainable management of water is essential. Overextraction of groundwater is a critical issue globally because aquifers are depleting and reducing water availability [1]. Nowadays, expansion of the agricultural sector has also become a major cause of mismanagement of this precious resource. It includes surface water as well as groundwater for irrigation purposes, but the main source of irrigation is groundwater [2]. Urbanization also affects water management, as water demand has significantly increased because of expanding infrastructure needs and higher standards of living [3]. Industrial growth has become a cause of water scarcity, as these processes require substantial amounts of water for different processes such as cooling, cleaning, and production [4]. Similarly, climate change is also responsible for water scarcity by changing precipitation patterns, increasing the frequency of droughts, and intensifying extreme weather events [5]. Sustainable water management (SWM) is a complex issue that encompasses scarcity, inefficiencies in distribution, and quality degradation. By 2025, half of the global population is predicted to reside in regions facing water scarcity [6].

Therefore, conventional methods of water management are no longer sufficient. In this digital era, geospatial technologies, big data analytics, the Internet of things (IoT), and artificial intelligence (AI) are gradually taking over conventional approaches in view of SWM. These factors enhance the capabilities of SWM by

B. Nagal (✉) · A. K. Prabhakar
Department of Civil Engineering, National Institute of Technology, Kurukshetra, Haryana, India
e-mail: 62200032@nitkkr.ac.in; akprabhakar@nitkkr.ac.in

S. Pathak et al. (eds.), *Intelligent Infrastructure and Smart Materials*,
https://doi.org/10.1007/978-3-031-92421-7_6

enabling predictive analytics and automated responses. Geospatial technologies include remote sensing (RS) and geographic information systems (GIS), which offer critical tools for SWM and help monitor water bodies, land use, and environmental changes using satellite imagery and RS data [7]. GIS technology helps in data gathering, live monitoring, precise mapping, and predictive modelling for groundwater systems [8]. The integration of GIS with RS methodologies facilitates a thorough understanding of the accessibility and SWM of this valuable resource [9]. GIS is also helpful for delineating areas with different groundwater potential (GWP), by combining RS data and using a multi-criteria decision-making (MCDM) approach for accurate assessment of water resources [10]. This also facilitates the process of delineating areas where groundwater has been replenished, identifying sites where artificial recharge is taking place, and assessing the quality of groundwater. This makes an important contribution to the assessment of groundwater, which further contributes to SWM [11].

Big data also play an important role in SWM, as it addresses the challenges of scarcity, quality, and distribution efficiency. In this technique, sensors collect real-time data on water levels, flow rates, and quality, enabling better monitoring and pollution tracking [12]. The IoT also provides a new mode in SWM, as it enables real-time monitoring of water levels, quality, and usage across various sectors. Sensors placed in water bodies, pipelines, and agricultural fields continuously collect data, which helps us detect leaks, monitor consumption, and manage distribution networks more effectively [13]. AI enables predictive analytics for water demand, identifies anomalies, and suggests optimal strategies for SWM. This includes machine learning (ML) models that predict rainfall patterns and groundwater levels. It can also systematize agricultural irrigation, ensuring precise water delivery on the basis of crop needs, thus conserving water while maximizing yields [14].

In addition, old infrastructure and inefficient water use practices lead to substantial losses. On a global scale, approximately 30% of water is estimated to be lost as a result of leaks and inefficiencies in the water distribution system [15]. Analysing water quality is also a major issue for managing water for different uses. Water sources are contaminated by industrial waste, agricultural run-off, and inadequate wastewater treatment. This creates problems with health risks and is harmful to ecosystems.

So the integration of these digital technologies into SWM is crucial, as traditional methods are no longer sufficient to manage the complexities of modern water challenges. Intelligent solutions provide the means to manage water resources more efficiently, sustainably, and equitably. In this digital age, implementing intelligent solutions for SWM is not just an option but also a necessity for a sustainable and resilient future.

2 Challenges in Water Management

Water management is of utmost importance for the survival of human civilizations, as evidenced by its importance in ancient societies such as Egyptians, Mesopotamians, and Indus Valley civilizations. In those years, efforts were made to develop advanced irrigation systems for agriculture and urbanization. At present, water management faces various challenges, including climate change impacts, population growth, urbanization, and agricultural demand including intensive farming practices. These contaminate water sources and deplete aquifers and surface water sources.

2.1 Climate Change Impacts

Climate change is a significant factor that affects water scarcity and various sectors, such as agriculture, industry, and domestic use, globally. This can lead to unpredictable precipitation patterns, prolonged droughts, and extreme weather events, exacerbating water scarcity [16]. The relationship between socioeconomic changes and climate variations also significantly influences water scarcity at both the global and regional levels [17]. Adaptive water supply planning is crucial in addressing the impacts of climate change on water scarcity. It is essential to take into account uncertainties in future water demand and climate conditions in order to efficiently mitigate these impacts [18]. Water supply management must also account for the contributions of land use/land cover changes and climate variability to ensure an optimal water supply [19]. The resilience of water utilities in facing climate change impacts is a growing concern, highlighting the importance of robust infrastructure and management strategies.

2.2 Population Growth

Population growth also affects water management, as water demand has increased manifold. The increasing population globally results in increased water consumption across various sectors, e.g. agriculture, industry, and domestic use [20]. As populations increase, the demand for water resources increases, thus intensifying pressure on existing water supplies. Planning for future water demand must incorporate population growth projections and adaptive approaches to ensure water availability across all sectors [21]. Furthermore, the advancement of innovative technologies and smart water management solutions can increase water use efficiency and improve the effects of population growth on water resources [22].

2.3 Urbanization

Urbanization significantly impacts water management by increasing water demand, availability, and quality. As cities expand and populations concentrate in urban areas, water demand increases, leading to increased stress on water resources [23]. Urbanization also affects the natural water cycle and increases impervious surfaces, which reduce infiltration and groundwater recharge (GWR) [24]. The transformation of land cover in urban areas is responsible for a decrease in water availability and increased run-off and directly decreases the GWR. The implementation of innovative techniques, e.g. rainwater harvesting, wastewater reclamation, and water pricing mechanisms, can mitigate the impacts of urbanization.

Urbanization represents a complicated challenge for water management. Comprehensive approaches that balance the water requirements of expanding urban populations are needed [25]. Understanding the complicated relationships among urbanization, water demand, and environmental repercussions makes it possible to create effective water management methods to protect urban water security.

2.4 Agricultural Demand and Intensive Farming Practices

Agricultural demand significantly impacts water management, primarily through intensive farming practices that deplete and contaminate water sources. These practices, which include high-volume irrigation and the extensive use of fertilizers and pesticides, are crucial for meeting global food production needs but pose serious challenges to sustainable water use [26]. Intensive irrigation practices lead to groundwater depletion, as farmers extract more water than replenish it naturally, risking the drying of crucial aquifers. The run-off from fields treated with fertilizers and pesticides carries nitrates, phosphates, and chemicals into water sources, leading to eutrophication and degradation of water quality [27].

To overcome these challenges, it is necessary to implement sustainable agricultural methods, such as precision farming, that increase water efficiency and minimize the need for chemical inputs. The utilization of advanced irrigation technologies, such as drip or sprinkler systems, can efficiently mitigate water loss. The implementation of integrated water resources management methodologies that are required across various levels can facilitate the development of more effective groundwater management methods [28]. It also enhances water conservation efforts and encourages the utilization of treated wastewater for irrigation purposes. Therefore, balancing agricultural water needs while guaranteeing water resource sustainability is complex but essential for water management.

2.5 Industrial and Agricultural Runoff

Industrial discharges frequently contain hazardous substances, including heavy metals, solvents, and toxic chemicals. These substances have the potential to persist in the environment and accumulate in water bodies, resulting in bioaccumulation in both aquatic and terrestrial organisms [29]. Agricultural run-off also transports fertilizers, pesticides, and animal waste, introducing excessive amounts of nutrients and organic matter into water systems. This phenomenon is known as eutrophication, which occurs when excessive algae growth depletes the oxygen levels in the water. This may also lead to the formation of dead zones that are uninhabitable for aquatic organisms [30].

The role of water management in addressing these pollution challenges is crucial and involves multiple aspects. The implementation of regulatory frameworks for pollutant discharges ensures adherence through continuous monitoring and the imposition of penalties. Furthermore, the imperative lies in the promotion and implementation of best management practices within the industrial and agricultural domains. These practices encompass cutting-edge technologies for treating wastewater, sustainable agricultural practices such as integrated pest management and precision farming techniques [31]. In addition, the establishment of infrastructure for wastewater treatment and reuse can help decrease the adverse effects of pollutants on natural water systems. Therefore, comprehensive water management policies that include preventive and corrective efforts are needed to reduce pollution and preserve water quality for future generations.

2.6 *Overextraction of Groundwater*

The excessive extraction of water, specifically from aquifers and surface water sources, poses a significant problem in water management. This has resulted in a loss of ecological, economic, and social concerns. Aquifers, which serve as crucial sources of potable water worldwide, are experiencing rapid depletion in numerous areas as a result of excessive extraction for agricultural, industrial, and domestic purposes [32]. The process of overextracting water results in a decrease in the water table, which in turn increases the difficulty and cost of accessing water. This can cause the land above to sink, resulting in damage to infrastructure and a reduction in the available surface area for agriculture and habitation. Similarly, the overconsumption of water from rivers and lakes to meet agricultural needs and support urban populations creates strain on these ecosystems. It also affects biodiversity and diminishes the ability of these systems to withstand natural variations and climate change [33]. A decrease in water level can modify natural flow patterns, deteriorate water quality by increasing the concentrations of pollutants, and reduce the recreational and aesthetic value of these bodies of water.

Efficient water management strategies to mitigate excessive extraction involve the implementation of sustainable practices for withdrawing water, regulatory measures such as the allocation of water rights, issuance of permits for groundwater extraction, and promotion of water-efficient technologies in agriculture and industry [34]. Furthermore, enhancing public awareness regarding the significance of water conservation and promoting for the use of treated wastewater can make a valuable contribution to SWM [35]. Therefore, the SWM is crucial to guarantee the long-term sustainability and accessibility of water resources.

3 Digital Water Management

Digital water management utilizes advanced technology to optimize water utilization, monitor water systems, and analyse data in view of SWM approaches. The implementation of this solution results in increased operational efficiency, minimized resource waste, and improved SWM capabilities.

3.1 Different Techniques for SWM

In this digital age, the role of modern technology in water management has become even more essential. Digital techniques for SWM involves different methods including advanced data analytics, IoT, AI, big data, blockchain, and geospatial technology as shown in Fig. 1. These technologies work together to improve digital water resource management (DWRM) and decision-making processes.

The integration of advanced technologies such as the IoT, AI, big data analytics, blockchain, RS, and GIS has transformed traditional water management practices. These enable more efficient, adaptive, and SWM of this precious resource. Currently, digital output instruments, such as meters and sensors, collect real-time data for detailed monitoring and analysis (Table 1).

At present, SWM systems utilize digital technologies to monitor multiple aspects of water quality and flow. These include a variety of applications:

- Multiparameter sensors for monitoring water quality and quantity.
- Wireless communication networks for data transmission.
- Data analytics software for real-time analysis and alert systems.
- Applications include water quality monitoring, leak detection, flood monitoring, and wastewater monitoring.

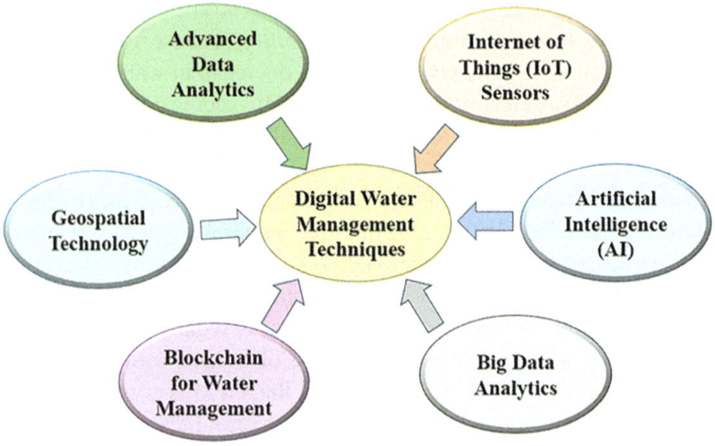

Fig. 1 Digital techniques for SWM

Table 1 Application of advanced technologies in SWM

Components	Purpose	Applications
Digital output instruments (meters and sensors)	To collect and transmit granular, real-time data for monitoring and analysis	**Rain gauges and flowmeters**: Quantitative assessment of precipitation and flow rates for environmental monitoring **Acoustic devices**: Utilize sound waves for identifying leaks and integrity testing **Video cameras**: Enhanced asset management through visual inspections **Smart water meters**: Precise measurement of water consumption and user data analytics **Pressure sensors**: Monitor and optimize pressure levels within pipelines for leak detection and pump performance efficiency
SCADA systems	To automate processing, control, and optimize the operations of water systems through real-time data acquisition and feedback control loops	**Pressure and pump management**: Automated control systems to maintain optimal pressure levels and operational efficiency **Plant control systems**: Real-time management of water and sewage treatment processes **Environmental system management**: Automated control of reservoir levels, water flows, and related environmental parameters
GIS	To facilitate the integration and spatial analysis of geographical and infrastructure data, enhancing the planning, management, and operational decision-making processes	**Asset mapping**: Geospatial mapping of infrastructure components for enhanced asset management strategies **Network models**: Development and integration of comprehensive network models for system-wide analysis and optimization **Environmental data management**: Spatial analysis of environmental datasets for sustainable management practices
Software	To enable sophisticated data management, analytical modelling, and operational reporting; crucial for strategic planning and operational excellence	**System integration**: Seamless integration with GIS and SCADA for cohesive water network management **Decision support systems**: Advanced algorithms for risk assessment and decision-making enhancement **Customer and billing management**: Automated systems for efficient customer data management and financial transactions **Hydraulic and hydrological modelling**: Utilize computational fluid dynamics and modelling to predict water flow and resource management for maximum efficiency**Cloud solutions**: Deployment of cloud-based platforms for scalable data management and enhanced accessibility

Supervisory control and data acquisition (SCADA) systems automate water system operations, enhancing efficiency through real-time data feedback. GIS tools analyse and manage spatial and infrastructure data, aiding in planning and operational decisions. Software solutions support data management and strategic planning, which are integrated with systems. These factors collectively enhance the precision, efficiency, and intelligence in SWM practices.

3.2 *Data Collection and Analysis Platforms*

These platforms serve as the foundation for data analytics in SWM, as they aggregate data from diverse sources, including sensors, meters, and SCADA systems, and apply advanced analytics techniques to extract actionable insights. This encompasses different techniques:

- Data collection and integration.
- IoT for real-time monitoring and visualization.
- Using ML techniques, predictive analytics, and modelling.
- AI for intelligent solutions.

These platforms enable water managers to access accurate, timely, and actionable information, thereby facilitating data-driven decision-making.

3.2.1 Internet of Things (IoT)

The IoT is transforming water management through the implementation of real-time monitoring and data collection systems [13]. IoT devices, such as sensors and smart meters, are deployed across water distribution networks, reservoirs, and agricultural fields to continuously gather data on water levels, quality, and usage [36]. These data are transmitted in real time to centralized systems, allowing for the immediate detection of leaks, anomalies, and inefficiencies. For example, smart water meters provide consumers with detailed insights into their water usage patterns, promoting water conservation and reducing waste.

In the agricultural sector, soil moisture sensors and automated irrigation systems ensure that crops receive the optimal amount of water, increasing water use efficiency and crop yields [37]. A representation diagram of the IoT is shown in Fig. 2. The IoT monitors the quality of water in a system utilizing environmental sensors to gather data on various parameters, including pH levels and temperature, from water sources. These data are transmitted through communication technologies, e.g. Wi-Fi or cellular networks, to a central system or cloud-based storage. After the data are stored, they undergo a processing phase to analyse and evaluate the quality of the water. An alert system is triggered when the data show signs of potential adverse effects, prompting the necessary responses [38].

In addition, the system provides support for data visualization, which facilitates the interpretation of data, as well as data sharing for collaborative efforts or regulatory compliance. The end-to-end process described here guarantees timely and efficient monitoring and management of water quality.

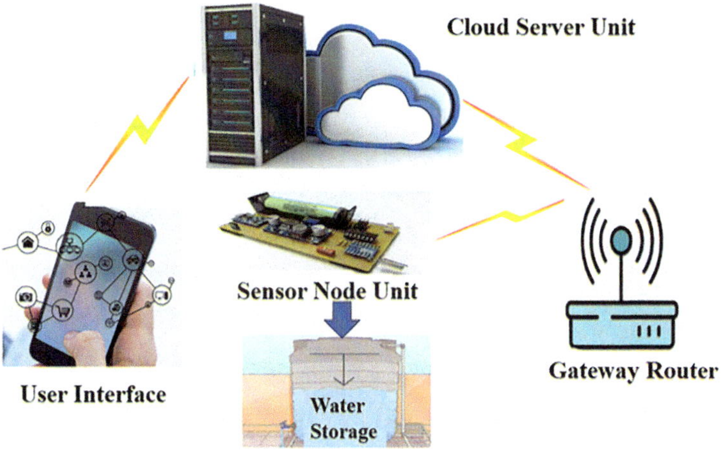

Fig. 2 Typical diagram of an IoT system

3.2.2 Artificial Intelligence (AI)

AI enhances the capabilities of water management systems by enabling predictive analytics and automated responses. AI algorithms can process data from various sources to forecast water demand, identify anomalies, and suggest optimal management strategies [39]. ML models predict rainfall patterns, groundwater levels, and potential droughts, which helps in the proactive management of water resources [40]. AI-driven systems can automate irrigation in agriculture, ensuring precise water delivery on the basis of real-time conditions and crop needs. These systems also support the acquisition and analysis of data in real time, allowing for predictive maintenance and dynamic optimization of water distribution. These methods are employed in AI-driven models such as neural networks and ensemble methods. These systems can forecast water demand, detect anomalies, and identify operational patterns [40].

AI analytics improve decision-making processes by providing actionable insights into water consumption trends and distribution dynamics, ultimately contributing to resource conservation and enhanced system performance [12]. The application of AI across water management domains, from distribution and precision agriculture to quality monitoring, drought prediction, and smart water grids, enhances water conservation efforts and supports SWM through improved efficiency, early detection, and data-driven decision-making.

3.2.3 Big Data Analytics

Big data analytics plays a crucial role in synthesizing information from diverse sources, providing a comprehensive view of SWM. This integration helps understand the impact of climate change on water availability and guides the development

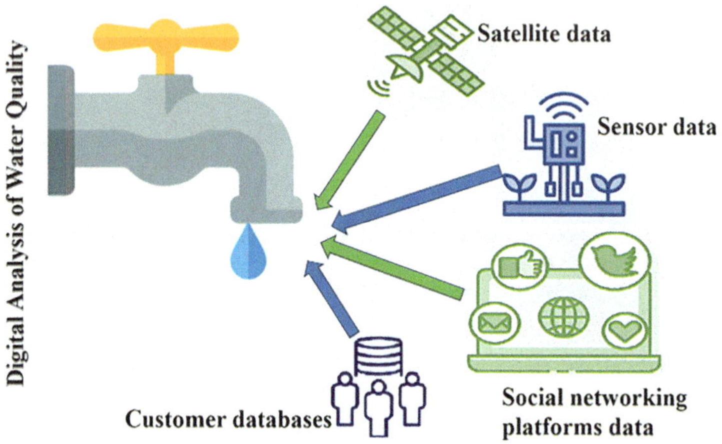

Fig. 3 Integrated data sources for water quality analysis

of resilient water infrastructure [41]. For analysing water quality using big data, it is required to integrate data from multiple sources, including satellite data, sensor data, social networking platforms, and customer databases (Fig. 3).

These diverse dataset inputs are used to assess and monitor water quality, ensuring comprehensive and accurate analysis. By analysing large datasets, water managers can identify trends, patterns, and correlations that were previously hidden, enabling informed decision-making. For example, analysing historical weather data and current usage patterns can help predict future water demand and optimize water supply accordingly [42]. Big data also supports the development of early warning systems for floods and droughts, enhancing preparedness and response strategies.

3.2.4 Geospatial Technology

Geospatial technology, including RS and GIS technologies, plays a critical role in comprehensive DWRM. Satellite imagery and remote sensors provide extensive data on water bodies, land use, and environmental changes, enabling accurate assessment and monitoring of water resources [7]. GIS integrates spatial and temporal data, allowing for detailed mapping and analysis of water distribution, quality, and usage patterns [8]. The GIS data model integrates different GIS maps that serve specific purposes. These factors help to delineate the GWP and groundwater storage (GWS), which ultimately helps in SWM. The sub-watershed maps identify specific drainage areas. In addition, specialized maps such as water quality, water level, pollution, rainfall, and utilization rating maps can be created with this (Fig. 4). Each of these maps provides detailed assessments and further enables comprehensive analysis and helps in DWRM.

These tools facilitate efficient planning, administration, and conservation initiatives, especially in areas with insufficient on-site monitoring data. For instance, GIS

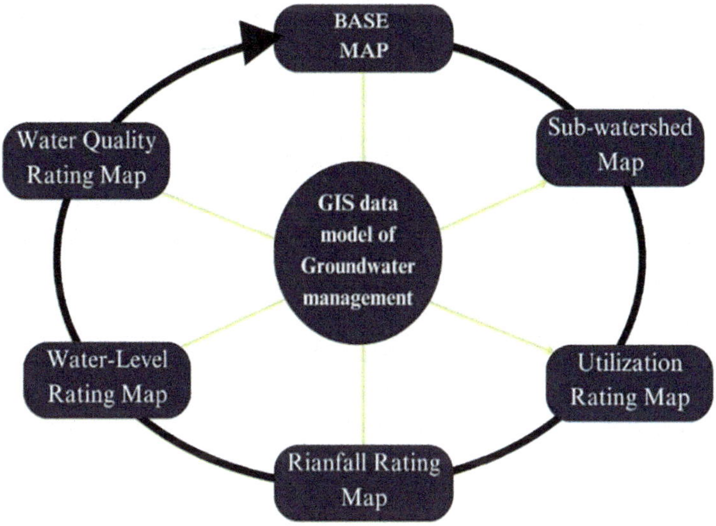

Fig. 4 Integration of spatial data for groundwater mapping

can be utilized to identify regions that are susceptible to flooding, record variations in water quality, and execute the most suitable sites for constructing new water infrastructure [9].

3.3 Modern Technology for Efficient Water Distribution

For the distribution of water, the incorporation of digital technologies has a great role. These include IoT sensors and AI-driven analytics, which facilitate accurate monitoring and control of water systems. These tools facilitate the identification of leaks, the optimization of irrigation schedules, and the prediction of water demand patterns. The combination of these approaches guarantees the SWM strategies, effectively addressing ecological and human requirements.

3.3.1 Leak Detection and Prevention

One of the most significant benefits of IoT-enabled water meters is their ability to detect leaks early. By continuously monitoring water flow, these meters can identify irregularities in the distribution system. A sudden increase in usage may indicate a leak. Once a potential issue is detected, an alert is sent to the utility or the consumer, enabling prompt action to prevent water loss. These data also provide real-time data that are instantly transmitted to centralized management systems. This real-time monitoring allows water utilities to track consumption patterns at a granular level.

These data are crucial for making informed decisions about water distribution, helping to ensure that supply meets demand while minimizing waste.

3.3.2 Personalized Water Consumption Insights

IoT-enabled water meters offer a level of transparency. Through mobile apps or online dashboards, users can access detailed information about their water consumption. This visibility empowers consumers to make more informed decisions about their water use, encouraging conservation efforts. The enhancement of water distribution efficiency in SWM involves the integration of advanced technologies such as AI, IoT, and data analytics.

To optimize water usage, reducing losses and ensuring equitable distribution are necessary. Integrating the IoT and AI into SWM is crucial for achieving sustainable water resource utilization, with future advancements and ethical considerations driving the evolution of more intelligent and equitable systems.

3.4 Precision Agriculture for Water Conservation

In the digital era, precision agriculture utilizes advanced technological frameworks and helps with SWM. These factors significantly enhance water conservation and distribution efficiency. This approach integrates real-time data from various sources. Precision agriculture facilitates targeted water application on the basis of crop-specific needs and environmental conditions, thus reducing water waste and improving irrigation efficiency. Big data technologies such as ML and deep learning significantly enhance precision agriculture by providing actionable insights and improving decision-making; addressing social and financial challenges remains crucial for their widespread adoption and effective implementation [43].

3.4.1 Soil Moisture Sensors

Soil moisture sensors are crucial in precision agriculture for advancing water conservation in this digital age for SWM. The transformative potential of AI and the IoT in modern agriculture emphasizes their role in enhancing efficiency, productivity, and sustainability through advanced technologies and real-time data integration [44]. These sensors deliver real-time measurements of the soil water content. These are integrated with digital platforms and analytical tools.

3.4.2 Automated Irrigation Systems

Automated irrigation systems in precision agriculture utilize sensors, ML algorithms, and real-time data analytics. These methods are used to optimize water utilization and enhance water conservation [45]. These systems incorporate soil moisture sensors and meteorological data to dynamically regulate irrigation schedules and volumes in accordance with the specific needs of crops and prevailing environmental conditions. This further reduces water wastage and enhances crop productivity. The potential of integrating ML models with digital farming solutions to optimize irrigation management emphasizes their role in promoting sustainable water use and reducing the challenges faced by farmers through remote monitoring and control [46].

4 Water Treatment and Reuse

The process of water treatment and its subsequent reuse has experienced substantial transformation, driven by advancements in technology and the utilization of data analytics. Traditional methodologies have been replaced by digital innovations, including real-time monitoring, predictive analytics, and automation. The synergy of the IoT, AI, and ML enables the precise control and monitoring of water quality and treatment processes [38]. Digital platforms help in decision-making, culminating in a great contribution in planning of SWM strategies. Digital tools and techniques with conventional techniques address water scarcity, contamination, and resource management challenges with enhanced precision and efficiency [47]. These factors also play crucial roles in advancing water treatment and reuse practices, ensuring a more sustainable future. This convergence of technology for SWM has the potential to revolutionize industry, enabling more effective and resilient water systems. By harnessing the power of digital innovations, water managers can optimize treatment processes, reduce environmental impact, and ensure a reliable water supply.

4.1 Need of Digital Water Treatment

Traditional water treatment technologies include sedimentation, filtration, and chlorination, which depend on physical and chemical mechanisms to eliminate impurities (Table 2). These methods are commonly employed because of their straightforwardness and efficiency in ensuring the provision of potable water. Digital water treatment technologies utilize advanced sensors, data analytics, and automation to improve the monitoring and treatment of water quality. These include various technological solutions, such as real-time water quality sensors, IoT-enabled systems, and AI-driven predictive maintenance.

Table 2 Traditional water treatment approaches

Physical processes	Sedimentation	Suspended solids settle to the bottom of a clarifier, facilitating removal
	Filtration	Porous media, such as sand or membranes, capture particulate matter and suspended solids
	Flotation	Air or gas bubbles adhere to suspended solids, enabling removal by skimming or overflow
Chemical processes	Disinfection	Oxidizing agents, such as chlorine or ozone, inactivate pathogenic microorganisms
	Coagulation and flocculation	The addition of chemical coagulants to neutralize electrostatic charges, facilitating the aggregation of particulate matter into larger, more readily removable flocs
Biological processes	Activated sludge	Microorganisms biodegrade organic matter and nutrients in the presence of oxygen.
	Trickling filters	A biofilm breaks down organic matter and nutrients as wastewater percolates through a porous medium

4.2 Digital Water Treatment

Digital innovations have fundamentally revolutionized water management, integrating advanced technologies seamlessly with established practices. The implementation of the IoT, AI, and ML algorithms improves water treatment processes and optimizes the resource distribution, resulting in enhanced operational efficiency [12, 40]. Furthermore, the application of data analytics and cloud computing facilitates remote monitoring and collaborative initiatives, providing water utilities [45]. Digital twins are computer-generated models that represent water systems. They enable the simulation and optimization of treatment methods and distribution networks.

Predictive analytics and ML algorithms analyse historical and real-time data to forecast future trends, identify potential anomalies, and optimize treatment processes. These technologies enable proactive decision-making, reduce the risk of water quality issues, and improve operational efficiency. The key aspects include the following:

- Predictive analysis for forecasting water demand, predicting water quality, and identifying potential failures.
- ML for developing predictive models, classifying and detecting anomalies, and optimizing treatment process control.

An advanced water treatment monitoring system incorporates sensors, a wireless smart-mesh IP sensor network, and a spectre cellular or ethernet gateway. The facility is equipped with sensors that monitor important parameters such as temperature, pressure, flow, and vibration. The sensors establish communication through intelligent edge nodes to a central gateway, which then transmits the data using ethernet or cellular networks for analysis as shown in Fig. 5. The system with a centralized architecture performs real-time data processing, facilitating automatic notifications for any deviations. This configuration guarantees effective, live monitoring, enabling

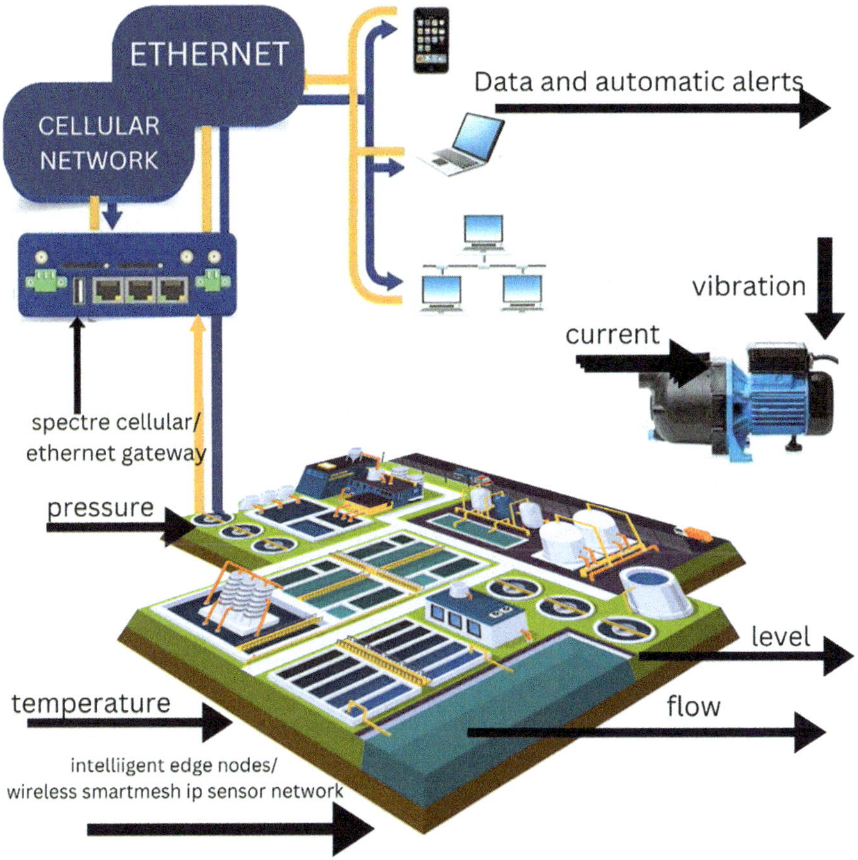

Fig. 5 Advanced water treatment monitoring system

operators to promptly identify and resolve potential problems, thereby maintaining the optimal performance and safety of the water treatment facility.

The integration of digital tools enhances the operational efficiency, resilience, and sustainability of SWM, ensuring a reliable and adaptable water supply for future generations. The adoption of a data-centric approach in water management provides decision-makers with extensive analytics, which facilitates more strategic SWM. These digital solutions also enhance operational efficiency, decrease operational expenses, and guarantee adherence to regulatory standards.

4.2.1 Automated Control Mechanisms

Automated control mechanisms regulate and optimize water treatment processes, distribution networks, and infrastructure in real time [48]. These mechanisms ensure optimal water quality, reduce energy consumption, and minimize environmental impact. The key components include:

- Sensors for monitoring water quality and quantity.
- Control algorithms for optimizing treatment processes and distribution networks.
- Actuators for implementing control actions.
- Communication networks for data transmission and system integration.

The integration of digital tools in water treatment has transformed industry, enabling real-time monitoring, predictive analytics, and automated control mechanisms. These advanced technologies enhance the operational efficiency, environmental sustainability, and overall resilience of water treatment techniques.

4.2.2 Advanced Reuse Systems

Advanced reuse systems include membrane bioreactors, advanced oxidation processes, and AI to treat and reuse water. These systems enable the removal of a wide range of contaminants, producing high-quality water suitable for various applications, including potable reuse. Research has shown that advanced reuse systems can achieve high levels of water quality, meeting, or exceeding regulatory standards.

4.2.3 Dynamic Response to Demand and Environmental Conditions

Advanced water reuse systems have the ability to adjust to variations in water demands, environmental conditions, and water quality parameters. This dynamic response enables the system to optimize water treatment and distribution in response to dynamic conditions. The incorporation of smart water grid systems, advanced reuse mechanisms, and adaptable response functionalities has transformed water reuse methodologies. This will facilitate effective, environmentally friendly, and robust water resource management. As research continues to advance, water reuse practices will become more effective, ensuring a reliable and sustainable water supply for future generations.

5 Community Engagement and Education

These initiatives help disseminate knowledge to communities regarding the significance of water conservation. These factors also promote active engagement in the preservation of water resources. Educational programs also enhance knowledge regarding water scarcity and promote SWM strategies. Through the active engagement of local communities, SWM strategies can be effectively tailored to address specific regional requirements by incorporating direct feedback. In addition, community-led initiatives such as watershed clean-ups play a crucial role in enhancing the overall efficacy of SWM endeavours, thereby guaranteeing a more sustainable approach to the utilization and conservation of this essential resource.

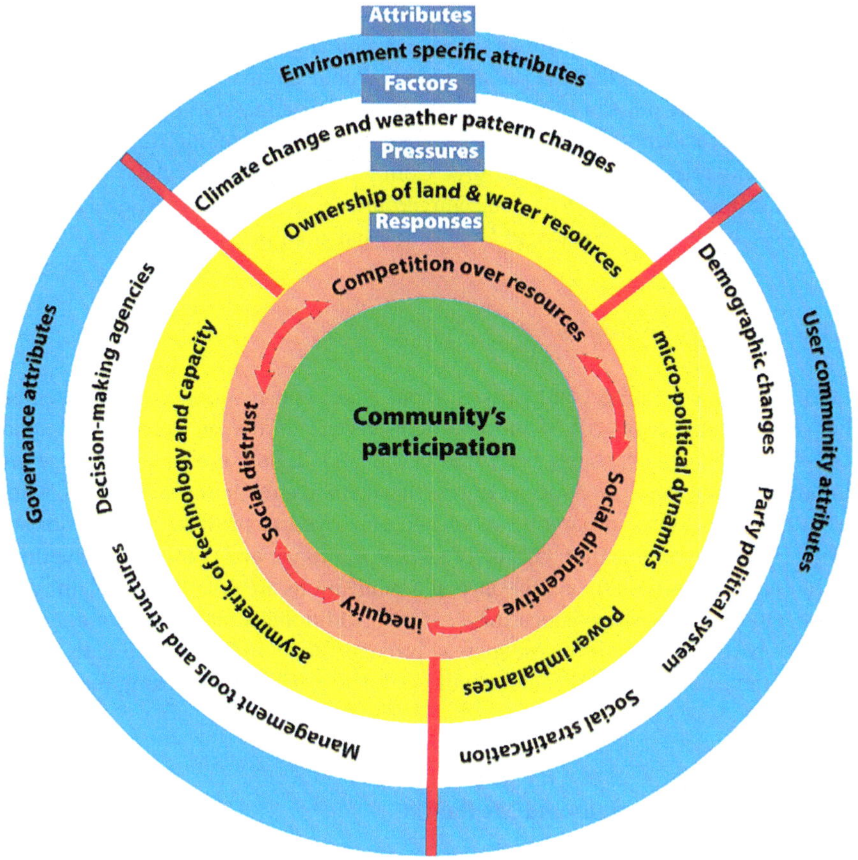

Fig. 6 Framework for analysing the factors influencing community participation

The participation of the community is influenced by a range of pressures, responses, and attributes as shown in Fig. 6. Factors such as climate change, competition for resources, and social distrust pose challenges for community engagement. The response includes the ownership of resources, social disintegration, and inequity. The attributes are associated with the environment, governance, and the user community. These attributes encompass factors such as technology, demographic changes, and decision-making agencies. The interconnections between these elements collectively determine and impact the level of community engagement [49].

5.1 Role of Community Engagement in SWM

Community engagement in SWM involves planning, decision-making, and implementing water conservation strategies. This participatory approach is essential for building local support and a sense of responsibility towards water resources. To effectively spread awareness and educate the public about SWM, strategies include organizing workshops and seminars. These provide interactive platforms for raising awareness and sharing the best possible ideas. Public campaigns using social media and traditional media can boost these messages. Participatory approaches such as encouraging residents to contribute to data collection and monitoring enhance the accuracy and relevance of SWM efforts. Community feedback mechanisms, such as surveys, allow the public to voice their problems and suggestions, ensuring that SWM strategies are responsive to local needs and conditions.

5.1.1 Building Awareness

Strategies for educating communities about water conservation and management help build awareness of SWM and water conservation. This involves targeting educational campaigns through both traditional and digital media to increase the importance of water conservation. Interactive methods, including workshops, webinars, and educational programs, should be employed to provide practical knowledge. For data visualization, GIS elaborates water usage and conservation methods [50]. Partnerships with local schools and organizations are crucial for widening the reach and impact of these educational efforts. Additionally, engaging communities through participation initiatives, such as projects, helps individuals to gain hands-on experience and a deeper understanding of and commitment to long-term SWM practices.

5.1.2 Public Campaigns and Media

Social media platforms and traditional media channels enhance the scope of audience engagement. Public campaigns and media serve as crucial tools in facilitating public engagement in SWM by enhancing awareness. The strategic utilization of diverse media channels, including television, radio, social media, and online platforms, enables the dissemination of information on water conservation practices, policy updates, and the significance of sustainable water usage on a large scale. Campaigns are typically formulated with the objective of engaging the general public by means of persuasive messages, visually appealing content, and interactive components. These findings further emphasize the importance of water-related matters and provide actionable measures for enhancement. Effective storytelling and precise communication can affect how the general public perceives and behaves.

In addition, media coverage pertaining to SWM subjects has the potential to stimulate community dialogues, galvanize back for local endeavours, and foster collaboration among residents and organizations. Through the utilization of various media platforms, public campaigns efficiently facilitate the dissemination of information and encourage active participation in SWM practices.

5.1.3 Citizen Science Initiatives

This facilitates the participation of residents in the process of data collection and monitoring. Citizen science represents a transformative approach to community engagement in SWM through public participation in scientific data collection and analysis. These initiatives enable the general public to participate in water quality monitoring and pollution tracking by utilizing standardized protocols and digital tools. Through the direct engagement of citizens, these programs improve the efficiency of data collection networks, offer real-time analysis, and increase the accuracy of monitoring efforts. In addition, citizen science facilitates increased public awareness and comprehension of SWM issues, as participants acquire practical knowledge of scientific methodologies and environmental problems.

5.2 Educational Initiatives for SWM

Introducing SWM topics within the school and university curriculum offers a theoretical framework. Different types of water management practices that may significantly impact Sustainable Water Management (SWM) through the curriculum are shown in Fig. 7. Through this, related knowledge is provided through experiential learning activities and applied projects. The collaboration between educational institutions and SWM organizations enhances these programs.

These findings establish a connection between academic learning and practical applications, as well as facilitating professional growth. Specialized training, facilitated through workshops, certification programs, and online courses, provides

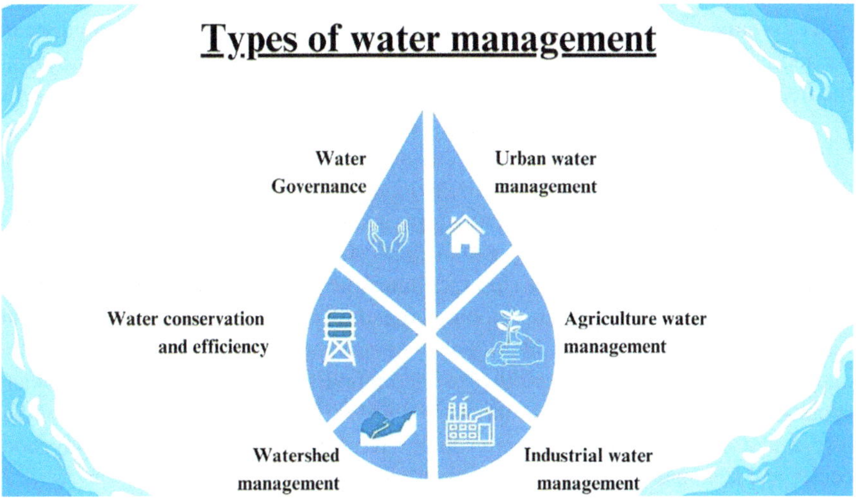

Fig. 7 Different types of water management practices in the curriculum of educational institutions

individuals with flexible educational opportunities, catering to the needs of students and professionals. Digital technologies, including IoT sensors and data analytics, help to enhance SWM in all manners. This comprehensive approach not only enhances educational outcomes but also fosters a more informed and technologically proficient community in the field of SWM.

5.3 *Workshops and Certification Programs*

Specialized training and workshops and certification programs for SWM provide targeted, specialized training designed to enhance practical skills and knowledge. Workshops typically involve intensive, short-term sessions that focus on specific SWM topics. These sessions often include interactive components such as case studies and hands-on exercises. Certification programs, on the other hand, offer a structured pathway for acquiring formal qualifications and demonstrating expertise in Sustainable Water Management (SWM) practices. These programs include covering theoretical concepts, practical applications, and industry standards. Both workshops and certification programs contribute to professional development by equipping individuals with the expertise needed to address complex SWM challenges and implement effective solutions.

6 Conclusion

The complexities of sustainable water management (SWM) encompass multiple factors, such as climate change, population growth, urbanization, and overexploitation of groundwater. Traditional water management methods are no longer adequate for addressing these complexities. The integration of digital technologies, including the Internet of Things (IoT), artificial intelligence (AI), big data analytics, and GIS, offers innovative methods to improve the efficiency, flexibility, and long-term viability of digital water resource management (DWRM). These technologies enable real-time monitoring of water systems, utilization of predictive analytics, and automation of responses, thereby enhancing proactive management. By incorporating modern digital technologies, water management can improve its capacity to withstand and adapt to challenges. This facilitates the sustainable utilization of water resources in the face of growing global demands. The future of water management relies on the deployment of intelligent systems that optimize resource utilization and provide the required adaptability to accommodate changing environmental and societal conditions. Thus, the adoption of digital solutions is not just an option but also a necessity to protect water resources for future generations, ensuring both ecological sustainability and societal resilience.

7 Future Directions

Sustainable water management (SWM) should focus on the continued integration and advancement of digital technologies to address emerging challenges. The subsequent key domains delineating potential in the future are as follows:

1. **Enhanced AI and ML Applications**: Advanced AI and ML models will be utilized to process larger and more complex datasets in real time. These models enhance predictions related to water demand, climate effects, and system anomalies in the field of SWM. These technologies will utilize proactive management strategies to efficiently control water resources. These will also optimize water networks to accommodate population growth and adapt to climate change.
2. **Development of Smart Water Grids**: Smart water grids, analogous to smart energy grids, represent a significant advancement in the field of SWM through the incorporation of the IoT, AI, and advanced analytics. The establishment of these interconnected networks would facilitate the implementation of real-time monitoring systems and automated leak detection algorithms. This would result in a reduction in water loss and the optimization of resource allocation.
3. **Blockchain for DWRM**: The implementation of blockchain technology has the potential to enhance digital water resource management (DWRM) by ensuring transparency, security, and decentralization in the management of water rights, transactions, and data sharing.
4. **Integration of Advanced Water Treatment Technologies**: Future SWM strategies should prioritize the development of advanced treatment technologies such as nanofiltration and bioelectrochemical systems. These technologies should be integrated with AI and the IoT for real-time monitoring. This will improve the quality of water, facilitate the safe reuse of wastewater, optimize operational efficiency, and reduce costs.

References

1. Bahinipati, C. S., & Viswanathan, P. K. (2019). Can micro-irrigation technologies resolve India's groundwater crisis? Reflections from Dark-Regions in Gujarat. *International Journal of the Commons, 13*, 848–858.
2. Mishra, A. K., & Singh, V. P. (2010). A review of drought concepts. *Journal of Hydrololgy, 391*, 202–216.
3. Awasthi, S., et al. (2022). Analyzing urbanization induced groundwater stress and land deformation using time-series Sentinel-1 datasets applying PSInSAR approach. *Science of the Total Environment, 844*.
4. Chowdhary, P., Bharagava, R. N., Mishra, S., & Khan, N. (2020). Role of industries in water scarcity and its adverse effects on environment and human health. In *Environmental concerns and sustainable development* (pp. 235–256). Springer Singapore. https://doi.org/10.1007/978-981-13-5889-0_12
5. IPCC. (2023). *Climate change 2022-impacts, adaptation and vulnerability*. NBER Working Papers. Retrieved from http://www.nber.org/papers/w16019

6. UNDESA. (2015). *International decade for action 'WATER FOR LIFE' 2005–2015. Routledge handbook of water and health.* Retrieved from https://www.un.org/waterforlifedecade/index.shtml. https://doi.org/10.4324/9781315693606-45
7. Shukla, A. K., Ojha, C. S. P., Shukla, S., & Garg, R. D. (2021). Water quality challenges in Ganga River Basin, India. In *The Ganga River Basin: A hydrometeorological approach* (pp. 1–19). Springer. https://doi.org/10.1007/978-3-030-60869-9_1
8. Shukla, A. K., Shukla, S., Surampalli, R. Y., Zhang, T. C., Yu, Y. L., & Kao, C. M. (2023). Modeling microconstituents based on remote sensing and GIS techniques. In *Microconstituents in the environment: Occurrence, fate, removal and management* (pp. 227–246). John Wiley & Sons Publications. https://doi.org/10.1002/9781119825289.ch10
9. Sharma, S., Pathak, S., & Kumar, S. (2024). A planning-support tool for spatial suitability assessment of harvesting sites for stormwater infrastructure. *Journal of Hazardous, Toxic, and Radioactive Waste ASCE, 28*(1), 04023044.
10. Nayak, D., & Shukla, A. K. (2022). Review of state-of-the-art research on river hydrological hazards, restoration, and management. In *River dynamics and flood hazards: Studies on risk and mitigation* (pp. 463–482). Springer Nature Singapore. https://doi.org/10.1007/978-981-19-7100-6_25
11. Chowdhury, A., Jha, M. K., & Chowdary, V. M. (2009). Delineation of groundwater recharge zones and identification of artificial recharge sites in West Medinipur District, West Bengal, using RS, GIS and McDm techniques. *Environmental Earth Sciences, 59*, 1209–1222.
12. Jayaraman, P., Nagarajan, K. K., Partheeban, P., & Krishnamurthy, V. (2024). Critical review on water quality analysis using IoT and machine learning models. *International Journal of Information Management Data Insights, 4*, 1–12.
13. Abebe, A. (2024). Internet of Things (IoT) enabled water distribution system for smart water management. *International Journal of Wireless Communications and Mobile Computing, 11*, 1–10.
14. Hussain, A., Deshwal, A., Priyadarshi, M., Pathak, S., Sambandam, G., Chand, S., & Shukla, A. K. (2024). Landfill leachate analysis from selected landfill sites and its impact on groundwater quality, New Delhi, India. *Environment, Development and Sustainability*, 1–26.
15. Alsaydalani, M. O. (2024). Hydraulic modelling for leakage reduction in water distribution systems through pressure control. *The Open Civil Engineering Journal, 18*, 1–14.
16. Gosling, S. N., & Arnell, N. W. (2013). A global assessment of the impact of climate change on water scarcity. *Climatic Change, 134*, 371–385.
17. Graham, N. T., et al. (2020). Humans drive future water scarcity changes across all shared socioeconomic pathways. *Environmental Research Letters, 15*, 1–10.
18. Molinos-Senante, M., Hernández-Sancho, F., Mocholí-Arce, M., & Sala-Garrido, R. (2014). A management and optimisation model for water supply planning in water deficit areas. *Journal of Hydrology, 515*, 139–146.
19. Zhu, K., Cheng, Y., Zhou, Q., Kápolnai, Z., & Dávid, L. D. (2023). The contributions of climate and land use/cover changes to water yield services considering geographic scale. *Heliyon, 9*, 1–23.
20. Okello, C., Tomasello, B., Greggio, N., Wambiji, N., & Antonellini, M. (2015). Impact of population growth and climate change on the freshwater resources of Lamu Island, Kenya. *Water, 7*, 1264–1290.
21. Younis, H. I., Kizhisseri, M. I., & Mohamed, M. M. (2023). Forecasting future water demands for sustainable development in Al-Ain City, United Arab Emirates. *Water, 15*, 1–29.
22. Palermo, S. A., et al. (2022). Smart technologies for water resource management: An overview. *Sensors, 22*, 1–23.
23. Pathak, S., Sharma, S., Banerjee, A., & Kumar, S. (2024). A methodology to assess and evaluate sites with high potential for stormwater harvesting in Dehradun, India. *Big Data Research, 35*, 100415.
24. Mautner, M. R. L., Foglia, L., Herrera, G. S., Galán, R., & Herman, J. D. (2020). Urban growth and groundwater sustainability: Evaluating spatially distributed recharge alternatives in the Mexico City Metropolitan Area. *Journal of Hydrology, 586*, 1–21.

25. Jato-Espino, D., & Pathak, S. (2021). Geographic location system for identifying urban road sections sensitive to runoff accumulation. *Hydrology, 8*(2), 72.
26. Mateo-Sagasta, J., Zadeh, S. M., Turral, H., & Burke, J. (2017). *Water pollution from agriculture: A global review*. The Food and Agricultural Organization.
27. Pericherla, S., Karnena, M. K., & Vara, S. (2020). A review on impacts of agricultural runoff on freshwater resources. *International Journal on Emerging Technologies, 11*, 829–833.
28. Meran, G., Siehlow, M., & von Hirschhausen, C. (2001). The economics of water. *New Perspectives Quarterly, 18*.
29. Gani, A., Singh, M., Pathak, S., & Hussain, A. (2024). Groundwater quality index development using ANN model of Delhi Metropolitan City, India. *Environmental Science and Pollution Research*, 1–16.
30. Gerlach, S. A. (2005). Industrial effluents. In *Marine pollution* (pp. 289–299).
31. Liu, Y., et al. (2017). A review on effectiveness of best management practices in improving hydrology and water quality: Needs and opportunities. *Science of the Total Environment, 601–602*, 580–593.
32. Turner, S. W. D., Hejazi, M., Yonkofski, C., Kim, S. H., & Kyle, P. (2019). Influence of groundwater extraction costs and resource depletion limits on simulated global nonrenewable water withdrawals over the twenty-first century. *Earth's Future, 7*, 123–135.
33. Yadav, M., Vashisht, B. B., Jalota, S., & Kumar, D. (2023). *Soil-water, agriculture, and climate change* (Vol. 113). Springer International Publishing.
34. Cosgrove, W. J., & Loucks, D. P. (2015). Water management: Current and future challenges and research directions. *Journal of the American Water Resources Association, 5*, 4823–4839.
35. Garcia, X., & Pargament, D. (2015). Reusing wastewater to cope with water scarcity: Economic, social and environmental considerations for decision-making. *Resources, Conservation and Recycling, 101*, 154–166.
36. Singh, M., & Ahmed, S. (2020). IoT based smart water management systems: A systematic review. *Materials Today: Proceedings, 46*, 1–8.
37. Kurundkar, S., et al. (2024). IoT-enabled smart irrigation system for precision farming using microcontroller. In *Data science and intelligent computing techniques* (pp. 945–952). Soft Computing Research Society.
38. Miller, M., Kisiel, A., Cembrowska-Lech, D., Durlik, I., & Miller, T. (2023). IoT in water quality monitoring-are we really here? *Sensors, 23*, 1–12.
39. Kamyab, H., et al. (2023). The latest innovative avenues for the utilization of artificial Intelligence and big data analytics in water resource management. *Results in Engineering, 20*, 1–11.
40. Machnoor, A., Ajayakumar, V., & Malagatti, M. (2024). Machine learning in water management. In *Science and technology: Recent updates and future prospects* (Vol. 1, pp. 59–75). B P International.
41. Gaffoor, Z., Pietersen, K., Jovanovic, N., Bagula, A., & Kanyerere, T. (2020). Big data analytics and its role to support groundwater management in the Southern African development community. *Water (Switzerland), 12*, 1–28.
42. Jamal, E., Moumen, A., Rissouni, Y., Chao, J., & Tahi, A. (2021). Big data analytic and IoT for water resources. In *Proceedings of the 2nd international conference on big data, modelling and machine learning* (pp. 433–439). SCITEPRESS—Science and Technology Publications.
43. Coulibaly, S., Kamsu-Foguem, B., Kamissoko, D., & Traore, D. (2022). Deep learning for precision agriculture: A bibliometric analysis. *Intelligent Systems with Applications, 16*, 1–18.
44. Alazzai, W. K., Abood, B. S. Z., Al-Jawahry, H. M., & Obaid, M. K. (2024). Precision farming: The power of AI and IoT technologies. *E3S Web of Conferences, 491*, 1–6.
45. Tace, Y., et al. (2022). Smart irrigation system based on IoT and machine learning. *Energy Reports, 8*, 1025–1036.
46. Et-taibi, B., et al. (2024). Enhancing water management in smart agriculture: A cloud and IoT-based smart irrigation system. *Results in Engineering, 22*, 1–15.

47. Siddique, I. (2024). Sustainable water management in urban areas: Integrating innovative technologies and practices to address water scarcity and pollution. *SSRN Electronic Journal, 8*, 172–178.
48. Hamed, M. O. A. (2018). Automatic control technologies to enhance water. In *Desalination and water treatment* (pp. 271–291). InTech.
49. Shunglu, R., et al. (2022). Barriers in participative water governance: A critical analysis of community development approaches. *Water, 14*, 1–24.
50. Yahya, B. M., Ahmed, K. A., & Salih, A. M. (2023). Water Resources Management and Applications using GIS: An overview. *The Asian Institute of Research Engineering and Technology Quarterly Review, 6*, 65–73.

Evaluation of Industrial By-Products as Partial Fine Aggregate Replacements in Concrete: A Comparative Analysis of ISF Slag, Fly Ash, and Nanosilica

Gaurav Bharti, B. Bharath, Amit Tripathi, Pushpendra Kumar Sharma, and Bishnu Kant Shukla

1 Introduction

Slag from the imperial smelting furnace (ISF) has been used in concrete technology, giving concrete its unique qualities and sustainability. ISF slag is a by-product of the imperial smelting process, which is mostly used to yield lead and zinc. It comprises a variety of oxides, including Fe_2O_3, CaO, SiO_2, and ZnO. ISF slag, which was once thought of as a waste product, has been shown to have potential in a number of industries, most notably building.

Studies have indicated that ISF slag has interesting properties for concrete manufacture. ISF slag concrete's strength and abrasion properties were investigated by Tripathi et al. (2013), who found that the material is viable for use in construction [1]. In a similar vein, Atzeni et al. (1996) investigated the application of granulated slag from lead and zinc processing in concrete technology, confirming its value in improving the characteristics of concrete [2]. Morrison et al. (2003) examined the use of ferro-silicate slag from ISF zinc production in place of sand, and they found that this improved the performance of concrete [3].

Adding industrial by-products to concrete, like ISF slag, is in line with larger initiatives to support environmentally friendly building methods. The advantages of employing fly ash as a substitute for sand were covered by Siddique (2003), Pappu et al. (2007), and Rajamane et al. (2007), who also highlighted the possibility of

G. Bharti · B. Bharath · A. Tripathi
Department of Civil Engineering, JSS Academy of Technical Education, Noida, India

P. K. Sharma
School of Civil Engineering, Lovely Professional University, Phagwara, India

B. K. Shukla (✉)
Department of Civil Engineering, JSS Academy of Technical Education, Noida, India

School of Civil Engineering, Lovely Professional University, Phagwara, India

S. Pathak et al. (eds.), *Intelligent Infrastructure and Smart Materials*,
https://doi.org/10.1007/978-3-031-92421-7_7

using ISF slag in similar applications [4–6]. Tripathi et al. (2016) assessed ISF slag's performance in concrete as a natural sand substitute and found it to be suitable [7]. Subsequent research by Singh and Siddique (2014) and Tan and Phan (2024) emphasised the durability and mechanical qualities of concrete including substitute aggregates, such as coal bottom ash and nanosilica [8, 9].

ISF slag is valuable in concrete applications, as evidenced by recent experiments [10–12]. Its application as a structural fill in reinforced soil structures was studied by Prasad and Ramana (2015), who showed that it is more adaptable than conventional concrete mixtures [13]. The possibility for wider adoption is reinforced by the insights provided by Dash et al. (2016) and Singh et al. (2017) regarding the use of industrial waste and ISF slag in concrete, respectively [14, 15]. Further research on substitute fine aggregates for concrete production by Krithika et al. (2019) solidified ISF slag's position as a practical choice [16].

Imperial smelting furnace slag, a residue from zinc and lead production, is another promising material in construction. Utilising ISF slag in concrete and other building materials helps in managing industrial waste effectively. By incorporating these industrial by-products, the construction industry can significantly reduce waste, lower carbon emissions, and enhance the performance and lifespan of building materials.

2 Materials

2.1 Fly Ash

GPC is an open-air produced inorganic polymer created from industrial and agricultural waste with an alumina and high silica content. The thermally activated natural minerals such as bentonite, wheat straw ash, and kaolinite, as well as industrial waste items including rice husk ash, fly ash, as well as alkaline activating solutions, give the GPC, which is a combination of alumina and silica. These arrangements polymerise these materials into subatomic bonds and structures to form a solidified fastener. GPC is often alluded to as "Alkali-Activated Materials" (AAMs); in any case, the two frameworks are synthetically unique. While geopolymerisation makes a three-dimensional stable polymer structure, enactment in AAM does not decisively prompt unsteady monomers. In comparison to GPC, the AAMs are more resilient while having less strength. The shape and appearance of fly ash particles are primarily spherical, ranging from 1 to 100 μm in diameter. These spherical particles contribute to the improved workability and reduced water demand in concrete mixes. Siddique (2003) explored the effect of replacing fine aggregates with Class F fly ash, highlighting the spherical nature of fly ash particles, which enhance the packing density and reduce the porosity of the concrete matrix [4]. Rajamane et al. (2007) also discussed the spherical morphology of fly ash [6] as shown in Fig. 1, which aids in the lubrication of concrete mixtures, thereby improving the compressive strength of the final product.

Fig. 1 Fly ash

Fig. 2 Imperial smelting furnace slag

2.2 *ISF Slag*

ISF slag is a by-product of the imperial smelting process used in zinc and lead production. ISF slag is characterised by its angular and irregular particle shape, which results from the rapid cooling of molten slag as shown in Fig. 2. This angularity can enhance the mechanical interlocking within concrete, potentially improving its strength and durability. Tripathi et al. (2013) investigated the strength and abrasion characteristics of ISF slag concrete [1], noting the rough and angular texture of ISF slag particles, which contribute to the mechanical properties of the concrete. Additionally, Yin et al. (2016) and Varadharajan et al. (2023) evaluated the chemical stability of ISF slag [17, 18], providing insights into its morphological characteristics which influence its performance in concrete applications.

2.3 Nanosilica

Nanosilica is a synthesised material known for its extremely fine particle size, typically in the range of 1–100 nm as shown in Fig. 3. The shape of nanosilica particles can vary, but they are often spherical or nearly spherical, contributing to their high surface area and reactivity. This nanoscale size and shape allow nanosilica to act as a filler material, enhancing the density and mechanical properties of concrete. [10]. Tran and Phan (2024) conducted an experimental study on the use of nanosilica in concrete [10], highlighting the uniform spherical shape of nanosilica particles that enhance the microstructure of the concrete matrix. Nigam and Verma [11] and Varadharajan et al. [19] examined the effect of fine sand and nanosilica on the mechanical properties of concrete, demonstrating that the incorporation of nanosilica improves the compressive strength and durability due to its fine particle size and shape.

Fig. 3 Nanosilica

3 Literature Review

Because of its flexibility and durability, concrete is a popular building material. For sustainable development, researchers are incorporating industrial by-products and alternative materials into concrete to increase its properties while reducing environmental impact. This literature review looks at studies on the effects of industrialised by-products such as ISF slag, fly ash, and nanosilica on concrete mechanical properties, durability, and performance. Tripathi et al. (2013) investigated ISF slag concrete strength and abrasion. ISF slag enhanced concrete's abrasion resistance, making it appropriate for wear-sensitive applications [1]. Atzeni et al. (1996) explored granulated slag from lead and zinc smelting for use in concrete technology. According to their findings, granulated slag might be used instead of standard aggregates without compromising the structural reliability of concrete. The study emphasised the environmental benefits of employing industrial by-products in concrete to minimise natural resource consumption [2]. Morrison et al. (2003) investigated ferro-silicate slag by ISF zinc production as a concrete sand substitute. Ferro-silicate slag concrete is stronger and more robust than conventional concrete [3]. This study showed that industrial slags may be utilised to manufacture concrete, which promotes sustainable building. Tripathi and Chaudhary (2016) assessed the performance of ISF slag as a concrete sand alternative [7]. They discovered that ISF slag might replace natural sand in concrete compositions, enhancing mechanical properties and the environment. The study looked at the environmental benefits of using ISF slag in concrete, particularly in terms of reducing the footprint of concrete manufacturing. Siddique (2003) found that substituting Class F fly ash for fine aggregate altered the mechanical properties of concrete [4]. Fly ash increased concrete workability while decreasing permeability, making it an interesting option for increasing concrete performance. Rajamane et al. (2007) estimated concrete compressive strength using fly ash as a sand substitute. Their prediction methodology for fly ash concrete compressive strength shown potential for high-strength concrete [6]. The study also emphasised the economic and environmental advantages of employing fly ash, a coal combustion by-product, in concrete making. Tripathi and Chaudhary (2016) directed a wide-ranging study on the use of imperial smelting furnace (ISF) slag as an ancillary for natural sand in concrete, published in the Journal of Cleaner Production. The research aimed to assess the performance of concrete mixtures incorporating ISF slag and evaluate its feasibility as an alternative to natural sand [7]. Singh and Siddique (2014) investigated the compressive strength, drying shrinkage, and chemical resistance of concrete with coal bottom ash as a sand substitute [8]. They discovered that bottom ash may replace sand in concrete, enhancing mechanical and chemical qualities. The study found that bottom ash may be used to create long-lasting, sustainable concrete. Sakalkale et al. (2014) and Tran and Phan (2024) examined nanosilica in concrete. According to the researchers, nanosilica enhanced the compressive strength and lifespan of concrete. Nanosilica filled the micro-pores in the concrete matrix, increasing its density and strength. According to studies, nanosilica has potential for enhancing concrete technology [9, 10]. Pappu et al. (2007) investigated Indian solid waste generation and building material recycling. Fly ash, slag, and bottom ash were discovered to be valuable in concrete manufacturing. The study emphasised the need of recycling solid waste in reducing pollution and promoting sustainable development [5]. The study looked at ways to recycle solid waste, which may help develop eco-friendly construction materials. According to the literature, industrial wastes such as ISF slag, fly ash, bottom ash, and nanosilica can be utilised to make concrete. Findings of recent researches highlight the need of adopting sustainable materials in concrete technology to promote environmentally friendly construction and high-performance building materials [20, 21]. To optimise sustainability and performance, future research should investigate the proper use of these by-products in real-world applications. Nanosilica reduced concrete permeability while increasing chemical resistance, making it more durable. Fine sand and nanosilica were shown to boost the compressive and tensile strengths of concrete, as per recent investigations [10, 11]. Fine sand improved particle packing, whereas nanosilica filled concrete matrix micro-pores, increasing strength and durability. Rao et al. (2024) investigated the strength of concrete using nanosilica powder and crushed tile waste. Their findings suggested that nanosilica and waste materials could produce long-lasting and cost-effective concrete [12]. Prasad and Ramana (2016) investigated the geotechnical properties of ISF slag as a structural fill in reinforced soil structures in "Geotextiles and Geomembranes". ISF slag has excellent strength, low compressibility, and good drainage, making it suitable for reinforced earth constructions [13]. The study discovered that ISF slag may be employed in geotechnical applications, hence increasing industrial by-product sustainability. Dash et al. (2016), in his review, investigated how fly ash, slag, and bottom ash impact the mechanical characteristics of concrete [14]. The authors emphasised that using industrial waste as a partial replacement for natural aggregates might improve concrete manufacturing sustainability. The study discovered that industrial waste materials might increase the compressive strength, durability, and workability of concrete, making them viable aggregate replacements. Singh et al. (2017) investigated the compressive strength of mortars incorporating ISF slag and marble powder. They discovered that ISF slag and marble powder increased mortar compressive strength. The study discovered that ISF slag's pozzolanic activity and marble powder's filling effect made the mortar matrix denser and more cohesive [15]. The findings indicated that using ISF slag and marble powder in mortars might create sustainable, high-performance construction materials. The literature review reveals that nanosilica, ISF slag, and other industrial waste materials may be utilised to produce concrete. These additives increase concrete's mechanical properties and durability while encouraging industrial waste recycling and reducing natural resource use. The findings underline the need of adopting sustainable materials in concrete technology for green buildings. Nanosilica increases concrete compressive strength and longevity, making it an ideal constituent for high-performance concrete. Similar to ISF slag, many industrial waste materials can increase the mechanical properties and sustainability of concrete. These findings can help the construction industry utilise more sustainable and high-performance concrete to address environmental and structural concerns. Industrial waste and new materials used in concrete production are a significant step towards sustainable development and building technology innovation. This comprehensive literature review examines the real advantages and applications of nanosilica, ISF slag, and other industrial waste materials. These materials require more research and development to fully realise their promise as sustainable and high-performance concrete. Sustainable construction technologies and the utilisation of industrial by-products have sparked interest in incorporating alternative materials into concrete mixtures. This literature review looks at how substitute fine aggregates and industrial slags impact concrete's mechanical properties, durability, and performance. The review proceeds in the order of the citation numbers. Kirthika et al. (2019) studied how clay in alternative fine aggregates impacts the mechanical properties and durability of concrete [16]. Clay with alternative fine particles decreased concrete workability and compressive strength. Treatment and clay content management can help to lessen the detrimental outcomes.

The study found that employing alternative fine aggregates to achieve concrete performance requires strict quality control. Yin et al. (2016) investigated the chemical stability and environmental effect of lead blast furnace (LBF) and imperial smelting furnace (ISF) slags in building materials. The chemical stability of LBF and ISF slags was satisfactory, with no heavy metal leaching [17]. This study shown that LBF and ISF slags can be long-term building materials if their chemical stability is studied and monitored. Granulated slag from lead and zinc processing was tested as a partial substitute for fine particles in concrete. Granulated slag's pozzolanic properties improved the compressive strength and durability of concrete [2]. The study found that employing industrial slags in concrete manufacturing might minimise the need for natural aggregates while also encouraging industrial by-product recycling.

Sugita et al. [22] investigated the mechanical properties and durability of concrete made with zinc slag as fine aggregate. Zinc slag performed exceptionally well as a fine aggregate in concrete, exceeding ordinary concrete. The study found that zinc slag can minimise the need for natural sand and promote industrial waste recycling in concrete production. Patil et al. (2016) investigated the mechanical properties and durability of cement concrete mixes containing imperial smelting furnace (ISF) slag as fine aggregate. ISF slag increased concrete compressive strength and durability, making it a viable substitute for natural fine aggregates [23]. The study also found that ISF slag decreases industrial waste while conserving natural resources. According to the research, alternative fine aggregates and industrial slags have a high potential in concrete manufacture. These additives increase concrete's mechanical properties and durability while also aiding industrial by-product recycling and reducing natural resource use. Kirthika et al. (2019) addressed clay-based fine aggregate issues [16]. Their findings emphasised the need for treatment and quality control in mitigating clay's detrimental effects on concrete performance. This research is critical for understanding alternative fine aggregates in concrete's limitations and solutions. Yin et al. (2016) claim that industrial slags in construction materials cause environmental difficulties [17]. LBF and ISF slags can be used in construction safely after being tested for long-term chemical stability. This research is critical to ensuring the environmental safety and sustainability of industrial slags in concrete. Atzeni et al. (1996) and Sugita (1988) report that granulated and zinc slag may be utilised as fine aggregates in concrete [2, 22]. Patil et al. (2016) investigated ISF slag as a concrete fine aggregate [23]. Their findings revealed that ISF slag increases concrete mechanical properties and durability. This study demonstrates that ISF slag may increase concrete performance and address industrial slag disposal issues. The literature review stresses the use of alternative fine aggregates and industrial slags in concrete making.

Power (2023) discusses employing metallurgical residues in construction [24]. Metallurgical wastes such as slag, dust, and sludge can be utilised in construction materials. These remnants increase the mechanical properties of concrete while also reducing industrial waste. According to Power [24], when properly managed and integrated, these residues can contribute to sustainable construction and improved building material performance. The chapter discusses current and future advancements in metallurgical waste in construction. Brand and Fanijo (2020), in their study, reviewed the impacts of steel furnace slag type on the mechanical characteristics, durability, and workability of concrete [25]. The investigation categorises BOF and EAF slag. The findings indicate that the kind and treatment of steel slag influences the performance of concrete mixes. BOF slag's high calcium oxide content increases concrete compressive strength and durability. However, free lime in BOF slag might cause expansion, necessitating cautious treatment. EAF slag increases concrete hardness and stability. According to the research, steel furnace slag may substitute natural aggregates, conserving resources and minimising waste. Haruehansapong et al. (2014), in their investigation on the influence of nanosilica particle size on cement mortar, concluded that smaller nanosilica particles boosted mortar compressive strength due to their higher surface area and reactivity [26]. Compressive strength gains plateaued or declined after the nanosilica content of the cement exceeded 3%. To improve cement-based material performance, nanosilica particle size and content must be adjusted. Mohammed et al. (2016) investigated nanosilica modified rubbercrete [27]. Rubbercrete is a mixture of waste tire crumb rubber and natural aggregates. Nanosilica was added to rubbercrete to increase mechanical properties and reduce rubber-induced strength loss. The study found that nanosilica boosted the compressive, flexural, and durability properties of rubbercrete. The pozzolanic reaction of nanosilica resulted in a denser matrix and improved rubber-cement paste bonding. This study demonstrates that nanosilica may enhance environmentally friendly concrete created from waste rubber, hence promoting sustainable building. Mallika Chekravarty et al. (2022) investigated the effects of nanosilica on the compressive, tensile, and flexural strengths of normal-strength concrete [28]. The study recommended nanosilica dosage of 2% by weight of cement, which enhanced concrete compressive strength. Chekravarty et al. [28] concludes that the incorporation of nanosilica considerably increases the mechanical properties of normal-strength concrete, making it a viable option for enhancing the performance and durability of concrete structures. The research highlights the potential of nanosilica as a beneficial additive in the construction industry, promoting the development of high-performance concrete. Nanosilica's filler effect and pozzolanic activity boosted concrete matrix density and cohesiveness, resulting in higher tensile and flexural strength. This study demonstrates that nanosilica can increase the performance of normal-strength concrete, allowing for the development of high-performance building materials. According to the reviewed study, metallurgical wastes, steel furnace slag, and nanosilica may all be employed in concrete. These additives increase concrete's mechanical properties and durability while simplifying industrial by-product recycling and reducing natural resource use. Power [24] discusses the use of metallurgical residues in construction, highlighting their potential to improve concrete performance and address industrial waste environmental issues. This chapter promotes industrial residue research in concrete applications. According to Brand and Fanijo (2020), incorporating new materials and industrial by-products into concrete, improves its mechanical properties and sustainability [25]. This literature study looks at how silica fume, nanosilica, copper slag, glass fibres, and fly ash impact concrete's mechanical properties, durability, and performance. The review is organised by citation numbers. Fallah-Valukolaee et al. (2022) offered a comparison of the mechanical properties and life-cycle evaluation of high-strength concrete employing silica fume and nanosilica as partial cement replacements [29]. The study looked at how silica fume and nanosilica increase the mechanical properties of high-strength concrete, as well as the environmental effects. Silica fume and nanosilica improved concrete's compressive, tensile, and durability. Nanosilica had a bigger impact because of its higher reactivity and smaller particle size, which raised the density of the concrete matrix. According to the life cycle assessment, using less cement might reduce the carbon footprint of concrete manufacture. This study demonstrates that silica fume and nanosilica increase the performance and sustainability of high-strength concrete.

Chithra et al. (2016) studied the synergistic impacts of colloidal nanosilica and copper slag on high-performance concrete's workability, mechanical properties, and durability [30]. Colloidal nanosilica improved the workability and cohesiveness of the concrete mix, whereas copper slag reinforced and extended its durability. The study found that these materials greatly increased compressive, flexural, and chloride penetration. The findings suggest that colloidal nanosilica and copper slag may be combined to produce high-performance concrete with improved mechanical and durability properties, hence increasing the use of industrial wastes in sustainable construction. Khan et al. (2018) performed experimental research on the mechanical characteristics of glass fibre-reinforced high-strength concrete, partially substituting cement with nanosilica to enhance the mechanical properties of the concrete [31]. Glass fibres boosted concrete's tensile and flexural strengths, whereas nanosilica increased its compressive strength and lifespan. Glass fibres combined with nanosilica resulted in a composite material with superior mechanical properties for high-performance structural applications. This study demonstrates that fibres and nanosilica may be used to generate high-strength concrete with better properties. Ahmad et al. (2024) conducted an assessment of fly ash with and without additional additions for stabilising expansive soils and concluded that expansive soils are prone to volume variations when moisture levels fluctuate [32]. According to the study, fly ash, alone or in combination with lime or cement, reduced the swell potential of expansive soils while increasing their compressive strength. Fly ash stabilised and reinforced the soil, preparing it for building. Fallah-Valukolaee et al. (2022) explored the advantages of silica fume and nanosilica in high-strength concrete [29]. Their research emphasised the importance of these ingredients in improving concrete's mechanical properties and sustainability. The life-cycle assessment revealed that silica fume and nanosilica minimise cement content, which is ecologically favourable. Chithra et al. (2016) investigated how colloidal nanosilica and copper slag improved high-performance concrete [30]. Their investigation revealed that these materials might produce concrete with improved workability, strength, and durability. Khan et al. (2018) investigated the use of glass fibres and nanosilica in high-strength concrete and discovered that they increased mechanical qualities [31]. The study discovered that fibres and nanosilica may be used to produce high-performance concrete for demanding structural applications. Ahmad et al. (2024) investigated fly ash's ability to stabilise expansive soils and enhance soil properties [32].

4 Physical and Chemical Properties of Fly Ash, ISF Slag, and Nanosilica

The physical and chemical features of industrial waste, such as particle size distribution, specific gravity, fine material concentration, and chemical composition, are critical in assessing their appropriateness and workability as substitutes for fine aggregates in concrete. These qualities reveal how well industrial wastes work in concrete mixes, guaranteeing that the final concrete retains the necessary structural integrity and performance attributes.

4.1 Shape and Appearance

Fly ash is a pozzolanic substance that does not have cementitious capabilities on its own. However, when finely split and exposed to moisture, it chemically interacts with lime to produce compounds having cementitious characteristics. This substance is produced by the burning of powdered coal. ISF slag, on the other hand, is black, vitreous, and granular, with hazardous elements including lead and zinc [1]. Nanosilica sludge is a wet industrial waste derived directly from nanosilica factory deposits.

4.2 Chemical Composition

4.2.1 ISF Slag

ISFS is mostly composed of Fe_2O_3, followed by SiO_2, CaO, ZnO, and Al_2O_3. Figure 1 and Table 1 show the typical chemical composition of ISFS employed in the present investigation. The concentration of Fe_2O_3 in ISFS is very high [33].

Table 1 Chemical composition of ISF slag

Chemical constituents	Composition (% by mass)
Fe_2O_3	35.76
CuO	0.26
SrO	0.05
ZrO_2	0.01
BaO	0.15
V_2O_5	0.06
Na_20	
ZnO	15.57
SO_3	2.73
Cr_2O_3	0.06
MgO	0.88
SiO_2	16.48
PbO	1.58
K_2O	1
TiO_2	0.39
CaO	18.19
Al_2O_3	5.04
P_2O_5	0.89
As_2O_3	0.11
MnO	1.14
NiO	0.02

Table 2 Chemical composition of fly ash

Constituents	SiO_2	Al_2O_3	Fe_2O_3	CaO	MgO	SO_3	Na_2O and K_2O
Fly ash class F (mass %)	54.9	25.8	6.9	8.7	1.8	0.6	0.6

4.2.2 Fly Ash

Fly ash is mostly composed of silica (SiO_2), alumina (Al_2O_3), calcium oxide (CaO), and ferric oxide (Fe_2O_3). Other minor components are magnesium oxide (MgO), sodium oxide (Na_2O), potassium oxide (K_2O), sulphur trioxide (SO_3), manganese oxide (MnO), titanium dioxide (TiO_2), and unburned carbon as shown in Table 2. The main ingredients can vary greatly: silica from 25% to 60%, alumina from 10% to 30%, and ferric oxide from 5% to 25%. Fly ash is classed as class F when the total of silica, alumina, and ferric oxide is 70% or higher, with reactive calcium oxide <10% [34–37].

4.2.3 Nanosilica

Nanosilica can be used to make new goods or as an admixture to improve natural resource efficiency and decrease environmental effect by eliminating nanosilica waste disposal [38–41] as shown in Table 3.

Table 3 Chemical composition of nanosilica

Compound	SiO_2	Al_2O_3	Fe_2O_3	CaO	MgO	Density (gm/cm^3)
Mass (%)	28.35	0.42	9.7	40.45	16.25	2.8

Table 4 Specific gravity of ISF slag, fly ash, and nanosilica

Materials	Specific gravity
ISF slag	3.88
Fly ash	2.72
Nanosilica	2.67

5 Tests Performed

5.1 Specific Gravity (SG) Test

Specific gravity is commonly described as the weight of a particular volume of material divided by the weight of an equivalent amount of water. Several investigations have recorded the SG of various industrial waste products, which is summarised in Table 4. Morrison et al. [3] reported that the SG of ISF slag is 3.88. Kosior-Kazberuk and Lelusz (2007) found a specific gravity of 2.72 for fly ash at 20% replacement [42]. Vigneshpandian et al. (2017) discovered that the SG of nanosilica is 2.67 [43].

5.2 Compressive Strength of Concrete

The concrete compression test employed a cube of 150 mm × 150 mm × 150 mm. All cubes were left to cure in the open air. At 7 and 28 days after curing, cubes of varied proportions were evaluated with a 3000 kN volume compression tester. The test comprised centring the test item in the testing equipment and delivering a uniform stress of 10 kg/cm^2 per minute. The loading continued until the values began to decline, indicating that the test component had failed. At this point, the machine came to a halt, and the current reading was recorded as the ultimate load. To compute the final cubic compressive strength, divide the ultimate load by the sample's cross-sectional area. Workability was tested using IS 2250-1981 [44].

5.3 Split Tensile Strength of Concrete

This indirect test method determines the tensile strength of a cylindrical specimen. In this test, a cylindrical test piece with a diameter of 100 mm and a length of 200 mm was allowed to cure for 28 days before being tested with a 3000 kN compression tester. To evade direct loading on the specimen, it was put beneath a wooden specimen. The test comprised progressively adding the stress until the test piece shattered, after which the measured value was recorded. The technique was carried out in line with IS 2250-1981 [44].

5.4 Flexural Strength of Concrete

Flexural strength, also known as modulus of rupture or transverse breaking strength, is a material factor that describes the stress in a material shortly before it gives or ruptures during a bending test. The transverse bending test, which is normally done with a three-point bending technique, is often used to assess this feature. This test requires bending a specimen with a rectangular or circular cross section until it breaks or gives. The flexural strength is the greatest stress that the material can bear throughout the test. This configuration provides for more homogeneous stress distribution and precise flexural strength measurement.

6 Results and Discussion

6.1 Compressive, Split Tensile, and Flexural Strength of Concrete

6.1.1 Comparative Analysis of Compressive Strength

The test findings for specimens containing varied amounts of ISF slag, fly ash, and nanosilica substitutes are compared. Table 5 compares the 28-day compressive, split tensile, and flexural strengths of mixes including ISF slag, fly ash, and nanosilica. Tables 2–4 offer these comparisons, which are based on research by Tripathi et al. (2013) for ISF slag, Siddique (2003) for fly ash, and Vigneshpandian et al. (2017) for nanosilica [1, 4, 43]. Figures 4, 5, and 6 illustrate these comparisons [37].

Table 5 Compressive strength (N/mm^2) comparison of ISF slag, fly ash, and nanosilica

Percentage of sand replacement (%)	ISF slag (0.4 w/c)	Fly ash (0.47–0.50 w/c)	Nanosilica (0.42 w/c)
0	48.25	26.4	31.42
10	54.00	28.2	31.8
20	52.70	30.8	33.02
30	55.46	34.9	38.60
40	56.17	38.9	39.10
50	51.50	40.0	40.22
60	51.28	–	–
70	50.74	–	–

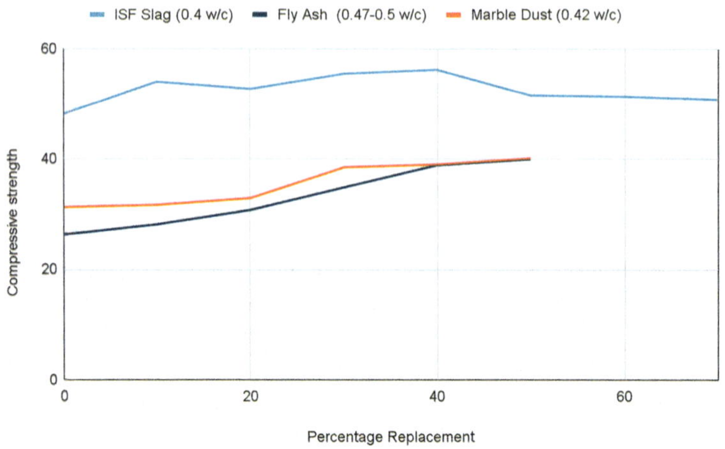

Fig. 4 Comparison graph of compressive strength of ISF slag, fly ash, and nanosilica

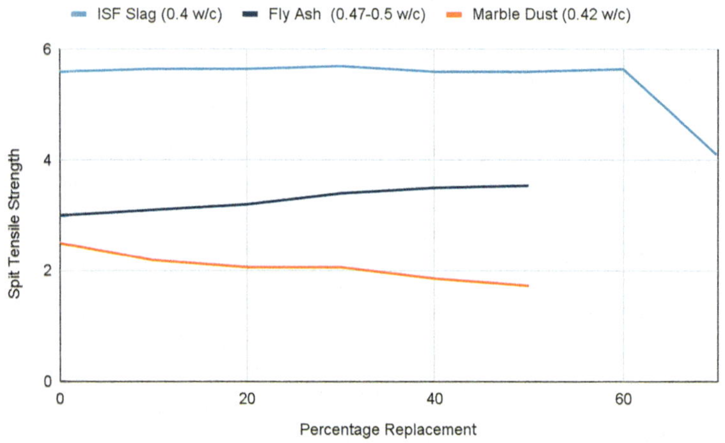

Fig. 5 Comparison graph of split tensile strength of ISF slag, fly ash, and nanosilica

Table 5 presents the compressive strength of concrete with varying percentages of sand replacement using three different materials: ISF slag, fly ash, and nanosilica. The water-to-cement (w/c) ratio is specified for each material: 0.4 for ISF slag, 0.47–0.50 for fly ash, and 0.42 for nanosilica. The compressive strength values are recorded for sand replacement percentages ranging from 0% to 70%. For ISF slag; the strength increases from 48.25 MPa at 0% replacement to a peak of 56.17 MPa at 40% replacement, then slightly decreases at higher replacement levels. For fly ash, the strength increases steadily from 26.4 MPa at 0% to 40.0 MPa at 50% replacement, with no data beyond this point. For nanosilica, the strength increases from 31.42 MPa at 0% to 40.22 MPa at 50% replacement, with no data available for higher replacement percentages. This table indicates that ISF slag and fly ash

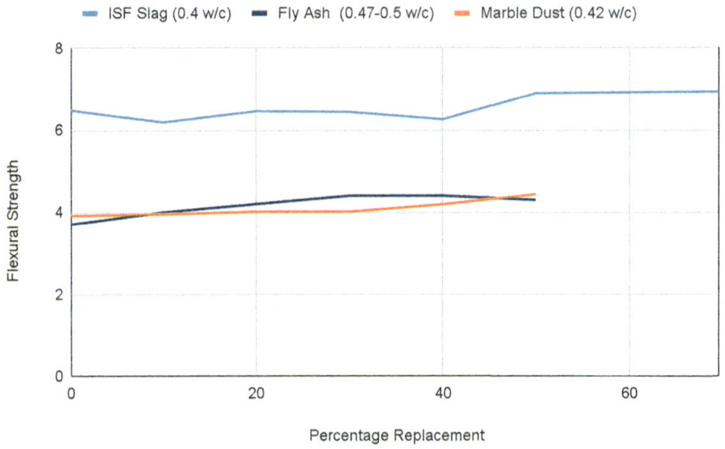

Fig. 6 Comparison graph of flexural strength of ISF slag, fly ash, and nanosilica

Table 6 Split tensile strength (in N/mm^2) comparison of ISF slag, fly ash, and nanosilica

Percentage of sand replacement	ISF slag	Fly ash	Nanosilica
0	5.6	3	2.51
10	5.65	3.1	2.2
20	5.65	3.2	2.069
30	5.70	3.4	2.069
40	5.60	3.5	1.865
50	5.60	3.54	1.737
60	5.65	–	–
70	4.10	–	–

improve the compressive strength of concrete up to certain replacement levels, while nanosilica shows a consistent increase within the measured range.

6.1.2 Comparative Analysis of Split Tensile Strength

The split tensile strength values are given for sand replacement percentages ranging from 0% to 70%. For ISF slag, the split tensile strength starts at 5.6 MPa for 0% sand replacement and increases slightly to 5.70 MPa at 30% replacement. It then stabilises around 5.60 MPa up to 60% replacement, with a significant drop to 4.10 MPa at 70%. For fly ash, the strength increases steadily from 3.0 MPa at 0% to 3.54 MPa at 50% replacement, with no data available beyond this point. For nanosilica, the split tensile strength decreases from 2.51 MPa at 0% to 1.737 MPa at 50% replacement, with no data available beyond this point.

Table 6 indicates that ISF slag maintains relatively stable split tensile strength up to 60% sand replacement, followed by a significant decrease. Fly ash shows a consistent increase in split tensile strength up to the measured limit, while nanosilica demonstrates a decreasing trend, suggesting that its effectiveness in enhancing split tensile strength diminishes with higher sand replacement percentages.

6.1.3 Comparative Analysis of Flexural Strength

The flexural strength values are provided for sand replacement percentages ranging from 0% to 70%. For ISF slag, the flexural strength starts at 6.48 MPa for 0% sand replacement, decreases slightly to 6.19 MPa at 10%, and then fluctuates slightly around 6.46–6.94 MPa up to 70% replacement, showing an overall increasing trend. For fly ash, the flexural strength increases from 3.7 MPa at 0% replacement to 4.4 MPa at 30% and 40%, before slightly decreasing to 4.3 MPa at 50%. There is no data available beyond 50% replacement for fly ash. For nanosilica, the flexural strength starts at 3.923 MPa at 0% replacement and generally increases to 4.444 MPa at 50% replacement, with no data available beyond this point.

Table 7 indicates that ISF slag maintains a relatively high flexural strength across various sand replacement percentages, with a general increase observed at higher replacements. Fly ash shows a consistent increase in flexural strength up to 40% replacement, followed by a slight decrease. Nanosilica shows a consistent increase in flexural strength up to 50% replacement, suggesting that it effectively enhances the flexural strength of concrete within the measured range.

- By reviewing various published articles, it was found that ISF slag, fly ash, and nanosilica as a replacement of sand shows encouraging results.
- The first result shows the comparison of compressive strength of concrete made using ISF slag, fly ash, and nanosilica as a percentage replacement of sand.
- The second result compares the split tensile strength of concrete using ISF slag, fly ash, and nanosilica as sand replacement.
- The third result shows the comparison of the flexural strength of concrete with ISF slag, fly ash, and nanosilica as percentage of sand replacement.
- By analysing multiple published articles, it was also concluded that the SG of ISF slag is more in comparison to that of fly ash and nanosilica.

Table 7 Flexural strength (in N/mm^2) comparison of ISF slag, fly ash, and nanosilica

Percentage of sand replacement	ISF slag (0.4 w/c)	Fly ash (0.47–0.50 w/c)	Nanosilica (0.42 w/c)
0	6.48	3.7	3.923
10	6.19	4.0	3.96
20	6.46	4.2	4.022
30	6.44	4.4	4.022
40	6.26	4.4	4.2
50	6.9	4.3	4.444
60	6.92	–	–
70	6.94	–	–

7 Conclusion

The present study investigated the use of industrial by-products, namely ISF slag, nanosilica, and fly ash, as partial replacements for fine aggregates in concrete. The findings demonstrate significant benefits in using these materials, particularly ISF slag, which due to its higher specific gravity, enhances the density and mechanical properties of concrete. Key conclusions include:

- ISF slag significantly improves the mechanical properties of concrete, with compressive strength reaching up to 56.17 MPa at a 40% replacement level.
- Fly ash can replace up to 50% of fine aggregate, with compressive strength peaking at 40 MPa.
- Nanosilica also performs well, with compressive strength reaching 40.22 MPa at a 50% replacement.
- Split tensile and flexural strength analyses indicate that ISF slag maintains relatively stable performance up to 60% replacement, while fly ash and nanosilica show optimal results at 50% replacement.
- Practical applications of such concrete include high-strength construction components and infrastructure projects where enhanced durability and mechanical properties are required.

Overall, the use of these industrial by-products not only addresses environmental concerns by reducing waste but also contributes to the production of high-performance concrete, thereby supporting sustainable construction practices.

References

1. Tripathi, B., Misra, A., & Chaudhary, S. (2013). Strength and abrasion characteristics of ISF slag concrete. *Journal of Materials in Civil Engineering, 25*(11), 1611–1618.
2. Atzeni, C., Massidda, L., & Sanna, U. (1996). Use of granulated slag from lead and zinc processing in concrete technology. *Cement and Concrete Research, 26*(9), 1381–1388.
3. Morrison, C., Hooper, R., & Lardner, K. (2003). The use of ferro-silicate slag from ISF zinc production as a sand replacement in concrete. *Cement and Concrete Research, 33*(12), 2085–2089.
4. Siddique, R. (2003). Effect of fine aggregate replacement with class F fly ash on the mechanical properties of concrete. *Cement and Concrete Research, 33*(4), 539–547.
5. Pappu, A., Saxena, M., & Asolekar, S. R. (2007). Solid wastes generation in India and their recycling potential in building materials. *Building and Environment, 42*(6), 2311–2320.
6. Rajamane, N. P., Peter, J. A., & Ambily, P. S. (2007). Prediction of compressive strength of concrete with fly ash as sand replacement material. *Cement and Concrete Composites, 29*(3), 218–223.
7. Tripathi, B., & Chaudhary, S. (2016). Performance based evaluation of ISF slag as a substitute of natural sand in concrete. *Journal of Cleaner Production, 112*, 672–683.
8. Singh, M., & Siddique, R. (2014). Compressive strength, drying shrinkage and chemical resistance of concrete incorporating coal bottom ash as partial or total replacement of sand. *Construction and Building Materials, 68*, 39–48.

9. Sakalkale, A. D., Dhawale, G. D., & Kedar, R. S. (2014). Experimental study on the use of nanosilica in concrete. *International Journal of Engineering Research and Applications, 4*(10), 44–50.
10. Tran, H. B., & Phan, V. T. A. (2024). Potential usage of fly ash and nano silica in high-strength concrete: Laboratory experiment and application in rigid pavement. *Case Studies in Construction Materials, 20*, e02856.
11. Nigam, M., & Verma, M. (2023). Effect of nano-silica on the fresh and mechanical properties of conventional concrete. *Forces in Mechanics, 10*, 100165.
12. Rao, A. U., Shetty, P. P., Bhandary, P. R., Tantri, A., & Yaragal, S. C. (2024). Assessment of fly ash and ceramic powder incorporated concrete with steam-treated recycled concrete aggregates prioritising nano-silica. *Emergent Materials, 7*(2), 443–472.
13. Prasad, P. S., & Ramana, G. V. (2016). Imperial smelting furnace (zinc) slag as a structural fill in reinforced soil structures. *Geotextiles and Geomembranes, 44*(3), 406–428.
14. Dash, M. K., Patro, S. K., & Rath, A. K. (2016). Sustainable use of industrial-waste as partial replacement of fine aggregate for preparation of concrete—A review. *International Journal of Sustainable Built Environment, 5*(2), 484–516.
15. Singh, Y., Vyas, A. K., & Kabeer, K. S. A. (2017). Compressive strength evaluation of mortars containing ISF slag and marble powder. *Materials Today: Proceedings, 4*(9), 9635–9639.
16. Kirthika, S. K., Surya, M., & Singh, S. K. (2019). Effect of clay in alternative fine aggregates on performance of concrete. *Construction and Building Materials, 228*, 116811.
17. Yin, N. H., Sivry, Y., Guyot, F., Lens, P. N., & van Hullebusch, E. D. (2016). Evaluation on chemical stability of lead blast furnace (LBF) and imperial smelting furnace (ISF) slags. *Journal of Environmental Management, 180*, 310–323.
18. Varadharajan, S., Kirthanashri, S. V., Singh, K., Irfan, D., Shukla, B. K., & Kumar, A. (2023, September). Replacement of fine aggregate with imperial smelting furnace slag for manufacture of ecofriendly concrete. In *AIP conference proceedings* (Vol. 2800, No. 1). AIP Publishing.
19. Varadharajan, S., Kirthanashri, S. V., Kumar, H., Irfan, D., Shukla, B. K., Kumar, A., Bishetti, P., & Bharti, G. (2023, September). Use of silica fume concrete to enhance mechanical properties and environmental benefits. In *AIP conference proceedings* (Vol. 2800, No. 1). AIP Publishing.
20. Maurya, N., Srivastav, Y., Rawat, S., Sharma, M., Srivastava, R., Sihag, P., & Shukla, B. K. (2024, February). Augmenting concrete performance and sustainability with recycled glass: A critical examination of material characteristics and construction applications. In *AIP conference proceedings* (Vol. 3050, No. 1). AIP Publishing.
21. Varadharajan, S., Kirthanashri, S. V., Tiwari, A., Jaiswal, A., & Shukla, B. K. (2023). Use of waste marble dust and waste glass powder and bacterial solution for manufacture of eco-friendly concrete. *Lecture Notes in Civil Engineering, 281*, 195–207.
22. Sugita, S., Shoya, M., & Sugawara, T. (1988). Studies on the properties of concrete with zinc slag as fine aggregate. *CAJ Rev*, 92–95.
23. Patil, S. B., Vyas, A. K., Gupta, A. B., & Patil, R. S. (2016). Imperial smelting furnace slag as fine aggregate in cement concrete mixes. *The Journal of Solid Waste Technology and Management, 42*(2), 128–136.
24. Power, G. (2023). Use of residues from the metallurgical industry in construction. In *Advances in mechanics of materials for environmental and civil engineering* (pp. 53–63). Springer International Publishing.
25. Brand, A. S., & Fanijo, E. O. (2020). A review of the influence of steel furnace slag type on the properties of cementitious composites. *Applied Sciences, 10*(22), 8210.
26. Haruehansapong, S., Pulngern, T., & Chucheepsakul, S. (2014). Effect of the particle size of nanosilica on the compressive strength and the optimum replacement content of cement mortar containing nano-SiO2. *Construction and Building Materials, 50*, 471–477.
27. Mohammed, B. S., Awang, A. B., San Wong, S., & Nhavene, C. P. (2016). Properties of nano silica modified rubbercrete. *Journal of Cleaner Production, 119*, 66–75.

28. Chekravarty, D. S. V. S. M. R. K., Mallika, A., Sravana, P., & Rao, S. (2022). Effect of using nano silica on mechanical properties of normal strength concrete. *Materials Today: Proceedings, 51*, 2573–2578.
29. Fallah-Valukolaee, S., Mousavi, R., Arjomandi, A., Nematzadeh, M., & Kazemi, M. (2022, December). A comparative study of mechanical properties and life cycle assessment of high-strength concrete containing silica fume and nanosilica as a partial cement replacement. In *Structures* (Vol. 46, pp. 838–851). Elsevier.
30. Chithra, S., Kumar, S. S., & Chinnaraju, K. (2016). The effect of colloidal nano-silica on workability, mechanical and durability properties of high performance concrete with copper slag as partial fine aggregate. *Construction and Building Materials, 113*, 794–804.
31. Khan, M. I., Sravanthi, M., & Prasanna, E. L. (2018). An experimental study of the mechanical properties of s glass fiber reinforced high strength concrete partially replacing cement with nano silica. *International Journal of Civil Engineering and Technology, 9*(4), 1398–1409.
32. Ahmad, S., Ghazi, M. S. A., Syed, M., & Al-Osta, M. A. (2024). Utilization of fly ash with and without secondary additives for stabilizing expansive soils: A review. *Results in Engineering, 102079.*
33. Van der Sloot, H. A. (1991). Systematic leaching behavior of trace elements from construction materials and waste material. *Studies in Environmental Science, 48*, 19–36.
34. Gamage, N., Liyanage, K., Fragomeni, S., & Setunge, S. (2011). Overview of different types of fly ash and their use as a building and construction material. In *Proceedings of the international conference of structural engineering, construction and management, Kandy, Sri Lanka.*
35. Shukla, B. K., Gupta, A., Gowda, S., & Srivastav, Y. (2023). Constructing a greener future: A comprehensive review on the sustainable use of fly ash in the construction industry and beyond. *Materials Today: Proceedings, 93*(3), 257–264.
36. Varadharajan, S., Kirthanashri, S. V., Maurya, N., Bishetti, P., Shukla, B. K., & Bharti, G. (2023). Utilization of fly ash in concrete: A state-of-the-art review. *Lecture Notes in Civil Engineering, 281*, 189–194.
37. Varadharajan, S., Kirthanashri, S. V., Dwivedi, A., Irfan, D., Shukla, B. K., Kumar, A., Bishetti, P., & Bharti, G. (2023, September). Use of fly ash and metakaolin to manufacture sustainable concrete. In *AIP conference proceedings* (Vol. 2800, No. 1). AIP Publishing.
38. Hameed, M. S., & Sekar, A. S. S. (2009). Use of waste and by-products as fine aggregate in concrete. *Nature Environment and Pollution Technology, 8*(3), 503–508.
39. Maurya, N., Srivastav, Y., Rawat, S., Ranjan, Y., Srivastava, R., Shukla, B. K., & Varadharajan, S. (2023). Reinforcing civil infrastructure with waste glass-enhanced concrete: A comprehensive review of properties, performance and applications. *Materials Today: Proceedings, 93*(3), 239–248.
40. Shukla, B. K., Bharti, G., Sharma, P. K., Sharma, M., Rawat, S., Maurya, N., Srivastava, R., & Srivastav, Y. (2024). Sustainable construction practices with recycled and waste materials for a circular economy. *Asian Journal of Civil Engineering*. https://doi.org/10.1007/s42107-024-01111-y
41. Varadharajan, S., Kirthanashri, S. V., Samim, M., Shukla, B. K., & Bharti, G. (2023). A review on influence of replacing industrial wastes with fine aggregate in concrete. *Lecture Notes in Civil Engineering, 281*, 183–188.
42. Kosior-Kazberuk, M., & Lelusz, M. (2007). Strength development of concrete with fly ash addition. *Journal of Civil Engineering and Management, 13*(2), 115–122.
43. Vigneshpandian, G. V., Shruthi, E. A., Venkatasubramanian, C., & Muthu, D. (2017, July). Utilisation of waste marble dust as fine aggregate in concrete. In *IOP conference series: Earth and environmental science* (Vol. 80, No. 1, p. 012007). IOP Publishing.
44. Jyothi, T. K., Raghunath, S., Ranganath, R. V., & Jagadish, K. S. (2019). Strength and elastic properties of tank-bed soil and lime–pozzolana-based geopolymer units and prisms. *Earthen Dwellings and Structures: Current Status in their Adoption*, 245–255.

Machine Learning Perspectives on Solar Panel Efficiency: The Impact of Pollutants and Environmental Factors

Richa Verma, Bhupender Parashar, Priyanka Kulshrestha, and Bishnu Kant Shukla

1 Introduction

The escalating global demand for electricity has caused the emergence of solar photovoltaic (PV) systems as a sustainable renewable energy solution [1, 2]. However, this unlimited renewable energy, which when harnessed by solar PV cells for direct energy production, faces efficiency challenges, and despite its potential, actual performance often falls short of projected capacities, with current efficiency hovering around 26% [3–6]. It is well established that the performance of PV system [7] is affected by environmental variables, and recently it is being studied how pollutants also wield significant influence. Thus, these factors add a level of complexity to the optimization of solar PV systems [8]. These panels, which are typically exposed to open environments, demonstrate fluctuations in the output power due to changes in such diverse operational factors [9]. Variables such as atmospheric temperature, radiation, relative humidity, surface temperature of the solar panels, and operational duration significantly influence the output power of PV panels. Output power and current showcase a direct relationship with solar radiation, whereas there is a logarithmic correlation with voltage [8]. Rising surface temperatures have a negative

R. Verma
Department of Physics, JSS Academy of Technical Education, Noida, India
e-mail: richaverma@jssuninoida.edu.in

B. Parashar
Department of Mathematics, JSS Academy of Technical Education, Noida, India
e-mail: parashar_b@jssaten.ac.in

P. Kulshrestha
Department of Home Science, Bhagini Nivedita College, University of Delhi, New Delhi, India

B. K. Shukla (✉)
Department of Civil Engineering, JSS Academy of Technical Education, Noida, India

S. Pathak et al. (eds.), *Intelligent Infrastructure and Smart Materials*,
https://doi.org/10.1007/978-3-031-92421-7_8

impact on panel output, which has been extensively studied, thus registering reductions in both output power and fill factor as surface temperature increases [10, 11]. Additionally, heightened humidity levels adversely affect PV panel performance [12, 13].

Hence, before installing PV panels in open settings, a thorough investigation into the impact of these operational parameters on panel output power is essential [14, 15]. For effective planning and implementation of solar power-based electric PV systems, it becomes necessary to collect comprehensive data on all operating conditions and factors. Before setting up a plant in a particular site, prediction models for output power have gained importance, which need to be as accurate as possible. They are based on recent observations and dependable data regarding operating parameters [16–18].

The integration of machine learning into data prediction has demonstrated real-time accuracy, which is required to make informed decisions and automated actions without human intervention [19]. Prediction models have considerably developed with time as the adaptability of machine learning has increased for both structured and unstructured data [20, 21]. With iterative learning processes, the models develop and evolve independently. In this scenario, it has been found that even if it is well known that output power is intricately related to operating conditions, the existing studies have yet not sufficiently examined this relationship, which is crucial for understanding how different factors interact to affect the performance of PV panels.

We conducted a wide-ranging study using machine learning techniques to comprehend the relationship between environmental variables, pollutants, and the performance of solar photovoltaic systems. Within this investigation, we initially tried to relate 20 atmospheric and environmental parameters, including solar flares and CMP outbursts observed in 2022, with electrical output parameters. Thereafter, 10 parameters were identified, indicating significantly greater correlations, either positive or negative, with the output compared to others. It is to be noted that six factors are connected to atmospheric conditions: air temperature, relative humidity, cloud opacity, GHI, DNI, and DHI, while the remaining four relate to pollutants: PM_{10} ($\mu g/m^3$), NO_x (ppb), SO_2 ($\mu g/m^3$), and benzene ($\mu g/m^3$).

Solar power generation potential in Rajasthan is the highest among all the states in India. As of 2023, operational solar power projects in this state produce approximately 16,060 MW of solar energy. This includes ground-mounted, rooftop, hybrid solar, and off-grid categories [15]. The present 1-year study investigates the impact of environmental variables and pollution parameters on the power output of a polycrystalline (2.64 kW) rooftop photovoltaic panel installation at the specified site in Jaipur, Rajasthan, India.

In comparison to conventional models based on physics and empirical correlations, machine learning has emerged as a sophisticated statistical method, as it is capable of establishing more precise associations between input parameters and output. Machine learning techniques are able to process large datasets, help to understand complex patterns, and then offer accurate predictions, which make them a valuable tool for understanding the influence of environmental parameters and pollutants on PV system performance [22]. The results obtained in this study will further enable the use of machine learning and emphasize its potential to improve accuracy in solar power modeling. It highlights the importance of selecting the right input parameters for the optimal performance of a solar energy system. Also, by using machine learning to study the data, astute correlations between environmental parameters, pollutants, and PV system performance could be identified.

The remainder of the chapter is structured as follows: section "Collection of Data" outlines how the 1 year data was collected and the challenges that were faced. The results and implications of the machine learning analysis on system performance are discussed in section "Methodology". Section "Machine Learning Modeling" concludes by highlighting the prospects of machine learning for optimizing solar PV systems. Therefore, the objective of this chapter is to use computational models effectively through various machine learning approaches. These tools are highly valuable for analyzing and predicting solar power generation. These models are designed to accurately forecast PV panel output power by accounting for diverse operating conditions, thereby advancing our understanding and predictive capabilities.

2 Collection of Data

The data was collected from the solar photovoltaic (PV) facility, namely, Ultra Vibrant Solar Energy Private Limited. These rooftop panels are situated at Fitpack Textiles Pvt. Ltd. in Jaipur, Rajasthan. The location is given by coordinates 26.76640°N, 75.8326°E, and the link is provided below:

https://maps.app.goo.gl/pHLiy3EW7UEeTTLp6

Figure 1 depicts the location of the plant in Jaipur city, Rajasthan. The selected site was considered to get good solar radiation all through the day. There are two inverters in the plant, with capacities of 20 and 100 kW. The data collected was from 2022. The output current and voltage were measured at every 30-min interval from 6 AM to 3 PM, from which the output power of the system was determined. Figure 2 illustrates the daily electrical output collected. In the month of December, a technical breakdown occurred, resulting in the plant malfunction, so there is minimum data for many days in December.

For 2022, the authors examined how the electrical output power of solar PV panels varies with respect to environmental factors such as solar radiation, atmospheric conditions, solar flares, and CMP outbursts. The atmospheric conditions that were studied include cloud opacity, air temperature, direct normal irradiance (DNI), diffuse horizontal irradiance (DHI), global horizontal irradiance (GHI),

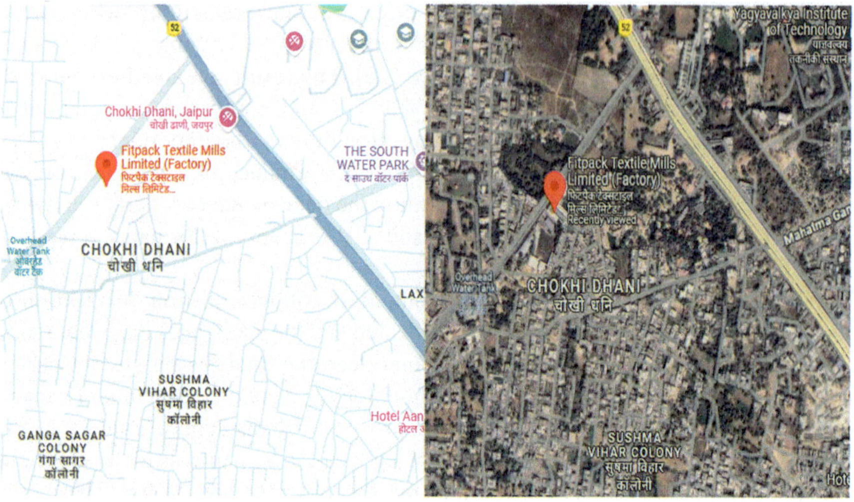

Fig. 1 Location details of Fitpack Textile Mills Limited (https://maps.app.goo.gl/pHLiy3EW7UEeTTLp6)

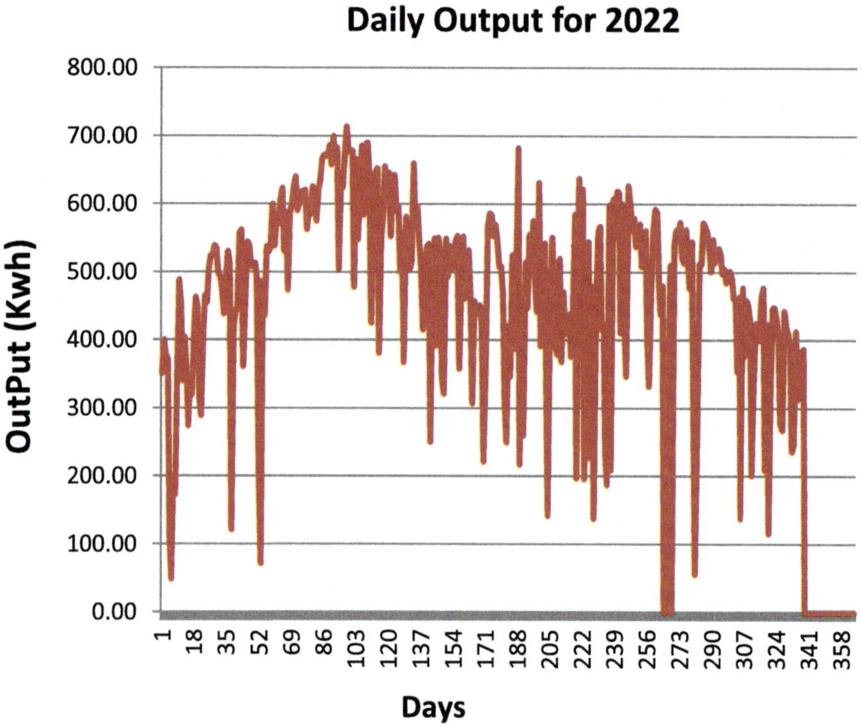

Fig. 2 Daily output from INV 1 and INV 2

Table 1 Monthly average atmospheric data (Solcast API)

Month	Air temp	Cloud opacity	DNI	DHI	GHI	Relative humidity	WD (degree)	BP (mmHg)
January	12.84	17.83	150.21	79.49	146.60	76.00	150.65	734.95
February	18.50	6.43	211.56	126.74	205.69	52.46	156.91	732.86
March	26.60	8.90	226.02	157.04	251.42	40.13	178.34	731.04
April	32.53	13.79	210.05	160.28	273.56	24.39	203.60	728.20
May	34.93	18.16	139.91	117.26	261.49	32.03	196.84	724.06
June	33.68	31.44	125.42	102.30	238.53	45.12	194.43	723.62
July	29.15	50.09	94.40	74.43	200.21	81.31	137.90	723.82
August	27.94	47.61	80.00	61.94	174.50	82.21	154.75	724.31
September	28.39	25.18	166.09	119.06	217.98	74.79	155.93	724.62
October	24.73	9.20	216.74	135.40	211.31	66.20	137.95	730.05
November	20.26	3.85	219.16	119.51	183.57	59.21	148.06	732.87
December	15.62	0.98	240.09	118.83	170.78	62.97	125.22	733.25

relative humidity, and wind speed, obtained from the data produced by Solcast API (https://solcast.com/Solcast PV). The average monthly radiation data obtained from Solcast API are presented in Table 1. The wind direction (WD) in degrees, barometric pressure (BP), and pollution factors were acquired from the Central Pollution Control Board (CPCB) located at the All India Central Control Room for Air Quality Management center in Adarsh Nagar, Jaipur. The parameters considered include PM2.5 (μg/m^3), PM 10 (μg/m^3), NO_x (ppb), NH_3 (μg/m^3), SO_2 (μg/m^3), CO (mg/m^3), ozone (μg/m^3), benzene (μg/m^3), MP-xylene (μg/m^3), and O-xylene (μg/m^3). Table 2 presents the monthly average data for pollutants obtained from CPCB.

Due to their high intensity, M- and X-class solar flares were chosen to estimate their impact on the solar output of the system. Solar flares and coronal mass ejection (CME) incidences in 2022 were obtained from the NASA website DONKI, which serves the research community by providing online data on space weather events.

https://kauai.ccmc.gsfc.nasa.gov/DONKI

Thus, in our study, we constructed an output power prediction model utilizing machine learning (ML). To achieve this, input parameters were gathered from the aforementioned sources to develop a robust ML model.

Table 2 Monthly average data for pollutants (CPCB India)

Month	$PM_{2.5}$ (µg/m³)	PM_{10} (µg/m³)	NO_x (ppb)	SO_2 (µg/m³)	NH_3 (µg/m³)	CO (mg/m³)	Ozone (µg/m³)	Benzene (µg/m³)	MP-Xylene (µg/m³)	O-Xylene (µg/m³)
January	65.92	106.09	56.41	11.14	49.25	0.92	37.45	4.96	1.51	1.84
February	50.32	112.55	45.84	11.65	39.12	0.66	47.96	3.45	0.66	1.33
March	60.41	125.44	59.51	12.95	47.86	0.80	52.01	4.78	3.42	2.59
April	62.49	152.76	65.96	14.09	41.84	0.90	41.54	4.07	4.92	2.46
May	82.33	186.78	38.81	10.93	26.35	0.72	52.80	2.91	4.09	1.90
June	46.90	124.06	29.59	11.07	19.72	0.76	53.76	1.47	2.41	1.19
July	18.57	39.42	29.79	9.52	31.49	0.75	38.56	0.69	1.38	0.73
August	19.51	49.80	28.25	8.86	29.35	0.66	31.33	0.66	1.79	0.81
September	32.54	67.06	37.36	12.60	32.56	0.80	36.68	1.49	3.43	1.99
October	56.56	102.74	58.44	14.48	40.19	0.86	50.87	2.50	7.53	4.19
November	65.95	114.15	89.34	10.35	60.03	1.13	44.09	5.12	14.86	8.05
December	55.25	113.41	83.93	14.88	52.88	0.96	40.13	5.21	11.61	6.41

3 Methodology

The aim of this research is to study the influence of environmental and atmospheric parameters on the performance and operation of PV systems in 2022 employing machine learning methods. The researchers organized and examined 20 atmospheric and environmental variables, including solar flares and CMP outbursts observed in 2022, to evaluate the performance of the rooftop solar installation. These variables, which affect the solar panel output, were then categorized into two groups: environmental factors and atmospheric parameters. A comprehensive analysis was conducted for each group, reaching into their individual sub-factors and relating and understanding their impact on the system output. The objective is to provide a realistic prediction of the expected production of energy from solar panels at the specified location under prevailing conditions.

Several challenges were encountered during data collection and analysis. There was insufficient data for the month of December 2022 due to plant breakdown. This hindered the analysis. Another factor, that is, wind speed, had to be excluded from the analysis due to discrepancies between wind speed data obtained from the Central Pollution Control Board (CPCB) of the Government of India and Solcast API. To enhance accuracy for this particular factor, it is recommended to either install or use wind speed measurement devices on location, especially considering the significance of this parameter. Additionally, as solar flares and CMP occurrences are sporadic, here the analysis highlights the days of their incidence to assess whether there is any impact on output fluctuations. These challenges notwithstanding, the study aims to provide valuable insights into the efficiency of PV systems in relation to all the above-mentioned factors.

3.1 Atmospheric Parameters

The atmospheric parameters considered are air temperature, cloud opacity, direct normal irradiance (DNI), diffuse horizontal irradiance (DHI), global horizontal irradiance (GHI), relative humidity, barometric pressure (in mm Hg), and wind direction (in degrees) [7, 9, 23]. The correlation analysis of these factors with the output is presented in Table 3.

(i) Air temperature: This represents the temperature of the air surrounding the solar panels. Our analysis shows that higher air temperatures can lead to increased solar panel efficiency due to improved conductivity, as there is a

Table 3 Correlation analysis of atmospheric factors

	Air temp	Cloud opacity	DNI	DHI	GHI	Relative humidity	WD (degree)	BP (mmHg)
Output power	0.36	0.43	0.49	0.61	0.74	0.51	0.25	0.06

positive correlation (0.36) between air temperature and solar output. Hence, higher air temperatures are coupled with increased solar power generation. Although we know that excessively high temperatures can also reduce efficiency, this has not been captured in the preliminary analysis.

(ii) Cloud opacity: This refers to the thickness or density of clouds in the sky. The more opaque or thicker clouds, the more sunlight will be blocked, and as the solar radiation reaching the solar panels reduces, the solar power output decreases. There is a negative correlation (−0.43) between cloud opacity and solar output.

(iii) Direct normal irradiance (DNI): This factor measures the intensity of solar radiation that reaches a surface directly, with no obstruction from clouds or other atmospheric elements. There is a positive correlation (0.49) between DNI and solar output, indicating that higher DNI levels are associated with increased solar power generation.

(iv) Diffuse horizontal irradiance (DHI): DHI refers to the quantity of solar radiation obtained per unit area by the surface without shadows and it comes equally from all directions after being scattered by atmospheric particles. There is a strong positive correlation (0.61) between DHI and solar output.

(v) Global horizontal irradiance (GHI): GHI refers to the entire solar radiation received from the entire hemisphere above a horizontal surface on the Earth's surface. This includes both direct and diffuse radiation. This factor is related to the total solar energy available for conversion by solar panels. There is a strong positive correlation (0.74) between GHI and solar output. As GHI increases, solar power generation increases.

(vi) Relative humidity: This is the ratio of the quantity of water vapor present in the atmosphere at that particular time and place to the maximum amount it could hold at the same temperature. Higher relative humidity levels indicate there is more moisture, which can reduce solar panel efficiency due to diffused light and also shadows created by clouds formed because of the moisture. There is a negative correlation (−0.51) between the two.

(vii) Wind direction (WD): A positive correlation (0.25) between wind direction and solar output indicates a moderate association between the two. This factor may have a role in the planning of the solar panel installation so as to maximize energy transfer and minimize wind-induced shading effects.

(viii) Barometric pressure (BP): This refers to the pressure exerted by the atmosphere at a given point. Changes in barometric pressure do point to forthcoming weather disturbances, which may indirectly influence solar panel performance through changes in cloud cover or other atmospheric conditions. There is a weak negative correlation (−0.06) between barometric pressure and solar output.

Overall, the strongest positive correlations with solar output are found for air temperature, GHI, DHI, and DNI, while negative correlations with cloud opacity and relative humidity suggest that these factors have the most significant influence on solar power generation.

Table 4 Correlation analysis of pollutants

	$PM_{2.5}$ (μg/m³)	PM_{10} (μg/m³)	NO_x (ppb)	SO_2 (μg/m³)	NH_3 (μg/m³)	CO (mg/m³)	Ozone (μg/m³)	Benzene (μg/m³)	MP-Xylene (μg/m³)	O-Xylene (μg/m³)
Output power	0.15	0.27	0.13	0.24	0.05	−0.02	0.23	0.13	−0.02	−0.03

3.2 Pollutants

The parameters that are considered here are $PM_{2.5}$ (μg/m³), PM_{10} (μg/m³), NO_x (ppb), SO_2 (μg/m³), NH_3 (μg/m³), CO (mg/m³), ozone (μg/m³), benzene (μg/m³), MP-xylene (μg/m³), and O-xylene (μg/m³). The correlation analysis of these factors with the output is described in Table 4.

(i) $PM_{2.5}$ and PM_{10}: Particulate matter (PM), including $PM_{2.5}$ (diameter ≤ 2.5 μm) and PM_{10} (diameter ≤ 10 μm), showed positive correlations with solar output. The moderate positive correlation for PM_{10} (0.27) suggests a stronger association compared to $PM_{2.5}$ (0.15). Particulate matter in the atmosphere can scatter sunlight, potentially influencing solar irradiance, and moderate levels of PM_{10} pollution may contribute to diffuse sunlight, enhancing solar panel performance.

(ii) NO_x and SO_2: Nitrogen oxides (NO_x) and sulfur dioxide (SO_2) exhibited positive correlations with solar output 0.13 and 0.24, respectively, indicating potential influences on solar irradiance. These pollutants, commonly emitted from combustion processes, can form aerosols that absorb or scatter sunlight. Additionally, chemical reactions involving NO_x and SO_2 may produce secondary aerosols, further impacting solar radiation.

(iii) NH_3 (μg/m³) and CO (mg/m³): Ammonia (NH_3) showed a weak positive correlation (0.05) with solar output, suggesting a minimal association. Carbon monoxide (CO) exhibited a very weak negative correlation (−0.02), implying a negligible negative association.

(iv) Ozone: This displayed a moderate positive correlation (0.23), indicating a stronger positive association with solar output [24]. In the lower atmosphere, ozone can absorb solar radiation. This can increase atmospheric stability and reduce cloud formation, resulting in clearer skies and ultimately enhancing solar irradiance levels.

(v) Benzene, MP-xylene, and O-xylene: Benzene exhibited a weak positive correlation (0.13) with solar output, indicating a slight positive association. Conversely, both MP-xylene and O-xylene showed very weak negative correlations (−0.02 and −0.03, respectively), suggesting minimal negative associations. These volatile organic compounds are commonly released from a variety of sources, such as vehicle exhaust, industries, and fuel combustion. Increased levels of these VOCs can lead to the formation of secondary pollutants such as organic aerosols [25, 26], which can affect atmospheric composition and optical properties, potentially influencing solar irradiance reaching solar panels.

Additionally, the spatial and temporal variability in the concentration of these VOCs, due to differences in emission sources, atmospheric transport, and chemical transformation processes, may contribute to localized variations in pollutant concentrations and atmospheric conditions. Higher levels of benzene pollution may indicate urban or industrial areas with higher levels of economic activity and infrastructure development. In contrast, the weak negative correlations for MP-xylene and O-xylene might suggest differences in emission sources or atmospheric processing compared to benzene, resulting in contrasting effects on solar output. In summary, the correlations between air pollutants and solar output underscore the complex interactions between atmospheric composition, aerosol concentrations, and solar radiation, and the correlations between air pollutants and solar output suggest that variations in atmospheric composition and aerosol concentrations affect the amount and quality of sunlight reaching solar panels.

3.3 CMEs and Solar Flares

Figure 3 depicts the daily output mapped with the solar flares and CME occurrences.

Coronal mass ejections (CMEs) are unpredictable and explosive events on the Sun that emit large amounts of energy and charged particles into space, potentially

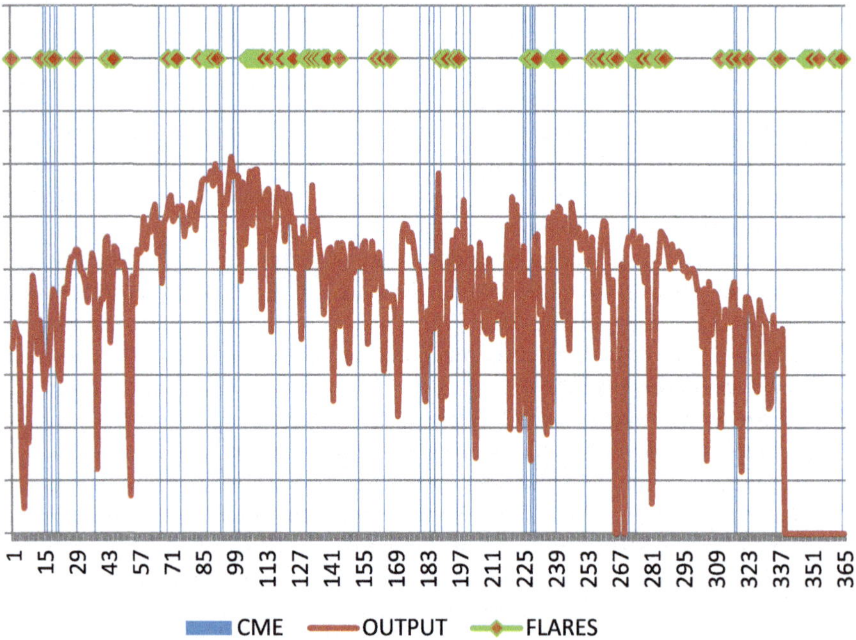

Fig. 3 Daily output mapped with the days of solar flares and CME occurrences

causing geomagnetic storms and affecting Earth's magnetosphere. Solar panel performance is primarily influenced by sunlight exposure and panel efficiency, with local weather conditions playing a significant role. Despite CMEs' broader effects on space weather and technology, their direct relationship with solar panel output is limited. In our study, we analyzed solar activity, including solar flares and CMEs, in relation to solar output. We mapped all the occurrences of solar activity, including solar flares and CMEs with the solar output. The data was studied monthly as well as daily. Only M- and X-class solar flares were considered, as they are the most potent and consequential in terms of their potential to affect space weather and Earth's systems [27]. However, we found a negligible correlation between solar panel output and these parameters. This lack of correlation may be attributed to the design of solar panels, optimized primarily for visible light conversion. Additionally, the Earth's atmosphere acts as a natural shield against harmful solar radiation associated with CMEs and solar flares, further mitigating their direct impact on solar panel output. In summary, there is no significant correlation between CMEs, solar flares, and solar panel output. This can be because of many reasons including the design and efficiency of solar panels, natural shielding mechanisms, spatial and temporal variability, and system resilience.

Hence, from the preliminary study, ten parameters, including six atmospheric variables (air temperature, relative humidity, cloud opacity, global horizontal irradiance (GHI), direct normal irradiance (DNI), and diffuse horizontal irradiance (DHI)), as well as four pollutants (PM10, NO_x, SO_2, and benzene), demonstrated significant correlations with solar output. By leveraging these parameters and employing machine learning tools, the aim is to predict and forecast solar panel output accurately.

4 Machine Learning Modeling

In the present study, it is observed that machine learning models are important for understanding and forecasting the complex interactions between environmental elements and the efficiency of solar panels. Through the examination of patterns, correlations, and interdependencies among these variables, machine learning techniques are able to build strong predictive models.

These models can generate more accurate solar panel output projections by utilizing current environmental data.

Moreover, it has been found that the effects of pollutants can be effectively combined with other relevant factors using machine learning techniques. This method can be used to create detailed prediction models that accurately forecast the performance of solar panels in a variety of environmental settings, including populated areas like Jaipur.

4.1 Regression Analysis for Solar Power Generation

Problem synopsis: Our target is to study and analyze the relationship between solar power generation through solar panels and various environmental factors.

Data collection: Data for this study was collected from a solar power plant, including factors such as air temperature, relative humidity, cloud opacity, direct normal irradiance (DNI), diffuse horizontal irradiance (DHI), global horizontal irradiance (Ghi), PM10 concentration, nitrogen oxides (NO_x) concentration, sulfur dioxide (SO_2) concentration, and benzene concentration.

Mathematical modeling: Linear regression is a common choice for such analyses.

Modeling tools: To understand the complex interaction between environmental elements and electricity generation, Python and machine learning technologies were used in this investigation of a solar power plant. Using the Pandas package ensured a clean dataset for analysis by facilitating easy preparation and data manipulation. The utilization of Matplotlib and Seaborn facilitated the visual examination of the variable distribution, thereby facilitating the detection of anomalies and trends in the data.

The predictive engine utilized in this study was the Scikit-learn linear regression model, which allowed for the quantification of the effects of air temperature, humidity, and pollutant concentrations on the generation of solar power.

The model was tested and trained to assess its predicted accuracy objectively. Python's flexibility was a major factor in making the entire process much more efficient. It offered data scientists and analysts a complete ecosystem, encompassing everything from data pretreatment to model training and evaluation. Due to Python libraries' comprehensive visualization features, the data were easier to grasp. The analysis was made apparent and reproducible by the way, the code snippets condensed its main points. The clarity and reproducibility of the analysis were further aided by succinct code snippets that highlighted key aspects of the research.

5 Results and Discussion

The current work demonstrates how well Python and machine learning technologies work together to extract useful insights from large, complicated datasets. Such approaches become essential for maximizing renewable energy sources as the energy environment changes, providing a window into the direction sustainable energy management will take. The role of atmospheric parameters, as given in Table 1, and pollutant parameters, as given in Table 2, was considered.

Based on the correlation coefficients provided in the two tables, it appears that pollution generally exhibits weaker correlations with the main output in the solar panel system compared to other factors such as environmental conditions (e.g., solar radiation, temperature, humidity, etc.).

In the analysis of pollutants, the correlation coefficients range from −0.02 to 0.27, indicating generally weak to moderate associations with the output variable. Conversely, in the previous analysis of environmental factors affecting solar output, the correlation coefficients ranged from 0.10 to 0.74, indicating moderate to strong associations with solar output. This implies that while pollutants might affect solar panel output to some extent, other environmental factors like solar radiation, cloud cover, and temperature likely exert a more substantial influence. Nevertheless, it is crucial to acknowledge that in urban environments such as Jaipur, pollutants still play a significant role.

5.1 Linear Regression Analysis Results

Table 5 presents the intercept and other parameters calculated in the regression model.

Interpretation of linear regression analysis results is summarized in Table 6.

The interpretation of the above results is as follows:

Intercept: In the linear regression model, the intercept indicates the predicted value when all the independent variables take zero value. In the present study of solar power plants, the intercept value is approximately 303.31, which suggests the level of power output in the absence of any external influences.

Table 5 Linear regression analysis results

Coefficients	Attribute
−4.61577	Air temperature
−1.09398	Relative humidity
−0.67747	Cloud opacity
−0.80244	Direct normal irradiance (DNI)
0.365301	Diffuse horizontal irradiance (DHI)
2.173438	Global horizontal irradiance (Ghi)
−0.57431	PM10 concentration
0.385641	Nitrogen oxides (NO_x) concentration
3.17368	Sulfur dioxide (SO_2) concentration
1.954178	Benzene concentration

Table 6 Linear regression analysis results

Significant factors obtained from modeling	Values
Intercept	303.30709857
Mean absolute error	44.46490070998011
Mean squared error	2811.6049440037077
Root mean squared error	53.024569248638954
R^2	0.843435017786182

Mean absolute error (MAE): The MAE is around 44.46. This statistic represents the average absolute deviation between the predicted and actual power generation figs. A smaller MAE indicates a more precise model, with this model's predictions deviating by about 44.46 units from the actual figures on average.

Mean squared error (MSE): The MSE is approximately 2811.60, which reflects the average of the squares of the errors between predicted and actual values. A smaller MSE signifies a more accurate model. Here, the model's predictions, on average, differ by about 2811.60 squared units from the actual values.

Root mean squared error (RMSE): The RMSE is approximately 53.02. It is the square root of the MSE and provides a more comprehensible measure of prediction error. A lower RMSE indicates better model accuracy, with this model showing an average deviation of about 53.02 units from the actual figures.

R^2 value: The R^2 value is 0.84. It suggests that approximately 84.34% of the variation in power generation is accounted for by the linear regression model. This high value indicates a strong link between the model's independent variables (like air temperature and humidity) and power generation, emphasizing the model's predictive strength.

6 Conclusion

The linear regression model exhibits a sensibly solid fit to the data, as shown by the low values of MAE, MSE, RMSE, and the high R^2 value. This recommends that the selected autonomous variables significantly explain the variation in solar power generation. However, deciphering these results within the context of the solar plant's operations and the environmental variables that affect power production is essential. For instance, solar power output naturally fluctuates based on factors such as environmental conditions, panel efficiency, and solar irradiance.

While the model takes these variations into consideration to a large extent, it might miss some unexpected anomalies or unprecedented events that could result in notable variations in power output. This model can be used to inform a variety of strategic and operational choices made by the solar power plant.

To optimize the impact on overall power production, maintenance activities can be planned for times when lower power output is anticipated. The model is useful as it forecasts additional power output under various scenarios, aiding in investment decisions such as purchasing new modern panels or upgrading existing facilities. Beyond its current applications, the model lays the groundwork for further research and development. While the high R^2 value provides a solid establishment, there is continuous room for improvement. Future studies could investigate more granular data, incorporate additional variables, or examine nonlinear models that might more accurately capture the complex dynamics of solar power generation. Combining real-time data feeds would permit the model to alter its predictions based on changing conditions dynamically.

To summarize, the regression model is an important device for analyzing and predicting solar power generation. Its reliability and robustness are evidenced by strong error measurements (MAE, MSE, and RMSE) as well as a high R^2 value. The model provides significant predictions of power generation and is also able to catalyze further innovations and enhancements in renewable energy generation. It provides both immediate operational benefits and long-term strategic insights, underscoring the significance of detailed measurable investigation in industrial settings.

Conflict of Interest Statement None declared.

Data Availability Statement All data generated or analyzed during this study are available upon reasonable request to the corresponding author.

References

1. Obaideen, K., Olabi, A. G., Al Swailmeen, Y., Shehata, N., Abdelkareem, M. A., Alami, A. H., Rodriguez, C., & Sayed, E. T. (2023). Solar energy: Applications, trends analysis, bibliometric analysis and research contribution to sustainable development goals (SDGs). *Sustainability, 15*(2), 1418.
2. Maka, A. O., & Alabid, J. M. (2022). Solar energy technology and its roles in sustainable development. *Clean Energy, 6*(3), 476–483.
3. Lin, H., Yang, M., Ru, X., Wang, G., Yin, S., Peng, F., Hong, C., Qu, M., Lu, J., Fang, L., & Han, C. (2023). Silicon heterojunction solar cells with up to 26.81% efficiency achieved by electrically optimized nanocrystalline-silicon hole contact layers. *Nature Energy, 8*(8), 789–799.
4. Richter, A., Hermle, M., & Glunz, S. W. (2013). Reassessment of the limiting efficiency for crystalline silicon solar cells. *IEEE Journal of Photovoltaics, 3*(4), 1184–1191.
5. Liberos, M., González-Medina, R., Garcerá, G., & Figueres, E. (2019). A method to enhance the global efficiency of high-power photovoltaic inverters connected in parallel. *Energies, 12*(11), 2219.
6. Nayak, P. K., Mahesh, S., Snaith, H. J., & Cahen, D. (2019). Photovoltaic solar cell technologies: Analysing the state of the art. *Nature Reviews Materials, 4*(4), 269–285.
7. Fouad, M. M., Shihata, L. A., & Morgan, E. I. (2017). An integrated review of factors influencing the performance of photovoltaic panels. *Renewable and Sustainable Energy Reviews, 80*, 1499–1511.
8. Ates, A. M., & Singh, H. (2021). Rooftop solar Photovoltaic (PV) plant–performance analysis and critical evaluation. *Journal of King Saud University-Science, 33*(3).
9. Song, Z., Wang, M., & Yang, H. (2022). Quantification of the impact of fine particulate matter on solar energy resources and energy performance of different photovoltaic technologies. *ACS Environmental Au, 2*(3), 275–286.
10. Sarmah, P., Das, D., Saikia, M., Kumar, V., Yadav, S. K., Paramasivam, P., & Dhanasekaran, S. (2023). Comprehensive analysis of solar panel performance and correlations with meteorological parameters. *ACS Omega, 8*(50), 47897–47904.
11. Yelmen, B., & Çakir, M. T. (2016). Influence of temperature changes in various regions of Turkey on powers of photovoltaic solar panels. *Energy Sources, Part A: Recovery, Utilization, and Environmental Effects, 38*(4), 542–550.

12. Qian, D., Zheng, Z., Yao, H., Tress, W., Hopper, T. R., Chen, S., Li, S., Liu, J., Chen, S., Zhang, J., & Liu, X. K. (2018). Design rules for minimizing voltage losses in high-efficiency organic solar cells. *Nature Materials, 17*(8), 703–709.
13. Boyle, L., Flinchpaugh, H., & Hannigan, M. P. (2015). Natural soiling of photovoltaic cover plates and the impact on transmission. *Renewable Energy, 77*, 166–173.
14. Mussard, M., & Amara, M. (2018). Performance of solar photovoltaic modules under arid climatic conditions: A review. *Solar Energy, 174*, 409–421.
15. Ummadisingu, A., & Soni, M. S. (2011). Concentrating solar power–technology, potential and policy in India. *Renewable and Sustainable Energy Reviews, 15*(9), 5169–5175.
16. Sharma, D. K., Singh, A. P., & Verma, V. (2014). A review of solar energy: Potential, status, targets and challenges in Rajasthan. *International Journal of Engineering Research, 3*(3).
17. Gómez, I., Molina, S., & Galiana-Merino, J. J. (2023). Evaluating the influence of air pollution on solar radiation observations over the coastal region of Alicante (Southeastern Spain). *Journal of Environmental Sciences, 126*, 633–643.
18. Mustafa, R. J., Gomaa, M. R., Al-Dhaifallah, M., & Rezk, H. (2020). Environmental impacts on the performance of solar photovoltaic systems. *Sustainability, 12*(2), 608.
19. Zazoum, B. (2022). Solar photovoltaic power prediction using different machine learning methods. *Energy Reports, 8*, 19–25.
20. Frederiksen, C. A. F., & Cai, Z. (2022). Novel machine learning approach for solar photovoltaic energy output forecast using extra-terrestrial solar irradiance. *Applied Energy, 306*, 118152.
21. Ramadhan, R. A., Heatubun, Y. R., Tan, S. F., & Lee, H. J. (2021). Comparison of physical and machine learning models for estimating solar irradiance and photovoltaic power. *Renewable Energy, 178*, 1006–1019.
22. Murugan, D. K., Said, Z., Panchal, H., Gupta, N. K., Subramani, S., Kumar, A., & Sadasivuni, K. K. (2023). Machine learning approaches for real-time forecasting of solar still distillate output. *Environmental Challenges, 13*, 100779.
23. Huld, T., & Gracia Amillo, A. M. (2015). Estimating PV module performance over large geographical regions: The role of irradiance, air temperature, wind speed and solar spectrum. *Energies, 8*(6), 5159–5181.
24. Dawo, C., Afroz, M. A., Iyer, P. K., & Chaturvedi, H. (2020). Effect of UV-ozone exposure on the dye-sensitized solar cells performance. *Solar Energy, 208*, 212–219.
25. Odum, J. R., Jungkamp, T. P. W., Griffin, R. J., Forstner, H. J. L., Flagan, R. C., & Seinfeld, J. H. (1997). Aromatics, reformulated gasoline, and atmospheric organic aerosol formation. *Environmental Science & Technology, 31*(7), 1890–1897.
26. Ghigo, G., & Tonachini, G. (1999). From benzene to muconaldehyde: Theoretical mechanistic investigation on some tropospheric oxidation channels. *Journal of the American Chemical Society, 121*(36), 8366–8372.
27. Schwenn, R. (2006). Space weather: The solar perspective. *Living Reviews in Solar Physics, 3*(1), 1–72.

Environmental Consequences of Rapid Industrialization: A Case Study of India's Industrial Clusters and the Urgent Need for Sustainable Policies

Sandeep Budde, P. S. Chani, and Sandeep Agrawal

1 Introduction

The Industrial Revolution, which led to substantial social stratification, paved the way for more facilities and the upliftment of the lower classes. Industries are the primary contributors to the gross domestic product (GDP) of the nation. The industrial accidents that have happened recently all over the world suggest that since the 1800s, when the industrial era first began, the developing world has experienced more misfortunes [1]. However, the actions taken to address these situations appear to be a new paradigm in every discussion, which has caused the sector to stagnate. Because land, labour, and capital are more readily available in urban areas than in rural ones, industrialization is one of the reasons that resulted in an increase in across major cities in India. Most metropolitan cities and other urban regions exhibit socioe-conomic stratification [2]. Growing industries have both positive and

S. Budde (✉)
Built Environment Lab, Department of Architecture and Planning, Indian Institute of Technology, Roorkee, India

Urban Environment Observatory, School of Urban & Regional Planning Program, University of Alberta, Edmonton, AB, Canada

Bharti Institute of Public Policy, Indian School of Business, Hyderabad, India
e-mail: sbudde@ualberta.ca

P. S. Chani
Built Environment Lab, Department of Architecture and Planning, Indian Institute of Technology, Roorkee, India
e-mail: p.chani@ar.iitr.ac.in

S. Agrawal
Urban Environment Observatory, School of Urban & Regional Planning Program, University of Alberta, Edmonton, AB, Canada
e-mail: sagrawal@ualberta.ca

S. Pathak et al. (eds.), *Intelligent Infrastructure and Smart Materials*,
https://doi.org/10.1007/978-3-031-92421-7_9

harmful effects on the environment and society. These implications include increased employment possibilities, population growth, an increase in the pollution index, greenhouse gas emissions that pose health risks, and an impact on communities and way of life. A process-oriented framework is required to quantify the adverse effects of industrial environments on their immediate surroundings, as many industrialized, emerging, and impoverished nations lack one. Concerns are heightened by the urbanization that accompanies industrial growth, necessitating the use of useful ecological assessment and mitigation techniques. This essential topic is examined from several perspectives in this chapter, with a focus on pollution caused by industrial gaseous release. With 39 cities ranking in the top 50 [3], India is the nation with the greatest air pollution in the world. But it is noticeably lacking from conferences that address the detrimental effects of industrialization on health. The report highlights the need for additional research, particularly in the light of India's rapidly increasing urbanization and industrialization [4–7]. The conclusion, which concentrates on industrial areas that pollute the environment, emphasizes the need for environmentally conscious legislation and environmental advocacy, especially in densely populated areas where industrial emissions constitute a serious threat to the environment and public health.

2 Background

Indoor pollution is blamed for about seven million fatalities globally, with China and India having the highest death tolls. According to estimates from the World Health Organization, over 4.7 lakh deaths in India alone are attributed to indoor pollution [8]. India is rapidly becoming an urban nation. Hence, more research in this field is required. Concerns regarding industrial areas and establishments are growing because the two countries, both India and China with the most significant death rates, are those whose economies are focused on manufacturing and services. The ESPOO Convention was established in Finland in 1991 and broadly addresses the impact of industrialization on health (ESPOO: UNECE Convention on Environmental Impact Assessment (EIA) in a Transboundary Context). Neither India nor China are members [9].

When compared to other countries, India has the most polluted cities worldwide. While the third interim WHO limit is only 15, there are claims that the PM2.5 concentration in some cities can approach 135 $\varepsilon g/m^3$. Also, according to the WHO, 15 of India's top 20 most polluted cities are isolated. The reasons for the increased particulate matter are growing industry and vehicle development [7]. The industrial GDP in future economic plans is increasing in the coming years, as per the Niti Ayog [10]; it caused the nation with the youngest population to have different economic and environmental policies, where the northern strip of the Indian subcontinent has the majority of the highly polluted cities. The most current draft EIA, which considered air-generated toxins in its risk assessment reports, demonstrated the necessity of environmental activism [11].

2.1 Structuring Your Chapter

"Industrial Area" refers to an area intended for industry growth. Industrial development is a primary driver driving India's economic growth. Industrial activities include SEZs, industrial estates, National Investment and Manufacturing Zones (NIMZs), and specialized investment regions. A comparable concept also has to be studied in India, where the United Nations Industrial Development Organization believes that 388 established industrial clusters exist [12].

The clusters were established under various Five-Year plans that outlined distinct industries across the Indian subcontinent. Among the most common types of clusters are those in the automotive, pharmaceutical, electrical, cement, logistical, and warehousing sectors. The old socio-economic structure of society has given way to a new social stratification due to industrialization, most notably in large cities. The expanding industries significantly impact the environment and society positively and negatively. Some of these repercussions are opportunities for employment, economic expansion, a rise in pollution and greenhouse gas emissions, health concerns, and an impact on communities and way of life. Tools and actions must be taken to assess and mitigate these adverse effects of industrial advancements and activities. As a result, assessments of the environment and livability are practical tools for improving the declining social and environmental sectors [13]. The expansion and effects of the Active Pharmaceutical Producing Industries (API) on the pharmaceutical industry are particularly covered in this chapter.

India has the most significant industrial economy, and environmental degradation leads to industrial pollution caused by businesses and their associated operations, especially by factories and industrial plants that dump and release waste products into the air and water, primarily harming the surrounding environment. The nation's industrialization has led to a recent surge in demand for IDAs in major cities in developing countries. These IDAs were proposed then and are now integrated into the city as part of densely populated settlements. Industrial waste is the primary source of pollution in the world, endangering both people and the environment. India's pharmaceutical sector has grown significantly over the past few decades thanks to a consistent population rise and its reputation as a low-cost manufacturing destination for many multinational drug companies. Notably, its bulk medicine production industry has seen tremendous growth since the 1970s and is centred mainly in Hyderabad, a city in southern India. However, this has produced rewards for shareholders and positive economic results for businesses headquartered in India and other countries. However, the effects of growing pharmaceutical production on the environment and residents, who live close to industries and industrial parks, have received very little attention.

2.2 *Industrial Estate Planning Approach in India*

Following globalization in 1991, India granted the ULBs various powers to allow unrestricted industrial development, which had dire repercussions in the following years. The framework illustrates the gaps in the approach between the various industries. Whereas the Central Pollution Control Board (CPCB) is the central organization responsible for overseeing the release and monitoring of pollutants near industrial areas, the Department of Commerce and Economics at the centre primarily disseminates timely growth and production data completed by various states nationwide.

The 73rd and 74th Amendment Acts, enacted in 1992 and 1993, gave urban local bodies authority over industrial areas. The State Industrial Infrastructure Corporation (SIIC), the Industrial Area Local Authority (IALA), which is led by elected officials from IDAs with industries within the industrial area, and the State Pollution Control Boards (SPCBs), which is responsible for monitoring pollution at the local level, are the three main entities that govern industrial regions. The flowchart in Fig. 1 illustrates this process. There is a difference between the land use planning authority, which approves the industry without considering pollutant dispersion and chimney height, and the CPCB and SPCB [14], which can investigate this issue but lack the authority to decide on land use and industrial unit planning within the industrial boundary. India's industrial area planning strategy is characterized by rigid departmental consensus and little technical assistance in pollution modelling.

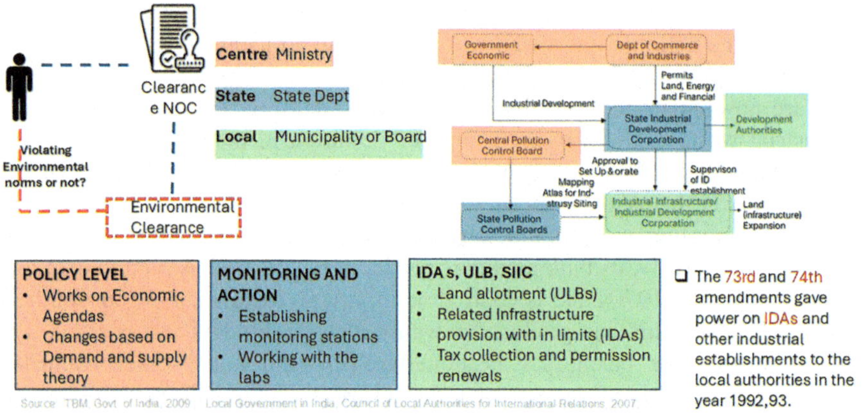

Fig. 1 Powers process between centre, state, and local authority in India (Source: TBM, Government. of India, 2009, Local Government in India, Council of Local Authorities for International Relations, 2007)

3 Methods and Materials

Most current literature addresses the topic in three main ways, emphasizing the cluster development concept and its local and global implications. It also discusses the industrial clusters throughout the Indian subcontinent and how their GDP is distributed among the nation's metropolitan areas. The second section of the background study addresses local and international laws and policies about environmental issues and pollution categories. The main discussion topics will be the EIA, its history across various countries, and how laws and policies have changed. Various international survey organizations provide categories and phases for risk assessment, and the processes and methodologies for risk assessment are laid out in depth throughout; the third area of requirement for the study is related to many macro-, meso-, and micro- category investigations that have been conducted, as well as pollutant modelling and approaches that have emerged to date.

3.1 Industrial Cluster Establishment and Growth

The idea of "industrial clustering" emerged in late marketplaces around the globe when the focus shifted from industrial cities to forming industries or groups of comparable sectors spread throughout the cities. In 1910, the English economist Alfred Marshall expounded on the elements influencing the formation of clusters. He elucidated the economic labour stratification that Jane Jacobs advocated in her book "Life and Death of Great American Cities," where emphasis is placed on the connections between urban regions and how the vitality of these clusters contributes to the life of a city. Economists like Giacomo Becattini had a different perspective on industrial clusters, believing that people are entwined with the dynamics of the area. [15, 16]

The first step in the evolution of industrial clusters is to concentrate on the economic activity linked by favourable axillary spill overs; these areas need to be analysed geographically rather than just economically. Let the notion of the industrial cluster be explained in the following four points, considering that the growth paradigm of these economic clusters varied depending on the economies of the nations in which they are located.

- Economies of scale mostly drove them at the time. These are mainly comparable sectors with actors typically associated with each other. As clusters are formed to improve economic activity, they have enormous dynamism in altering their economic activity. These changes are based on changes in business models, industries, and technologies.
- Secondly, they are selected based on their geographical location.
- Thirdly, these industrial clusters collaborate in rivalry, producing high production that indirectly affects the surrounding environment: (e.g. pharmaceutical

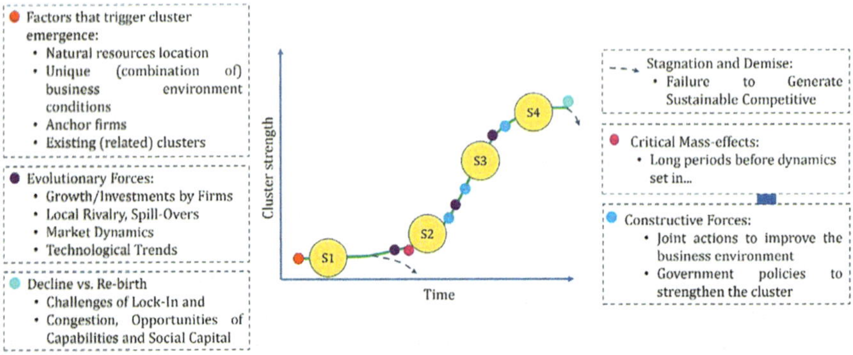

Fig. 2 The evolution of clusters (*Source: Authors compilation* [15])

industries depend on active pharmaceutical ingredients (API) and help in setting up more auxiliary units).

- Fourthly, in terms of how they evolve, they exhibit an S-curve, with changes to their strength occurring over time. Over time, this form of development is evolved in terms of type, technology, and industry category. However, as explained below, the state of these industrial institutions must be viewed in the context of a similar theory as shown in Fig. 2.

3.2 *Different Geographers' Coin Cluster Definitions*

"Clusters are a geographically proximate group of interconnected companies and associated institutions in a particular field linked by commonalities and complementarities. Clusters encompass an array of linked industries and other entities important to competition…including governmental and other institutions—such as universities, standard-setting agencies, think tanks, vocational training providers and trade associations [17] …geographically bounded concentration of similar, related or complementary businesses, with active channels for business transactions, communications and dialogue, that share specialized infrastructure, labour markets and services, and that are faced with shared opportunities and threats" [18, 19].

Regional clustering has been used to describe industrial districts of small crafts firms, high-technology centres, agglomerations of financial and business service firms in cities, company towns, and large branch plants and their supply chains". "…clusters at least must be characterized along relevant dimensions if appropriate policies are to be devised … (these include) …density. breadth, depth…activity base…growth potential…innovative capacity" [20, 21].

The industrial clusters in the developing nations are seen differently when compared to the developed nations wherein the factor-driven economies, the cluster

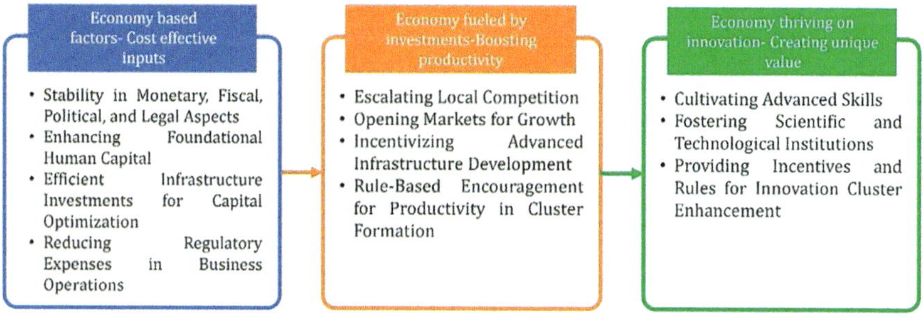

Fig. 3 Stages of economic development (*Source:* Authors' compilation from Ref. [21])

development process starts with a different theoretical input according to the comparative advantage of nations. According to the literature, factor-driven economies established industrial clusters with efficient basic infrastructure. During that time, industries were given incentives in terms of regulatory benefits and ease of doing business policies as discussed in Fig. 3 [22].

China and India are becoming the world's largest producers of pharmaceuticals. While China has emerged as the primary provider of the active pharmaceutical ingredients (APIs) needed to build medications, India also produces a sizable portion of APIs. It has developed a speciality in the processing of drugs, which it then exports as completed goods to all markets worldwide [23]. This research examines the latter, concentrating on the significant risk to public health provided by pollution from antibiotic manufacturing facilities in India, which is thought to be a factor in the rising rates of drug resistance both domestically and internationally. Due to travel and trade with India, antibiotic-resistant genes have the potential to spread globally, which poses a severe threat to public health.

The literature collectively suggests that India's economic growth has been accompanied by increasing inequality between rural and urban areas and states. Aneja (2020) finds that the tertiary and secondary sectors are more responsible for raising income inequality among states, while the primary sector offsets this gap [19]. Both highlight the divergence in per capita income among states in the post-reform period, with different states following different steady-state paths. Mazundar (2010) attributes the rising inequality to the peculiar sectoral composition of India's growth, particularly the dualistic nature of the manufacturing sector [19]. Overall, they indicate that economic growth in India has not been evenly distributed, leading to disparities in income and inequality between different industries and states as explained in Fig. 4.

This is partly because, since liberalization, private money has made up a larger portion of all investments, and that private capital is more likely to be concentrated in urban and coastal areas. This leads to further concentration of wealth in already more prosperous regions. An assessment of the performance of Software Export Zones (SEZs) shows that most of them were established in districts that were more industrialized than the national average and close to metropolitan cities [24]. Since

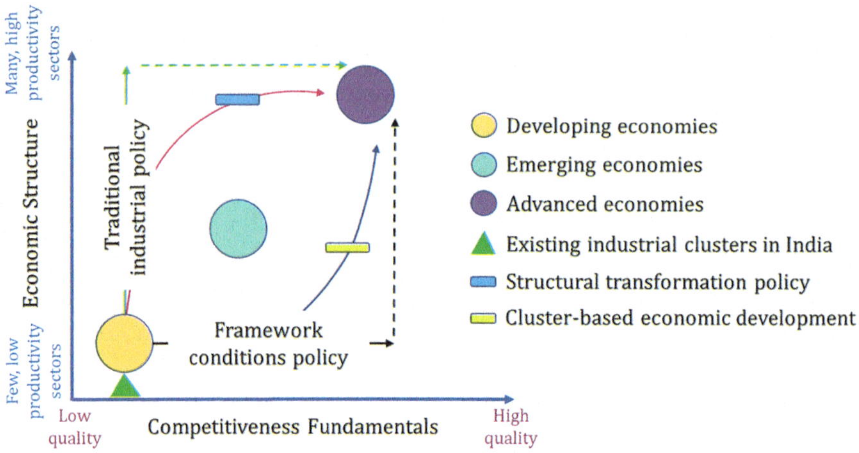

Fig. 4 Comparison of development approaches (Ketels [15])

the corridor policy is designed around important transportation corridors that already handle the majority of the nation's freight traffic—beginning with the Western Dedicated Freight Corridor—it also carries the risk of perpetuating inequity trends. Not enough consideration has been given to how these measures can amplify already existing inequities in the Indian economy. While the federal government controls some parts of industrial growth, such as transportation infrastructure like highways and railroads, some industries, and corporation and income taxes, state governments determine a large portion of the trajectories of industrial development. There are differences in the experiences of states regarding industrial success because states are in charge of all other industries, land, and the development and provision of industrial infrastructure. In addition to historical institutional development trajectories, sociocultural conditions, and unequal access to natural resources, policy decisions also play a role in this [25].

India's pharmaceutical industry clusters are now key factors influencing the country's reputation in the world market. Two notable examples are the Genome Valley in Hyderabad and the Pharmaceutical Cluster in Gujarat, which feature high concentrations of biotechnology and pharmaceutical enterprises, research facilities and academic centres. These clusters provide an atmosphere that is favourable for R&D, manufacturing, and technology transfer while also encouraging creativity, teamwork, and economic progress. In addition to boosting pharmaceutical exports and the country's economy, these clusters' cooperative ecosystem makes sure that there is a consistent supply of reasonably priced medications available to suit both local and global health-care demands.

These clusters face difficulties even though they have had tremendous success. Infrastructure growth, regulatory compliance, and the need for ongoing innovation are important issues that require attention. Strong legislative frameworks and enabling legislation are necessary to guarantee the quality and safety of products. Concurrently, it is necessary to make ongoing expenditures in transportation,

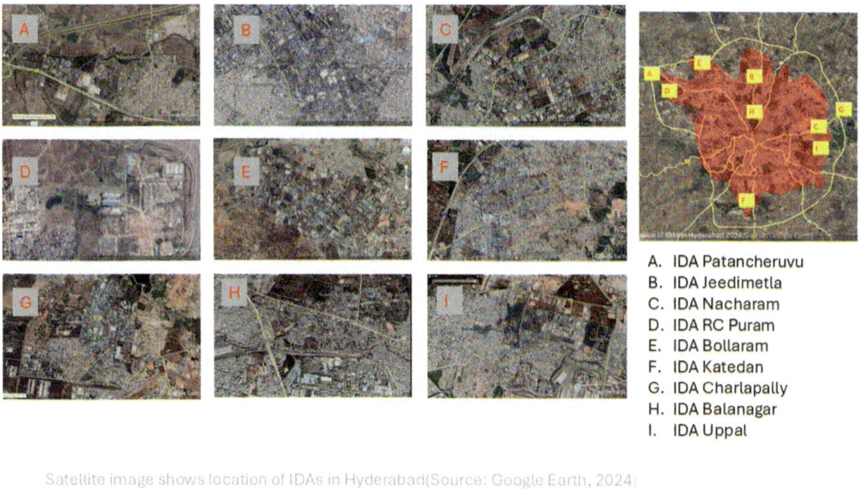

Fig. 5 Industrial development area across Hyderabad city

logistics, and utilities in order to meet the industry's expanding needs. In addition, maintaining the growth trajectory of these pharmaceutical clusters in India depends critically on industry participants and research institutions working together to cultivate an innovative culture. The industrial development area across Hyderabad city is shown in Fig. 5.

In the intricate economic landscape of Hyderabad, which is known as the drug capital of India which produces 40% of India's bulk drug production, specialized business clusters emerge as dynamic entities, each possessing a distinctive nomenclature and focus area such as NIMZ (National Investment and Manufacturing Zone). These clusters, characterized by their unique names and industry specializations, constitute the city's economic nuclei. Let us embark on a nuanced exploration of these clusters, delving into their idiosyncrasies, phases of development, and their profound influence on the urban fabric of Hyderabad. Table 1 explains the cluster wise stages of evolution.

3.2.1 Economic Synergy: S1

The initial establishment of Industrial Development Areas (IDAs) is proposed based on the economic Five-Year plans that existed in India during the time. IDA Bollaram, which is located 11 km from the central business district established in the year 1979, is one of the pharmaceutical clusters where all major production units are located. Urbanization trend clusters are a phenomenon where businesses collaborate to increase productivity though there is availability of auxiliary pharmaceutical units for profitability. This synergy is similar to teamwork in which mutualistic relationships between production units and APIs produce benefits for both parties.

Table 1 Industrial clusters and their establishment types

S. no	Name of the cluster	Type of cluster	Year of establishment	Distance from CBD (km)	Stage of cluster	Availability of APIs	Urbanization Trend (1980-2021)
A	IDA Patancheruvu	Pharma	1976	30	S3	Yes	↑
B	IDA Jeedimetla	Pharma	1975	14	S2	Yes	↔
C	IDA Nacharam	Pharma	1985	9	S3	Yes	↔
D	IDA RC Puram	Pharma	1983	30	S4	Yes	↑
E	IDA Bollaram	Pharma	1979	11	S1	Yes	↓
F	IDA Katedan	Pharma + food processing	1970	24	S2	No	↓
G	IDA Charlapally	Metallurgical + miscellaneous	1985	16	S3	No	↑
H	IDA Balanagar	Pharma + institutional	1971	10	S4	Yes	↔
I	IDA Uppal	Pharma + food processing	1965	10	S4	No	↔

↑ = High, ↔ = Saturated, ↓ = Low

This cluster had experienced maximum air pollution levels, thus effecting the urbanization trend.

3.2.2 Spatial Distribution: S2

IDA Jeedimetla and IDA Katedan were established in 1975 and 1970, respectively, following the implementation of India's first industrial policy and the establishment of Indian Drug and Pharmaceutical Limited (IDPL)—a major drug production unit operated by the Government of India since 1961 under the leadership of Pt. Jawaharlal Nehru. This initiative significantly boosted the growth of pharmaceutical manufacturing units in these areas. Moreover, the strategic location of the IDPL clusters on the outskirts of the city, along with their proximity to the CBD (Secunderabad), facilitated the distribution of major industries within these zones. This, in turn, encouraged the settlement of numerous auxiliary Active Pharmaceutical Ingredient (API)-producing industries.

3.2.3 Inter-cluster Collaboration and Competitive Dynamics: S3

IDAs located in Patancheruvu, Nacharam, and Charlapally, which are located at a distance from the CBD, helped in linking with more auxiliary units such as food processing industries depending on the packaging industries and pharmaceuticals industries dependent on the tablet processing industries. Similar to natural niche competition, intercluster cooperation and competitive dynamics are important in these economic ecosystems. Companies in a cluster work together in a manner similar to ecological mutualism, which produces gains in productivity as well as shared benefits. On the other hand, rivalry amongst clusters is comparable to niche competition, in which several economic players fight for supremacy and resources in the market, thus maximizing production and degrading the local environment.

3.2.4 Evolutionary Trajectories and Economic Phases: S4

IDAs, which are influenced by the multiple nuclei effect, started changing the morphological characteristics of the industries within the specific cluster due to excessive urbanization happening around these clusters. The locations such as Uppal, Balanagar, and RC Puram fall under this category where the existing pharmaceutical units are closed down and the pharmaceutical unit count is declining progressively and change in unit size and type helped in new technology upgrading within these clusters. This evolutionary journey follows a pattern resembling an S curve.

In conclusion, scientific approaches can be used to analyse Hyderabad's economic landscape, which is typified by specialized company clusters. These clusters show a complex yet intriguing interplay of economic entities, ranging from geographic optimization strategies to elements that will remain integral to Hyderabad's scientific and economic development as the city continues to grow. In summary, rather than being purely economic entities, Hyderabad's named and classified industrial clusters are dynamic ecosystems that shape the city's industrial identity. Every cluster contributes something distinct and strong to Hyderabad's economy, from pharmaceutical giants to metallurgical conjectures.

4 Conclusion

In conclusion, India's complex interplay of benefits and challenges is revealed by the delicate connection between industry, urbanization, and environmental effects. Without a doubt, the growth of numerous industrial clusters has stimulated economic expansion, produced job opportunities, and raised the GDP of the nation dramatically. However, in light of the current challenges and the industrial accidents that have occurred both in India and globally, a more concerted and forceful approach is needed to mitigate the negative impacts of industrial activity.

The dual nature of industrialization is highlighted by the socio-economic division observed in urban environments, particularly in metropolitan regions. While it generates employment, it exacerbates issues such as pollution, greenhouse gas emissions, and hazards to public health. A major barrier to sustainable development is the absence of a standardized framework for calculating and mitigating these negative impacts.

The fact that India is the most polluted nation in the world draws attention to the serious issue of indoor pollution and the necessity of conducting in-depth studies and actively participating in international conventions that address the negative health effects of industrialization, especially from the pharmaceutical industrial clusters, which create the issues of superbugs which become immune to antibiotic medicines (Kumar, 2012). As the nation increasingly urbanizes, practical approaches and policies are badly needed to assess and mitigate the negative social and environmental repercussions of industrial growth.

The importance of environmental activism and sustainable legislation is emphasized in the study's conclusion, particularly in industries with significant ecological footprints like the pharmaceutical business. Comprehensive collaboration is needed to strike a balance between economic growth and environmental preservation. India can only guarantee a more equitable and sustainable future for its growing population by adopting a coordinated plan to solve the challenges posed by industrialization.

References

1. Mishra, P. K., et al. (2009). Bhopal gas tragedy: Review of clinical and experimental findings after 25 years. *International Journal of Occupational Medicine and Environmental Health, 22*, 193–202.
2. Darton, R. C. (2015). Assessing and measuring environmental impact and sustainability. *Assessing and Measuring Environmental Impact and Sustainability*. https://doi.org/10.1016/B978-0-12-799968-5.00014-2
3. Zeb, N., Khokhar, M. F., Pozzer, A., & Khan, S. A. (2019). Exploring the temporal trends and seasonal behaviour of tropospheric trace gases over Pakistan by exploiting satellite observations. *Atmospheric Environment, 198*, 279–290.
4. Pandey, J. S., Kumar, R., & Devotta, S. (2005). Health risks of NO2, SPM and SO2 in Delhi (India). *Atmospheric Environment, 39*, 6868–6874.
5. Mishra, R. K., Agarwal, A., & Shukla, A. (2021). Predicting ground level PM2.5 concentration over Delhi using landsat 8 satellite data. *International Journal of Remote Sensing, 42*, 827–838.
6. Ghosh, S. (2001). Sustainable energy policies for clean air in India
7. Guttikunda, S. K., Nishadh, K. A., & Jawahar, P. (2019). Air pollution knowledge assessments (APnA) for 20 Indian cities. *Urban Climate, 27*, 124–141.
8. Drisse, M.-N. B., Goldizen, F. & Adair-Rohani, H. (2017). Atlas on children's health and the environment.
9. Timo, K. (2007). The convention on environmental impact assessment in a transboundary context (Espoo convention). In *Making treaties work: Human rights, environment and arms control* (pp. 218–242). https://doi.org/10.1017/CBO9780511494345.011

10. Niti (2014). Chapter: 2 Five year plans and SSIs in Indian Economy.
11. EIA 2020 Draft. 804 (2020).
12. UNIDO (2019). http://www.dcmsme.gov.in/clusters/clus/smelist.htm.
13. Fayyaz, A., Lund-Thomsen, P., & Lindgreen, A. (2017). Industrial clusters and CSR in developing countries: The role of international donor funding. *Journal of Business Ethics, 146*, 619–637.
14. Health E. (2014) S. M. & Industries, P. Central Pollution Control Board (Cpcb), p. 4.
15. Ketels, C. (2017). *Cluster mapping as a tool for development* (pp. 1–52).
16. Saadatyar, F. S., Al-Tabbaa, O., Dagnino, G. B., & Vazife, Z. (2020). Industrial clusters in the developing economies: Insights from the Iranian carpet industry. *Strategic Change, 29*, 227–239.
17. Porter, M. E. (2000). Location, competition, and economic development: Local clusters in a global economy. *Economic Development Quarterly, 14*, 15–34.
18. Gupta, A., & Shukla, A. K. (2024). Optimal approaches in global warming mitigation and adaptation strategies at city scale. *Discover Sustainability, 5*, 272. https://doi.org/10.1007/s43621-024-00497-8
19. Sharma, S. K., Deeksha, R. P. K., Shukla, S., & Shukla, A. K. (2024). Overview of sustainable development goals. In P. K. Rai, A. Singh, B. Beingachhi, & S. K. Singh (Eds.), *Natural resources management and sustainable livelihoods in the mountainous region. Advances in geographical and environmental sciences*. Springer. https://doi.org/10.1007/978-981-97-2100-9_14
20. Singh, A. (2010). Clusters in india.
21. Shukla, S., Deeksha Chand, S., Rai, P. K., Shukla, A. K., Misra, R. K., & Singh, U. (2024). Global and Indian perspectives on sustainable development, remote sensing, and climate change: An overview. In *Natural resources management and sustainable livelihoods in the mountainous region: Evidence, gap and future strategies*. Springer. https://doi.org/10.1007/978-981-97-2100-9_6
22. Kuchiki, A., & Tsuji, M. (2005). *Industrial clusters in Asia*. Palgrave Macmillan. https://doi.org/10.1057/9780230523647
23. IBEF. (2013). Pharmaceuticals 2013. *Pharmaceuticals*, 1–36.
24. Aneja, R., & Banday, U. J. (2020). Regional economic growth and inequality in India: A sector-wise decomposition analysis. *Arthaniti: Journal of Economic Theory and Practice, 20*(1), 1–16. https://doi.org/10.1177/0976747920910825

Comparative Assessment of Window Technology to Achieve Energy Efficiency: A Review

Manish Kumar Chitranshi

1 Introduction

Commercial and residential buildings account for some 20–40% of energy consumption, including around 30% of global carbon dioxide (CO_2) production in developed countries [1]. This consumption rate is a reflection of the increase in energy usage in growing economies which is directly correlated with the estimated annual population growth of 0.9% [2]. Climate change is making it increasingly difficult to achieve thermal comfort because of the increasingly extreme conditions and variations. The energy load is consumed by air conditioning which maintains the thermal comfort of a building. The study also discussed advances in windows technologies here we detailed both traditional and advanced technologies used in windows. The technology of traditional window (also known as static windows) includes passive technologies that can enhance its performance (both optically and thermally), such as insulated glass [3], tinted glass [4], anti-reflective glass [5], self-cleaning glazing, and low-E-glazing [6]. In contrast, the second types of windows, known by a range of names, such as active windows, dynamic windows, or advanced windows employs technologies such as liquid crystal glazing [7], electrochromic glazing [8], thermochromic glazing [9, 10], photovoltaic glazing [11], and gasochromic glazing [12].

M. K. Chitranshi (✉)
Amity School or Architecture & Planning, Amity University Madhya Pradesh, Gwalior, Madhya Pradesh, India
e-mail: mkchitranshi@gwa.amity.edu

S. Pathak et al. (eds.), *Intelligent Infrastructure and Smart Materials*,
https://doi.org/10.1007/978-3-031-92421-7_10

2 Performance in Optical and Thermal Condition

Windows serve a variety of functions in buildings, including admitting natural light, ventilation, passive solar gain, and beauty. The following essential parameters can be used to describe the thermal and vision characteristics of a window: "visible transmittance, SHGC & U-value", Density "Convection", contact "Conduction "and waves "Radiation" are the three modes of energy moves or heat transfer that impact the evaluation of the aforementioned characteristics. In what follows we will consider these key aspects. The *U*-value measures the insulating effectiveness of a material; this term is also known as both the total thermal transmittance and the total heat transfer coefficient. On the basis of temperature difference, the *U*-value is measured in W/m^2 K. The average thermal transmittance values in different building elements are shown in Table 1 [13]. The fabric's *U*-value will almost certainly increase both the energy efficiency of a building and the comfort of its occupants [14]. Heat transfer through window is primarily a result of conduction through the glass material. As a result, as shown in Table 2, different *U*-values can be obtained thriough different glass pane configurations and employing different combinations of materials between the panes of glass. In addition, it is also possible to transfer more heat through the glass panes, spacers and frames for windows. Further, in multi-pane windows, radiation occurs in three areas, on both the windows interior and external surfaces, also the window frames.

2.1 *Visible Transmittance*

The wavelength spectrum which is transmitted through a certain kind of glazing material is known as visual transmittance (VT). In the case of transparent windows, this measurement ranges from 90% and 10% for glazing that is extremely reflective.

Table 1 The *U*-values of different building materials [13]

Element	*U*-value (W/m^2 K)
Roof	0.16
Floor	0.25
Windows	2.00
External wall	0.30

Table 2 U-factor for different kinds of windows glazing pane [13]

Glass configuration	Cavity material	*U*-value (W/m^2 K)
Non-coated double-pane	33 mm aerogel (granular)	0.44
Non-coated double-pane	22 mm aerogel (monolithic)	0.65
Non-coated triple-pane	Argon with 16 mm	0.79
Non-coated double-pane	Argon with 15 mm cavity	1.20
Non-oated double-pane	Air cavity 15 mm	1.40
Non-coated double glass	12 mm cavity	2.80
Non-coated single glass 6 mm	–	5.70

For glazing with a reflective coating, the VT is determined by the type of vehicle, including the number of panes, the glazing, and the existence of coatings. Each of these can have an impact on transparency [14]. A high VT indicates that there is more daylight in a space and, in most cases, there is a reduction in the power consumption load of the building. For example, a typical two-layered glass window has a visual transmittance of 78%, which may reduce energy costs. Findings show that greater reductions can be obtained if low-E coatings [15] and colored films are used [16].

2.2 G-Factor

The G-Factor is also known as the solar heat gain coefficient (SHGC). This measures the percentage of solar radiation which comes through the window. This includes both the heat directly transmitted as well as the heat absorbed, and subsequently released, into the space [17]. This factorial value is used to evaluate the heat gain as a result of direct exposure to the sun's radiation, and the shading devices requirements of transparent material such as glass [18], and it is crucial in lowering the need for heating loads. SHGC values typically range from 0 to 1.

3 Dynamic Glazing Technologies for Solar Innovation

This type of window uses electrochromic technology to darken the pane of glass using moveable or switchable shade and ventilation systems which collect energy. This includes both optical and mechanical variables. In most cases, thermal characteristics are climate-dependent. Examples of these technologues include electrochromic and photochromic, thermochromic [19–21], latent Photo chromatic materials (PCMs), electrochromic, and switchable film [16] [21].

3.1 Phase-Change Film Produces Smart Window

Polychromic materials (PCMs) absorb near-infrared light from the Sun and convert it into heat for the building. This material can absorb the invisible wavelengths of the Sun's light, emitting it as heat; it can also be 'switched' so that it reflects those wavelengths of light instead. The application of this technology minimizes the degree of heat transfer. Recent studies have shown that double-pane windows with PCM fillings can reduce solar heat gain in buildings while simultaneously raising the temperature by 3 °C on the inside surface of the window.

3.2 *Building Integrated Photovoltaic Panels (bıpv)*

In the case of non-segregated solar photovoltaic (PV) material, it is difficult to maintain a high power yield while retaining transparency of window glass [22]. So far, research has taken a particular interest in the use of semi-transparent PV cells glazing innovation. Titanium oxide (TiO_2) film is used in Grätzel cell, conductive glazing materials with a typical range of 16–20% [23]. Recent advancements in window technology aim to convert solar energy into electricity. They offer an attractive alternative because of their reduced environmental impact, lower cost, and efficient use of space.

Making energy-efficient windows in accordance with local regulations Energy regulations urge the improvements in all the *U*-value factors, including the cost, weight, and dimensions of exterior glazing components. Finally, a summary of window technology based on the market is provided. The global window manufacturer companies are the primary sources of information on *U*-value. The use of substitute gases and electrochromic materials have resulted in significant cost savings, and there has been great advances in the use of vacuum-sealed glass and mostly driven as a result of their efforts to limit heat transmission by conduction. In the glazing samples, there was a lot of radiation.

4 Technologıes for Steady Glazıng

This section covers static (or traditional) glazing technologies. Recently, this sort of glazing has become the focus of research, particularly in terms of the thermic and visual behavior of glass. These include solutions such as pane windows, soda-lime-silica glass, safety glass, low-E windows, and film suspension and vacuumed glazing.

4.1 *The Thickness of the Glass*

An examination was made of the impact of glass thickness on the thermal performance of a single-glazed window. It was observed that the glass pane U-factor was significantly reduced for thicknesses of more than 12 mm. However, the loss of heat transmission comes at a cost to both the weight of the window and the level of visible light transmission.

4.2 Glazing with Many Panes

During Mid 20's double-glazed windows have been widely produced in UK [24]. The gap between the panes creates a thermal barrier which can increase the insulating properties of the window and thereby improve a buidling's energy efficiency [25]. The model employed in this study suggests that the use of mirror glass will bring energy savings which will produce, in a hot, dry climate, a payback period of approximately 9–10 years.

In a double-glazed window, the gap between the panes acts as a thermal barrier, thereby improving the performance of the window components [26]. This has been shown to equate to a reduction of 50% in heat loss when compared with traditional single glazing, while visual transmittance and SHGC Coefficient. Both remain relatively high [16]. Furthermore, heat transmission rate of a double-glazed window is two and a half times that of a single-pane window, which can help to improves thermal performance by around 50–70% in warm climates [28].

4.3 Tinted Glass

The use of tinted glazing panes aids in the preservation of material transparency the window's interior layer, which absorbs some of the heat while blocking daylight, [29]. The tinting of glass occurs during the floating process, when metallic components are added to the glass. The window transmittance and visibility are reduced as a result of this operation and coloring [30], maximizing the level of absorption of solar energy while simultaneously lowering sun glare by up to 50% for single panes of glass [16]. This glass absorption property can be tuned to spectral selectivity, meaning that the tinted windowpane can be manufactured to absorb only certain wavelengths of the solar spectrum in the near-infrared while passing on only sunlight. The description of tinted glass description is based on various factors, including color, width, insulating characteristics, fraction of solar heat, and the fraction of visible spectrum. These are summarized in Table 3 [30].

4.4 Coatings with a Low Emissivity

Long infrared radiations are reflected by a low e-surface such as a heat-reflecting mirror. This was created to minimize the levels of infrared and ultraviolet light that comes through the window pane, without any reduction in the amount of sunlight coming into the internal spaces. In this way, the insulating characteristics of glass are achieved through the decrease in heat transfer between the interior and the exterior.

Table 3 Tinted glass specifications for various factors

Colour	Width in mm	U-value	G-value	Visible transmittance
Bronze	4	5.8	0.68	0.60
	6	5.7	0.60	0.49
	8	5.6	0.53	0.40
	10	5.6	0.48	0.33
Green	4	5.8	0.63	0.79
	6	5.7	0.55	0.73
	8	5.6	0.50	0.68
	10	5.6	0.46	0.63
Grey	4	5.8	0.66	0.55
	6	5.7	0.57	0.43
	8	5.6	0.50	0.34
	10	5.6	0.45	0.26

Low-E glass, also known as "hard coat" Low-E, is created by applying a low-emissivity layer of indium tin oxide is applied in its molten stage and when it is just beginning to harden [31]. The manufacturing process of low-emissivity coatings consist of two methods, the sputtered method and pyrolytic techniques, in which atoms are expelled from solid targets by subjecting them to intense particle bombardment, to the already manufactured glass [16].

A study from Canada has reported saving of 13–17% in energy costs by using low cost soft-coat low-emissivity compared to hard-coat [32]. In the same study it was also found that hard-coat solutions also gives the best results when used on buildings in the United Kingdom residences.

Low-emissivity coatings can also possibly be tuned to certain optical spectrums and electromagnetic waves, in much the same way as reflective glass, in order to enhance their performance in terms of both thermal regulation and the amount of daylight passing through the window. Low-e coatings can lower the fraction of solar heat gain in order to make up for existing heating loads [33], which can be something of a disadvantage [34]. However, the advantages in terms of sunlight and glare reduction make these coatings a viable solution for emittance optimization and radiation control.

4.5 Anti-reflective Coatings

A recent technology for increasing daylight transmission is anti-reflective coatings, which can boost apparent transmittance by a good ratio range from 0.7 to 0.9 in multi pane-windows. Low-E The use of Tin (IV) Oxide as a coating had no effect on the total U-factor while boosting the G-value by a certain percentage.

4.6 *Self- Cleaning Coatings*

In the field of self-cleaning coatings, water is used in order to cut down on the use of human labor. This procedure has been utilized specifically for windows that have a high degree of condensation. PV panels, for example, are a power-generating solution. Hydrophilic coatings, by contrast, include a thin transparent layer. To create a static contact angle, a coating of titanium dioxide (TiO_2) is applied between the glass and the temperature of the water is <90°. When exposed to light, this chemically breaks down glazed debris. The 'photocatalysis process' describes how light interacts with materials [37, 38].

4.7 *Gases That Can Be Used as a Substitute*

The double-glazing business employs a variety of gases to fill the gap between the panes [27]. Window manufacturers make considerable use of noble (inert), because they are both noncombustible and unreactive Xenon is one type of gas. It is the most efficient of the noble gases, but its scarcity makes it more expensive to produce than krypton gas (U-value of 0.64 W/m^2), which has a lower U-value. Due to its medium-use, Xe has a relatively low usage. Compared to the previous model, the performance is better, but it is more expensive than argon solutions [13]. For gas-insulated multi-pane windows, the U-factor of W/m^2 ranges from 1.0 to 2.2, 0.5 to 0.8 W/m^2 and K, respectively [39]. The following section gives a description of the various methods which can be employed. Argon-filled solutions have been shown to reduce window thermal conductivity by around two-thirds when compared with air-based combinations in previous experiments and modelling [40]; they are also popular because they are odorless. Its non-toxic characteristics have allowed for its incorporation into colder climates, where these technologies are used in residential houses. Krypton-based gap fillers have reduced the overall U-factor by a factor of two. In theoretical comparisons, argon gaps account for 17.18% of the total, despite the fact that argon has greater manufacturing cost, it has become a more popular gas, and a mass production option. Alternatively, a xenon gap filler has been evaluated as an insulating material and found to be the most effective. All of their U-values (calculated at 0.28 W/m^2) K), despite the fact that element has a future similar to krypton, with higher production costs than krypton and argon technologies combined [14]. Multi-pane windows have recently gained popularity as a solution as an energy-efficient solution when replacing window glazing in old buildings. There is a genuine requirement for a thin and light solution. They are a non-degradable, environmentally friendly, and limited availability [27].

4.8 Vacuum-Sealed Windows

Vacuum-sealed glazing is regarded as a very promising solution for reducing heat gain and heat loss by convection and transfer in the air within the gap that has been filled. This can be attributed to a reduction in the value of gas particles involved in thermal performance. The manufacturing industry is currently working on the development of evacuated glazing [41, 42], which produces a lower U-factor through lowering heat transfer via supporting elements. Vacuum windows with commercially available technology can achieve *U*-values as low as 0.7 W/m^2. In vacuum glazing, heat transfer is limited and restricted by the vacuum. Conduction-based supporting elements like edge sealing and window frame materials are selected carefully with a view of reducing the overall U Value of the window assembly [43]. Commercially available self-cleaning hydrophobic coatings are divided into two categories: hydrophobic and hydrophilic coating. Water droplets will roll down hydrophobic coverings, having been repelled by them. The two surfaces on the glass have a high static contact angle; this is achieved by the use of moldable waxes, chemical compounds and chemical vapour infiltration. At the time of writing, the manufacturing sector is now working on producing evacuated glazing [41, 42], which will reduce heat transfer through the supporting elements to produce a lower *U*-value. Under the commercially available technology, vacuum windows can reach *U*-values as low as 0. In such a system, heat is transferred through contact between the panes in vacuum glazing. A support pillar array is based on conduction and edge sealing. As a result, it is necessary to develop methods for incorporating novel elements into the manufacturing process. To lower their *U*-values, components are put through their paces openings [43]. One of the new materials, aerogel, was developed in the 1930s. It is a highly insulating silica-based substance comprised of 4% silica and 96% air [45]. It has a molecular structure resembling foam and retains air. It reduces heat convection while permitting a high level of sunlight *U*-value. The benefits to the window industry of this material has been shown to inhibit heat transmission via conduction, with temperatures reaching a U-factor as low as 0.05 W/m^2 K being possible.

5 Conclusıon

This study provides a thorough examination of commercially available window products. These were evaluated for their insulating characteristics, the fraction of solar heat transfered, and the visible spectrum of light. On the basis of the available literature and thorough examination, the following conclusions can be made:

- The interaction of the exposed surface with the surroundings is directly linked to the energy loss through the building envelope. Traditional windows have relatively high *U*-values comparing to other elements because they are composed of materials with a high heat transfer rate (walls, roofs and floors). The focus of present and future research seems to be on applying static and dynamic technologies to decrease heat exchange on window elements.

- Since, it creates the sample weighted and massive, increasing glazing width has no beneficial effect on lowering the U-factor of pane windows. Similar effects on heat transfer may be seen when tinted windows are used alone; these effects solely affect the VT and the rate of solar heat transfer.
- However, glass treatments work well to reduce the impact of solar heat absorption through openings in structures. It was found, for instance, that using minimal, antiglare, and photocatalytic films can lower the U-factor by 0.48 (for low-emissivity materials) without affecting the transparency of the glazing.
- When compared to the air-filled pockets factor (0.62 W/m^2 K), gas-filled cavities may dramatically reduce the U-factor of windows (by 22.2% when using argon, 33.3% when using krypton, and 41.1% when using xenon, respectively).
- Double-sash windows will substantially minimize the thermal conduction of windows when compared with other window technologies (such as double, triple and vacuum windows). However, in comparison with conventional multi-pane glazing with an argon filling (1.2 W/m^2 K), the use of vacuum insulation glazing with 0.7 W/m^2 K may reduce the U-factor by 42%. This significantly enhances the insulating properties of the double-glazing technology which is presently in use.
- Although dynamic windows may control the VT and thermal conduction of windows, it is important to take into account the methods of production, material availability, and environmental factors when choosing the best option for each situation.
- By drastically reducing thermal value and heat transfer within the pane window, vacuum glazing provides a highly effective method to minimize heat transmission on windows. According to this analysis, appreciative circumstances will exist for the development of this technology in the future.
- PV material might prove an attractive in terms of power saving, with both a reduction in negative effects on the lifespan of windows and higher output yields.
- According to [14], based on the review, vacuum-glazing technologies with applications for aerogel may become available in future, once issues regarding the strength and stability of the material have been resolved. A similar approach to achieving building with net zero energy consumption for future building envelopes may be to incorporate building-integrated PVs directly into window working models.

References

1. Perez-Lombard, L., Ortiz, J., & Pout, C. (2008). A review on buildings energy consumption information. *Energy and Buildings, 40*(3), 394–398.
2. TWB. Retrieved from: https://data.worldbank.org/indicator/SP.POP.GROW?end=2021.

3. Crannell, H. (2011). Performance tests of large thin vaccum windows. *Nuclear Instruments and Methods in Physics Research Section A: Accelerators, Spectrometers, Dectectors and Associated Equipment, 629*(1), 387–391.
4. Casini, M. (2018). Active dynamic windows for buildings: A review. *Renewable Energy, 119*, 923–934.
5. Sertel, T., Sonmez, N. A., Cetin, S. S., & Ozcelik, S. (2019). Influences of annealing temperature on anti-reflective performance of amorphous Ta2O5 thin films. *Ceramics Interrnational, 45*(1), 11–18.
6. Sadooghi, P., & Kherani, N. P. (2019). Influence of slat angle and low-emissive partitioning radiant energy veils on the thermal performance of multilayered windows for dynamic facades. *Renewable Energy, 143*, 142–148.
7. Baliyan, V. K., Jeong, K. U., & Kang, S. W. (2019). Dichroic-dye-doped short pitch cholesteric liquid crystals for the application of electrically switchable smart windows. *Dyes and Pigments, 166*, 403–409.
8. Piccolo, A., & Simone, F. (2015). Performance reqirements for electrochromic smart window. *Journal of Building Engineering, 3*, 9–103.
9. Liang, R., Sun, Y., Aburas, M., Wilson, R., & Wu, Y. (2018). Evaluation of the thermal and optical performance of thermochromic windows for office buildings in China. *Energy and Buildings, 176*, 216–231.
10. Peng, J., Curcija, D. C., Thanachareonkit, A., Lee, E. S., Goudey, H., & Selkowitz, S. E. (2019). Study on the overall energy performance of a novel c-Si based semitransparent solar photovoltaic window. *Appiled Energy, 242*, 854–872.
11. Yang, J., Jo, H., Choi, S. W., Kang, D. W., & Kwon, J. D. (2019). Adoption of wide-bandgap microcrystalline silicon oxide and dual buffers for semitransparent solar cells in buliding-integrated photovoltaic window system. *Journal of Materials Science & Technology, 35*(8), 1563–1569.
12. Feng, W., Zou, L., Gao, G., Wu, G., Shen, J., & Li, W. (2016). Gasochromic smart window: optical and thermal properties, energy simulation and feasibility analaysis. *Solar Energy Materials and Solar Cells, 144*, 316–323.
13. Jelle, B. P., Hynd, A., Gustavsen, A., Arasteh, D., Goudey, H., & Hart, R. (2012). Fenestration of today and tomorrow: A state-of-the-art review and future research opportunities. *Solar Energy Materials and Solar Cells, 96*, 1–28.
14. Cuce, E., & Riffat, S. B. (2015). A state-of-the-review on innovative glazing technologies. *Renewable and Sustainable Energy Reviews, 41*, 695–714.
15. Solovyev, A. A., Rabotkin, S. V., & Kovsharov, N. F. (2015). Polymer films with multilayer low-E coatings. *Materials Science in Semiconductor Processing, 38*, 373–380.
16. Carmody, J., Selkowitz, S., Lee, E. S., & Arasteh, D. K. (2004). *Window systems for high-performance buildings*. W. W. Norton & Company. 978-0-393-73121-7.
17. Singh, M., Chandel, S. S., & Nagar, N. (2010). An empirial model for angle-dependent g-values of glazings. *Energy and Buildings, 42*(3), 375–379.
18. Oliveri, L., Fagnani, F., Peretti, P.-L., Cattaneo, C., De Carli, M., & Muscio, E. (2015). G-value indoor characterization of semi-transparent photovoltaic elements for building integration: New equipment and methodology. *Energy and Buildings, 101*, 84–94.
19. Kamalisarvestani, M., Saidur, R., Mekhilef, S., & Javadi, F. S. (2013). Performance, materials and coating technologies of thermochromic thin flims on smart windows. *Renewable and Sustainable Energy Reviews, 26*, 353–364.
20. Chen, X., Li, Q., Yang, Y., & Xu, X. (2012). Smart window coating based on nanostructured VO2 thin film. *Optik, 123*(13), 1187–1189.
21. Ismail, K. A. R., Salinas, C. T., & Henriquez, J. R. (2008). Comparison between PCM filled glass windows and absorbing gas filled windows. *Energy and Buildings, 40*(5), 710–719.
22. Meng, W., Peng, J., Yang, H., & Luo, Y. (2018). Performance evaluation of semi-transparent CdTe thin film PV window applying on commerical buildings in Hong Kong. *Energy Procedia, 152*, 1091–1096.

23. Skandalos, N., & Karamanis, D. (2015). PV glazing technologies. *Renewable and Sustainable Energy Reviews, 49*, 306–322.
24. Liu, C., Wu, Y., Li, D., Zhang, Y., Wang, Z., & Li, X. (2017). Effect of PCM thickness and melting temperature on thermal performance of double glazing units. *Journal of Building Engineering, 11*, 87–95.
25. Aguilar, J. O., Xamán, J., Olazo-Gómez, Y., Hernández-López, I., Becerra, G., & Orozco, A. (2017). Thermal performance of a room with a double glazing window using glazing available in Mexican market. *Appiled Thermal Engineering, 119*, 505–515.
26. Aydin, O. (2000). Determination of optimum air-layer thickness in double-pane windows. *Energy and Buildings, 32*(3), 303–308.
27. Selkowitz, S. E. (1978). Thermal performance of insulating window systems.
28. Arıcı, M., & Karabay, H. (2010). Determination of optium thickness of double-glazed windows for the climatic regions of Turkey. *Energy and Buildings, 42*(10), 1773–1778.
29. Li, C., Tang, J., Chen, T., Tian, Q., & Zhang, Z. (2015). Experimental and therotical study on the effect of window films on building energy consumption. *Energy and Buildings, 102*, 129–138.
30. Rezaei, S. D., Shannigrahi, S., & Ramakrishna, S. (2017). A review of conventional, advanced, and smart glaxing technologies and materials for improving indoor environment. *Solar Energy Materials and Solar Cells, 159*, 26–51.
31. Mempouo, B., Cooper, E., & Riffat, S. B. (2010). Novel window technologies and the code for sustainable homes in the UK. *International Journal of Low-Carbon Technologies, 5*(4), 167–174.
32. Laouadi, A., Gaur, A., Desjarlais, M., Chen, M., Moffatt, M., Riffat, S. B., Goudey, H., Arasteh, D., Selkowitz, S., & Fazio, M. (2008). Field performance of exterior solar shadings for residential windows: Winter results. In *IBPSA-Canada eSim Conference, Quebec, 2008, May*.
33. Kumar, K. S., Kavi Kumar, K. S., Koteswararao, H., Tang, T., & Bhowal, S. (2018). Experimental and therotical studies of various solar control windows glasses for the reduction of cooling and heating loads in buildings across different climatic regions. *Energy andd Buildings, 173*, 326–336.
34. Zhang, W., Lu, L., & Xu, X. (2019). Thermal and daylighting performance of glass window using a newly developed transparent heat insulated coating. *Energy Procedia, 158*, 1080–1085.
35. Rosencrantz, T., Bülow-Hübe, H., Karlsson, B., & Roos, A. (2005). Increased solar enegry and daylight utilisation using anti-reflective coatings in energy-efficient windows. *Solar Energy Materials and Solar Cells, 89*(2–3), 249–260.
36. IQ Glass UK. *Anti-reflective-glass-indicative-performance*. Retreived from https://www.iqglassuk.com/storage/documents/IQ-Glass-Anti-Reflective-Glass-Performance.pdf.
37. Parkin, I. P., & Palgrave, R. G. (2005). Self-cleaning coatings. *Journals of Materials Chemistry, 15*(17), 1689–1695.
38. Arabatzis, T., Nikolopoulos, F., Ioannidis, T., Tsoutsou, C., Papadopoulos, A., Li, W., Xu, Z., & Zhang, Z. (2018). Photocatalytic, self-cleaning, antireflective coating for photovoltaic panels: Characterization and monitoring in real conditions. *Solar Energy, 159*, 251–259.
39. Weller, B., Unger, S., Tschernitz, S., Hegger, M., & Knaack, U. (2012). *Glass in building*. Glass in Building, Birkhauser.
40. Lolli, N., & Aste, N. (2016). Aerogel vs. argon insulation in windows: A greenhouse gas emissions analysis. *Building and Environment, 101*, 64–76.
41. Fang, Y., & Arya, F. (2019). Evacuated glazing with tempered glass. *Solar Energy, 183*, 240–247.
42. Zhang, W., Lu, L., & Chen, X. (2017). Performance evaluation of vaccum photovoltaic insulated glass unit. *Energy Procedia, 105*, 322–326.
43. Cuce, E., & Cuce, P. M. (2016). Vaccum glazing for highly insulating windows: Recent developments and future prospects. *Renewable and Sustainable Energy Reviews, 54*, 1345–1357.
44. Surekha, K. S., Aliofkhazraei, M. (2014). Anti-abrasive nanocoatings.
45. Kistler, S. S. (1931). Coherent expanded aerogels and jellies. *Nature, 127*(3211), 741.

Recent Advancements Towards Leveraging the Role of Artificial Intelligence in Agriculture and Smart Farming

Abdul Gani, Mohit Singh, Shray Pathak, and Athar Hussain

1 Need for AI in Agriculture

Agriculture is an unskilled labour activity, and as the population and agricultural output requirements grow, automation will become more significant. Artificial intelligence (AI) greatly benefits agriculturalists in terms of technology, machinery, and its purposes. Supply and quality of the crop are confirmed through predictive analytics and better plant and harvest production management systems. Farmers use satellite imagery and meteorological data to calculate land area and track the general health of crops in real time [1]. Machine learning, AI, and Big data technologies may be used by firms to estimate prices, assess output and yield, and discover pest and disease incursions. They may counsel farmers on seasonal fluctuations, crop kinds to sow for maximum profit, minimum pesticide use, and forecast pricing tendencies [2]. Since AI significantly reduces labour costs and resource scarcity, it will be a useful tool for helping farmers handle the increasing complexity of modern farming. Many sectors are utilising AI technology to increase production and efficiency. AI technologies are assisting individuals in all industries in overcoming traditional barriers. Finance, transportation, universal health care, and agriculture are some of the industries that use AI applications [3].

The worldwide population is increasing while at the same pace as urbanisation is accelerating. The global population is projected to go beyond 9 billion by the year 2050, requiring an increase of 70% in farming production to encounter the demands

A. Gani · M. Singh (✉) · A. Hussain
Department of Civil Engineering, Netaji Subhas University of Technology, New Delhi, India
e-mail: abdul.gani.phd21@nsut.ac.in; mohit.singh.phd23@nsut.ac.in;
athar.hussain@nsut.ac.in

S. Pathak
Department of Civil Engineering, Indian Institute of Technology Ropar, Rupnagar, Punjab, India

S. Pathak et al. (eds.), *Intelligent Infrastructure and Smart Materials*,
https://doi.org/10.1007/978-3-031-92421-7_11

[4]. Only 10% of this enhanced yield may come from underused areas, with the balance being met by the current production increase. Farmers need a strategy to improve output since they are under pressure to meet rising demand. More individuals will require feeding. Because there is a constrained amount of fertile soil, farming also requires ingenuity [5]. So, there is a need for the development of measures to advance farmers in reducing or, at least, managing their risks. The employment of AI in agriculture on a worldwide scale is one of the most intriguing possibilities. Numerous food manufacturers are now struggling to control the hazards and dangers that bugs and other ailments offer to their crops. Changing climate, monoculture, and extensive pesticide use exacerbate these concerns. These factors combine to present farmers with a new challenge [6]. Farms and farmers are under significant strain since agriculture relies primarily on environmental energies for most of its supplies. This stress is worsened by the unpredictability of rain, a shortage in labour, and an annual desire for higher yields. This means that the farming business will need to dramatically scale up in the next years, and agriculture efficiency will be required to nearly double for us to fulfil our needs.

2 Role of AI in Agriculture

AI-based solutions help to increase efficiency across all industries and address problems that many firms, including those in the agricultural sector, encounter, such as weeding, irrigation, crop management, crop yield, crop planting, and soil composition sensing [7]. In order to provide high-value AI applications in the above-mentioned field, agricultural robots are being created [6]. With the world's population increasing, the agriculture field is in trouble, but AI can provide a better answer. AI-based solutions have enabled farmers to produce more yield with less input while also increasing output quality and assuring a speedier go-to-market for produced commodities.

2.1 Image Detection and Identification

The use of autonomous unmanned aerial vehicles (UAVs) for tasks including identification and tracking, search and rescue, body recognition and geolocation of people, and detection of forest fire has gained popularity recently. Due to their flexibility, unbelievable imagination technology that covers everything from shipping to photography, the ability to be piloted with a remote controller, and the devices' dexterity in the air, drones or UAVs are becoming more and more popular for reaching great heights and distances and carrying out a variety of applications [5, 6].

2.2 *Workforce and Skills*

Farmers may now collect massive amounts of information from both public and government websites, analyse it all, and get solutions for a variety of complex issues thanks to artificial intelligence. It also helps us to use irrigation with intelligence, improving farmer production. Soon, it will be known that farming is a combination of biological and technological abilities, which will not only result in improved quality for all the farmers but also lower costs and reduced loads [8].

2.3 *Maximise Production*

The greatest productivity levels for all crops are determined by seed quality and variety selection. New technologies have benefited in crop selection and even improved the ability to choose the hybrid seed options most suited to farmer needs. It has been applied by studying how the seeds react with variation in soil type and weather conditions. Plant diseases can be reduced by gathering this information [9]. We can now better link market dynamics, annual results, and client demands, enabling farmers to maximise agricultural returns.

2.4 *Farming-Related Chatbots*

Conversations with users are automated by chatbots, which are linguistic virtual assistants. With the help of chatbots powered by AI and machine learning, we can now understand natural language and have more individualised conversations with individuals. Agriculture has taken use of this potential by assisting farmers in gaining answers to unresolved questions, providing counsel, and making other suggestions. They are primarily prepared for shopping, travel, and media [10].

3 Impact of AI in Agriculture

Eventually, AI systems are computerised algorithms that accept information as inputs and interpret it to produce valuable output. As a result, they are interpreted as data-processing systems [11]. As a consequence, it is beneficial to be aware of how advantage is produced from big data analysis when considering how value may be produced from AI in various contexts. Artificial intelligence can be used in all of these diverse types of insights, and each will provide benefits to farmers [12]. Drawing conclusions value from insights frequently begins with qualitative and diagnostic analytics before sufficient information is produced to recognise how to implement prescriptive or predictive analytics. The different capabilities of artificial intelligence emerging in agriculture are explained in detail below.

3.1 Collecting Agricultural Data

Much of the early benefits that farmers will receive from artificial intelligence will be at the diagnostic and descriptive sides of the analytics range. Frankly, AI will deliver a sufficient understanding of a farm's condition, allowing the farmer to examine it, think about it, and make management decisions. Many of the precision-agriculture technologies presently in use on farms, such as animal activity monitors and high-resolution field maps, have close or exact counterparts [13]. Contemporary algorithms are developing and going beyond them. While the human eye can frequently perform similarly well, sadly it is not always possible; the application of artificial intelligence allows this capability to measure more common and extensive surveillance [14]. AI that can identify data from sound or mobility, or perhaps even anything that can be recorded by a sensor and transformed into a digital code, may also be used to report on agricultural operations.

Algorithms for artificial intelligence can also go beyond what people can feel, such as chemical signs that could suggest poisons. Furthermore, modern AI systems are being instructed to discover information by merging numerous data sources. When paired with soil moisture meter data from AI, aerial photography might greatly minimise the number of soil-moisture sensors required to build a high-resolution soil moisture map [15]. A significant recent agricultural use of AI has entailed data collecting using equipment such as mobility trackers on, or even in, farm animals, allowing assessment of their health, and behaviours. One of the oldest instances is the use of pedometers to sense oestrus in cattle like cows and buffalo, which may both save the labour associated with observing oestrus start and enhance the odds of attaining the desired calf sex through better positioning of insemination [3].

AI technology is also being utilised to enhance the extraction of information from both satellite and aerial pictures. Convert photos into relevant information, such as relevant data on a farmer's fields, these often need high computing expenses and subject expertise [15, 16]. Computational and AI will substantially automate these operations over the next decade, making it feasible to gather and record data on every property that can be recognised or extrapolated from such information data.

Another use of AI in agricultural production is to assist farmers in obtaining the information they want most efficiently. Language-translation AI, for example, may offer information to farm workers in the language that works best for them, such as assisting in the transmission of agricultural advice to farmers operating in places where people speak multiple languages. Such skills might considerably increase the transmission of solid agricultural advice, potentially leading to more widespread adoption of improved agricultural methods [12]. Chatbots, in which an engaging agent is merely a machine, might also assist farmers in obtaining the information they want.

3.2 Enhanced and New Alerts

After the ability to record some pertinent data has been demonstrated, it is very typical to try to extend that to creating computer performance that automates steps or does additional analyses as we move from descriptive to prescriptive analytics-based modelling. The simplest of them is to notify people when something that can be reported is discovered. For instance, animal motion detectors that indicate their behaviours are often accompanied by the creation of resources that deliver alerts of critical patterns or behaviours [17]. The potential of overreliance on computational alarms is one of its drawbacks. As with human senses, there will always be a chance that something will elude the machine's sensors. As a result, human inspection and testing will always play a vital role.

A significant characteristic of processes, including AI, seems to be that their effectiveness can drift over time due to changes in the qualities of what they are attempting to predict. It is therefore critical to put in place dependable procedures to evaluate the precision of such warning systems and to recalibrate them regularly [16]. This function is unlikely to be performed by farm labourers, underlining the requirement for new roles for professionals such as agricultural consultants or automated approaches.

3.3 Enhanced Diagnostics

As a result of being capable of properly monitoring and measuring agricultural systems, we can explore and comprehend the causes of numerous occurrences. One difficulty with natural systems is that they will not always perform as predicted because they are impacted by established or unknown causes in ways we do not comprehend. Modern large-scale agriculture methods have solved this issue in part by simplifying agricultural systems themselves. AI-assisted data gathering will allow for more accurate evaluations of how various elements interact to determine agricultural system features [2, 15]. Improved data acquisition for diagnostic analysis implies the need for humans to interpret the results and create meaningful visions.

When there is enough data, more powerful AI systems may occasionally be trained to effectively forecast complicated occurrences, compensating for a large variety of interrelating circumstances, in ways that we do not completely comprehend. However, one important advantage of AI in such situations is that it may demonstrate multiple parameters that can be employed to gain predictive precision when it was earlier unknown whether it existed [9]. This is a compelling reason to explore how the various components interact to produce the observed occurrence.

3.4 Enhanced Predictions

The capacity to forecast the growth of product and agricultural attributes across time and geography has increased as information, biological knowledge, and AI algorithms have improved. Predictions regarding how unrelated types of animals and plants will behave in different situations will be a big area of development. This is the traditional analysis of productivity as a function of genotypic, environment, and management (GEM) influences, a method that agricultural scientific institutions have roughly explained to propose varieties. Utilising the help of AI, this method will not only be solved more precisely than earlier but it could also be provided to farmers [18]. Alternative subject of research has been the production of present projections of animal or crop status as it develops, considering parameters such as nutritional status, developmental stage, and size. This necessitates the development of a GEM method that receives continuous inputs, containing unreliable values such as climate forecasts and updates them over time so that the present prediction is always up to date, comparable to a climate prediction [18]. Farmers would be able to optimise their operations with such a skill. Forecasts of the expected abundance of ready-for-harvest plants during the year, for contrast, may drive a farmer to plant extra now to make up for a predicted deficit.

Precise instantaneous forecasting is data concentrated, needing not just adequately accurate local information on the surroundings and plant or cattle but also up-to-date animal or crop status. As a result, the work required to produce such estimates is doubtful to always be justified. They are unquestionably useful in substantial-scale operations aimed at ensuring a regular pattern of crop delivery to consumers, which covers a substantial portion of farming production. The utilisation of information to train prescriptive and predictive models can raise questions about the use and dangers of partial, skewed, and erroneous datasets. Auspiciously, guidance on how to deal with such data difficulties is evolving as well; often, algorithms are added into models to describe prejudices or gaps in the datasets, allowing valuable information to be recovered and used.

Further than the local farms, bigger three-dimensional projections of animal and crop growth and output are useful for farming product customers or those impacted by crop production changes. However, I believe that developments in the availability of data and artificial intelligence will assist agriculture insurance to be significantly more suited to individual conditions. Aside from actual production, AI projections will increasingly be utilised to offer indicators of the overall health of the agricultural system and possible hazards that agriculturalists may wish to investigate more or address [13].

Another difficulty with delivering forecasts is confirming that the decision-maker chooses the best choice. When deciding what to do, it is critical to evaluate the ambiguity in forecasting, and attempts are ongoing to enhance the characterisation and representation of confidence in predictive model projections. It is acknowledged, however, that individuals fight to clarify such uncertainty in a way that leads to the best option. It will have a significant impact on which projections are made directly to farmers and how they are made. The temptation will be to couple or perhaps replace forecasts with the outcomes of various conceivable actions in light of various options, which will necessitate even more data and AI development.

4 Application and Approach of AI in Agriculture

The future of agriculture and the agricultural business is strongly reliant on innovative solutions and technical innovations that utilise non-traditional computer techniques to boost yields and better utilise resources. Crop simulations and judgement-making tools are increasingly being utilised in agriculture to boost output and resource utilisation. AI has huge potential to change farming by combining innovative techniques to anticipate farm yields. For a better understanding of technologies related to AI implemented in agriculture, the following context is divided into various agricultural unit management and explained in brief below:

4.1 Management of the Soil

A vital part of farming operations is managing the soil. Crop productivity will rise while soil resources are conserved thanks to a detailed understanding of various soil types and conditions. It describes the application of techniques, strategies, and treatments to improve soil functionality [8, 19]. A common soil survey technique can be used to look at pollutants in urban soils. Treatment of soil with compost and manure increases soil aggregation and porosity. Higher aggregation is a sign of the existence of organic materials, which are essential for preventing the formation of soil crusts. Alternative tillage strategies can be used to reduce soil physical deterioration. Organic materials must be used if you want to improve the condition of your soil. Several soil-borne diseases that require treatment via management of soil regularly have a significant impact on vegetables and other consumables of crop production [20, 21]. The estimation of the sustainability of soil managing techniques includes sensitivity to soil degradation, with consideration of the fact that soils differ in their ability to resist change and recover.

Table 1 provides a summary of the use of AI in soil management strategies. Through the creation of a set of workable management choices, evaluation of each alternative, and selection of the alternative that best meets the user-weighted set of criteria, the application of management-oriented modelling (MOM) decreases nitrate leaching [22, 23]. MOM employs hill climbing as a strategic search approach, with best-first as a tactical search method, to determine the shortest route from start

Table 1 Summary of use of AI in soil management strategies

Technique	Merits	Demerits	References
Fuzzy logic	Can categorise soil based on the dangers connected with it	Big information is required	[24]
MOM	Reduces nitrate leaching while increasing output	It takes time. Only nitrogen is allowed	[23]
DSS	Sedimentary and erosion yield are reduced	Training requires large amounts of data	[25]
ANN	It is possible to measure soil nutrients following erosion	Its estimation is limited to NH_4	[20]
ANN	Time saving, cost-effective, and 92% accurate	Required big data	[8]
ANN	Reports soil texture successfully.	It neither improves nor solves the problem of poor soil texture.	[9]
ANN	Capable of predicting soil moisture	Because weather conditions are rarely predictable, the forecast will fail over time	[14]
ANN	The soil texture is predicted	Training requires large amounts of data. Has limitations in terms of implementation	[26]
ANN	Can predict the average monthly soil temperature	Only takes temperature into account when determining how well soil performs	[27]
ANN	It is possible to forecast the soil activity of enzymes. Predicts and characterises the structure of the soil appropriately	Only a few soil enzymes are measured. It is more concerned with categorisation than with increasing soil performance	[28]

nodes to targets. Engineering knowledge is acquired in three steps for the construction of the Soil Risk Characterization Decision Support System: system implementation, conceptual design, and knowledge acquisition.

4.2 Management of Crops

Sowing is the first step in crop management, which continues with crop growth management, harvesting, crop storage, and crop shipping. It is characterised as programmes that increase the growth and productivity of agricultural goods. Crop productivity would unquestionably increase with a solid understanding of crop classes depending on season and soil type. Farmers must incorporate crop management practises to address water shortages brought on by the soil, the weather, or insufficient irrigation [29]. The use of flexible decision-tree-based crop management systems should be encouraged. The time, severity, and predictability of droughts are important considerations when making agricultural decisions.

In a broad sense, a global crop management interface for all facets of agriculture is offered by the universal crop management system. Mc Kinion and Lemmon proposed the first artificial intelligence approach in crop management in their article "Expert Systems for Agriculture"in 1985. Boulanger designed an additional expert system for corn crop security in his thesis. Roach presented the POMME expert system in 1987, which was designed for apple farm control [30]. The crop prediction approach is used to anticipate the best crop by detecting numerous soil factors and atmospheric conditions. There are many variables to take into account, including humidity, rainfall, temperature, depth, copper, phosphate, sulphur, manganese, organic carbon, potassium, calcium, magnesium, and iron, nitrogen, and soil type. Demeter is a remotely operated rowing machine with two video cameras and a GPS for navigation [31]. An entire field's worth of harvesting tasks can be planned, and then carried out by harvesting crop rows, rotating to cut subsequent rows, moving around in the field, and spotting unforeseen obstacles. The input and the output parameters were binary coded in order to train and test the network.

The separate software and hardware elements of the machine, which include the self-driving car, the manipulator, the final component, and the two computers with machine vision for sensing and 3D image processing of the fruit and the surroundings, and, finally, a controller strategy that creates collision-free motions for the manipulator throughout harvesting, comprises the use of AI in cucumber harvesting [22]. For each site, field-specific precipitation data and meteorological factors can be employed. The efficiency of rice yield projections is affected by adjusting ANN parameters. In model optimisation, fewer data groups necessitated fewer hidden connections and small learning rates. The best prototype so far had an accuracy of 94% with six inputs and two output classes. The unrivalled network produced had five hidden layers that were trained up to 300,000 echoes and averaged 85.9% precision. Table 2 shows the use of various AI techniques in crop management strategies are summarised below.

4.3 Disease Management

Plant diseases can pose a significant threat to the consumers, environment, world economy, and farmers. In India alone, diseases and pests destroy 35% of crops, costing farmers a great deal of money. The misuse of pesticides puts human health at risk because some of them are toxic and biomagnified [40]. To produce a good crop, agriculture must manage the disease. Different plant and animal diseases can occur for a variety of reasons, including hereditary factors, the texture of the soil, rainfall, dry climate, temperature, wind, and other environmental factors. Controlling the effects is a big difficulty, especially in large-scale farming, due to these factors as well as the unpredictable nature of some disorders' underlying influences. These impacts can be avoided by observing the crop, recognising illness, and giving proper treatment. Significant knowledge and skill are required to spot an incapacitated plant and afterwards take the necessary procedures to recover it [41]. Computerised systems are utilised all around the world to study the state and then provide solutions for controlling it. For disease identification, image sensing and analysis are

Table 2 Summary of use of AI in crop management strategies

Technique	Merits	Demerits	Reference
ANN	Can forecast crop sensitivity to moisture in the soil and salinity	Only takes soil temperature and texture into account	[12]
ANN	Rice yield predicted precisely	Time-consuming, and only available in a certain environment	[32]
ANN	Crop nutrition disorders are detected with greater than 90% accuracy	There were just a few symptoms evaluated	[33]
Fuzzy logic and ANN	Decreases the quantity of insects that damage crops	This shows a lack of ability to differentiate between crop and weed	[34]
Robotics	Crop harvesting success rate of 80%	Choosing accuracy and speed is slow	[35]
Fuzzy cognitive map	For improved crop management, predict cotton crop quality and yield	It moves quite slowly	[18]
ANN	Crop yield is predicted	Only takes weather into account while determining crop production	[36]
Prolog	Discards farm equipment that isn't used as frequently	Location specific	[37]
Calex	Can create crop management activity scheduling recommendations	Time-consuming	[38]
Robotics-Demeter	Capable of harvesting up to 40 hectares of crop	Expensive: Consumes a lot of energy	[39]

performed to guarantee that images of the leaves are divided into external sections like the background, unaffected area, and affected area of the plant leaf. The contaminated portion of that plant leaf is then harvested and delivered to the lab for further examination. This aids in the identification of insects and the detection of nutritional deficits [42].

Table 3 summarises the machine learning and artificial intelligence technologies in illness management that have been published in the literature. A farmer should use an integrated disease control and management approach that combines physical, chemical, and biological measures to successfully control illnesses and reduce losses [43]. Achieving them takes time and is not very cost-efficient, hence the necessity for an AI approach to disease control and management. The explanation block (EB) provides a clear perspective of the reasoning followed by the expert system's kernel. In the platform for forming intelligent conclusions for agricultural disease management, a unique technique to rule advancement utilising fuzzy logic is implemented. A text-to-speech (TTS) translator is used to provide text-to-talking user interface functionality. It offers a very engaging and effective user experience on the web for real-time interactions. A rule-based and forward-chaining empirical method was employed in the construction of the system, which aids in illness detection and treatment recommendations.

Table 3 Summary of use of artificial intelligencein disease management

Technique	Merits	Demerits	Reference
GIS, ANN	95% accurate prediction	Some farmers in remote areas will be denied access Internet based	[44]
Using rules-based expertise in disease detection	Diseases are detected more quickly, allowing for speedier treatment Because of its preventative strategy, it is cost-effective	It takes time To ascertain whether the pest has built up an immunity to the prophylactic intervention, ongoing monitoring is necessary	[45]
Fuzzy	For farms, Xpest offers pest-related information Additionally, Internet services help to sustain it, forecasting with great accuracy	Dependent on Internet	[46]
ANN	Has a prediction rate of more than 90%	The ANN does not eradicate infections or lessen their impact	[10]
Web-based expert system	High performance	Internet and web based	[47]
Fuzzy logic	Excellent precision Counters quickly to crop disease characteristics	It has limited use because it requires Internet access Its efficacy cannot be determined since just four seed harvests were evaluated	[48]
Web GIS, fuzzy logic	Eco-friendly and economical	Inefficiency as a result of the dispersed distribution It takes time to locate and distribute data A mobile browser determines the location of the data	[49]
FL and TTS converter	Rapidly resolves plant pathogenic issues	Increased speed Internet is required Its multimedia interface is a voice program	[50]
Database	In the testing set, accurate findings were found	Database inefficiency when used on a wide scale	[51]
Genetic algorithm, ANN, computer vision system	Worked at a high rate. Capable of multitasking	Dimension-based identification, which may have an impact on excellent species	[51]

4.4 Weed Management

The control of weeds, which is additionally the most difficult problem in pest control, is one of the most difficult difficulties for a farmer. Weed regularly affects predicted profit and production for farmers. Among all insecticides, the application of herbicides outnumbers all others, including the usage of insecticides [52–54]. No agriculturalist likes to apply pesticides, but no agriculturalist wants to watch weeds drawing up every drop of nutrients and water meant for the crops. The usage of herbicides hurts both human health and the environment.

A survey reveals that weed infestations reduce productivity by 50% for dry beans and corn harvests. The production of wheat is reduced by 48% as a result of weed competition. These losses could occasionally reach 60% [55]. According to one study, yield reductions in sesame crops range between 50% and 75%. Table 4 summarises the applications of AI in weed management. The variation in yield losses can be ascribed to the time of crop exposure to weeds and the geographical heterogeneity of weeds. Weed impacts the ecology negatively and advantageously [56, 57]. Weed effects include submerging during hurricanes, some types of weeds can have their way through rampant fires, some cause irreparable liver damage if metabolised, and they core strength out plants or crops by competing for water, nutrients, and sunlight, depending on the nearly equivalent Weed Science Society of America (WSSA) report. Some plants are toxic, causing allergic responses or potentially endangering public health.

Table 4 Summary of the use of AI in weed management

Technique	Merits	Demerits	References
UAV	The weed identification rate is extremely high in a relatively short amount of time	It is quite costly and needs a great deal of human knowledge	[58]
Support vector machine, ANN	Intercepts crop degradation quickly, allowing for immediate site-specific solutions	Only recognises low quantities of nitrogen	[59]
ANN, learning vector quantisation	The weed identification rate is very high with a quicker turnaround time	The manner of data input employed has an impact on the AI's performance	[60]
Digital image analysis, GPS	Has a rate of success of more than 60%	Its achievement took 4 years to acquire, and it is quite a time intensive	[61]
Sensor machine learning, robotics	It eliminates the time and gets rid of weeds that are resistant to herbicides	Expensive. The continuous use of heavy machinery directly reduces soil's ability to produce	[62]
Saloma expert system for estimation, forecasting, and weed management.	The high pace of adaption and prediction	Big data is required	[63]
ANN, GA	Excellent performance Reduces the need for trial and error	Big data is needed	[64]
UAV, GA	Can monitor weeds rapidly and effectively	Weed management is minimal to non-existent Expensive	[65]
ANN	Cost-effective, improved performance	New data presents an adaptation difficulty	[66]

4.5 Management of the Soil and Irrigation

Irrigation is a labour-intensive procedure in farming. Problems with soil and irrigation are critical for agriculture. Inadequate management of soil and irrigation consequences in agricultural losses and polluted features. Trained AI devices that are knowledgeable of historical weather patterns, crop types, and soil conditions may mechanise the irrigation process and improvement entire production [67, 68]. Because irrigation uses around 70% of the world's freshwater resources, programming this procedure can not only conserve water but also assist agriculturalists in dealing with their water challenges. Organisations such as CropIn, which uses AI to maximise per-acre value, and Intello Labs, which uses deep learning to analyse photos, are developing AI-based tools and sensors to monitor soil health [69, 70].

This section highlights specific studies in the realm of soil management and irrigation that have been helped by AI approaches. It used agriculturalists' knowledge to construct fuzzy system to award yields based on the adequacy of land maps produced with the system's assistance. The neural network system for assessing topsoil moisture in paddy was computed [71]. Some neural network approaches for rainfall forecasting were examined using four independent inputs. The study discovered that the neural network's radial basis function performs the best when compared to previous models.

4.6 Yield Prediction and Management

Agricultural yield prediction is particularly beneficial for crop cost predictions as well as marketing strategies. Additionally, in the age of accuracy agronomy, predictive models may be used to examine relevant factors that have a reliable effect on yield. With the emergence of innovative technologies such as AI, satellite pictures, machine learning, and sophisticated analytics, an environment for self-sustaining, effective, and precise agriculture is emerging [72]. The combination of this modern technology is assisting farmers in achieving maximum average output as well as improved constraints over the number of food grains, confirming they remain profitable.

ANN model employs a backpropagation technique to anticipate yield from topsoil restrictions. Statistical climatic data, optimal planting seasons, and real-time moisture absorption data and soil moisture may be recognised using AI algorithms to make projections and offer farmers feedback on the best planting time [73].

4.7 Industry Perspectives

Worldwide AI in agronomy was valued at USD 608.9 million in 2018 and is expected to grow at a CAGR of 25.4% from 2019 to 2025 [74]. AI practices utilised in agriculture can help increase production and productivity. As a result, agricultural organisations are using AI technology in the form of predictive analytics-based solutions. AI-based strategies and solutions profit from insect control, monitoring soil, delivering better crops, and enhancing agriculture-related duties included in the all-inclusive food stock [75]. AI is becoming more prevalent in the farming business, to enhance yield significance and accuracy as it assists in farm data analysis.

The continuously growing population necessitates the application of AI in the farming economy. The necessity for a green rebellion propelled by the Internet of Things, artificial intelligence, and data is defined by the inadequate supply of arable land and the requirement for increasing food production due to food security. Applications aided by AI supply various fields in the farming industry, such as optimisation and predictive analysis, insect infestation detection, soil monitoring, and diagnosing plant illnesses [76, 77].

5 Conclusion

Food consumption is expanding daily, and without the use of innovative agricultural practices, it will be difficult to meet the rising need. AI-opted technologies forecast weather conditions, evaluate crop sustainable development, and properly assess farms for the existence of illnesses or pest problems, as well as unsatisfactory plant nutrient uptake, using data including temperature, rainfall patterns, wind velocity, and solar radiation in conjunction with ML algorithms and images captured by satellite systems and drones. Agriculturalists who have Wi-Fi access can utilise AI applications to acquire an AI-tailored farming plan. Agriculturalists can meet the worldwide demand for increased food supply and competitiveness by implementing AI-driven solutions that boost productivity and revenue without depleting irreplaceable natural resources. AI provides great benefits in terms of utilising fewer resources while increasing output.

Farmers may utilise AI to gain real-time insights from their crops, detecting regions that require irrigation, fertilisation, or pest treatment. AI-implemented manufacturing and yield management: Artificial intelligence-assisted selecting, wrapping, and sorting improve food production, packaging, and classification. Farmers gain from its help in grasping agricultural data-driven insights such as solar radiation, rainfall, wind speed, and temperature. AI technologies may be able to alleviate farmers' difficulties, such as changing climate and bug and weed invasions that reduce harvests.

Artificial intelligence has been employed in crop selection and to assist farmers with fertiliser choices. The machine interacts among itself to determine which crop is acceptable for harvesting as well as the fertilisers that encourage exponential yield using the dataset that the operator has acquired and supplied to the program. Deep learning has broad applicability, and its use in business has advanced significantly. Deep learning has a benefit over machine learning in that it provides profundity to machine learning. Many important ways exist to provide food producers with greater harvests and effective field management. As foodstuff is the most basic necessity of each human being, this ultimately contributes to the country's general growth.

AI will be applied in agriculture to enhance the entire agricultural cycle. The use of AI will help farmers achieve their goal of a healthier crop by making better judgements in the field. Considering food serves as the most vital requirement for human beings, so superiority of statistics may be used more resources effectively and efficiently to foresee danger and evaluate situations, and hence take appropriate action before starvation persists as a humanitarian catastrophe, which in return can aid in the entire growth of the globe.

References

1. Waqas, M., Hawkesford, M. J., & Geilfus, C.-M. (2023). Feeding the world sustainably: Efficient nitrogen use. *Trends in Plant Science, 28*(5), 505–508. https://doi.org/10.1016/j.tplants.2023.02.010
2. Misra, N. N., Dixit, Y., Al Mallahi, A., Bhullar, M. S., Upadhyay, R., & Martynenko, A. (2022). IoT, Big Data, and artificial intelligence in agriculture and food industry. *IEEE Internet of Things Journal, 9*(9), 6305–6324.
3. Balaska, V., Adamidou, Z., Vryzas, Z., & Gasteratos, A. (2023). Sustainable crop protection via robotics and artificial intelligence solutions. *Machines, 11*(8), 774. https://doi.org/10.3390/machines11080774
4. Singh, S., & Jain, P. (2022). Applications of artificial intelligence for the development of sustainable agriculture. In *Agro-biodiversity and agri-ecosystem management* (pp. 303–322). Springer Nature Singapore. https://doi.org/10.1007/978-981-19-0928-3_16
5. Talaviya, T., Shah, D., Patel, N., Yagnik, H., & Shah, M. (2020). Implementation of artificial intelligence in agriculture for optimisation of irrigation and application of pesticides and herbicides. *Artificial Intelligence in Agriculture, 4*, 58–73. https://doi.org/10.1016/j.aiia.2020.04.002
6. Kagita, M. K., Kaosar, M., & Thilakarathne, N. (2022). Survey on AI-based IoT and drone-equipped smart agriculture. In *Artificial intelligence of things in smart environments: Applications in transportation and logistics* (p. 101).
7. Shaikh, T. A., Mir, W. A., Rasool, T., & Sofi, S. (2022a). Machine learning for smart agriculture and precision farming: Towards making the fields talk. *Archives of Computational Methods in Engineering, 29*(7), 4557–4597. https://doi.org/10.1007/s11831-022-09761-4
8. Chang, D. H., & Islam, S. (2000). Estimation of soil physical properties using remote sensing and artificial neural network. *Remote Sensing of Environment, 74*(3), 534–544. https://doi.org/10.1016/S0034-4257(00)00144-9
9. Elshorbagy, A., & Parasuraman, K. (2008). On the relevance of using artificial neural networks for estimating soil moisture content. *Journal of Hydrology, 362*(1–2), 1–18. https://doi.org/10.1016/j.jhydrol.2008.08.012
10. Wang, X., Zhang, M., Zhu, J., & Geng, S. (2008). Spectral prediction of Phytophthora infestans infection on tomatoes using artificial neural network (ANN). *International Journal of Remote Sensing, 29*(6), 1693–1706. https://doi.org/10.1080/01431160701281007
11. Berckmans, D. (2008). Precision Livestock Farming (PLF). *Computers and Electronics in Agriculture, 62*(1), 1. https://doi.org/10.1016/j.compag.2007.09.002
12. Dai, X., Huo, Z., & Wang, H. (2011). Simulation for response of crop yield to soil moisture and salinity with artificial neural network. *Field Crops Research, 121*(3), 441–449. https://doi.org/10.1016/j.fcr.2011.01.016

13. Anderson, W. K. (2010). Closing the gap between actual and potential yield of rainfed wheat. The Impacts of Environment, Management and Cultivar. *Field Crops Research, 116*(1–2), 14–22. https://doi.org/10.1016/j.fcr.2009.11.016
14. Zhao, Z., Chow, T. L., Rees, H. W., Yang, Q., Xing, Z., & Meng, F.-R. (2009). Predict soil texture distributions using an artificial neural network model. *Computers and Electronics in Agriculture, 65*(1), 36–48. https://doi.org/10.1016/j.compag.2008.07.008
15. Mhlanga, D. (2023). The role of FinTech and AI in agriculture, towards eradicating hunger and ensuring food security. In *FinTech and artificial intelligence for sustainable development: The role of smart technologies in achieving development goals* (pp. 119–143). Cham. https://doi.org/10.1007/978-3-031-37776-1_6
16. Rejeb, A., Rejeb, K., Zailani, S., Keogh, J. G., & Appolloni, A. (2022). Examining the interplay between artificial intelligence and the agri-food industry. *Artificial Intelligence in Agriculture, 6*, 111–128. https://doi.org/10.1016/j.aiia.2022.08.002
17. Akhtar, M. N., Ansari, E., Alhady, S. S. N., & Abu Bakar, E. (2023). Leveraging on advanced remote sensing- and artificial intelligence-based technologies to manage palm oil plantation for current global scenario: A review. *Agriculture, 13*(2), 504. https://doi.org/10.3390/agriculture13020504
18. Papageorgiou, E. I., Markinos, A. T., & Gemtos, T. A. (2011). Fuzzy cognitive map based approach for predicting yield in cotton crop production as a basis for decision support system in precision agriculture application. *Applied Soft Computing, 11*(4), 3643–3657. https://doi.org/10.1016/j.asoc.2011.01.036
19. Gani, A., Pathak, S., Hussain, A., Ahmed, S., Singh, R., Khevariya, A., Banerjee, A., Ayyamperumal, R., & Bahadur, A. (2023). Water quality index assessment of river Ganga at Haridwar stretch using multivariate statistical technique. *Molecular Biotechnology*, 1–24.
20. Behrens, T., Förster, H., Scholten, T., Steinrücken, U., Spies, E., & Goldschmitt, M. (2005). Digital soil mapping using artificial neural networks. *Journal of Plant Nutrition and Soil Science, 168*(1), 21–33. https://doi.org/10.1002/jpln.200421414
21. Gani, A., Singh, M., Pathak, S., & Hussain, A. (2023). Groundwater quality index development using the ANN model of Delhi Metropolitan City, India. *Environmental Science and Pollution Research*, 1–16.
22. Gani, A., Hussain, A., Pathak, S., & Banerjee, A. (2024). An empirical investigation on the elimination of heavy metals using bioremediation method for selected plant species. *Physics and Chemistry of the Earth, Parts A/B/C, 134*, 103568.
23. Li, M., & Yost, R. S. (2000). Management-oriented modeling: Optimizing nitrogen management with artificial intelligence. *Agricultural Systems, 65*(1), 1–27. https://doi.org/10.1016/S0308-521X(00)00023-8
24. López, E. M., García, M., Schuhmacher, M., & Domingo, J. L. (2008). A fuzzy expert system for soil characterization. *Environment International, 34*(7), 950–958. https://doi.org/10.1016/j.envint.2008.02.005
25. Montas, H., & Madramootoo, C. A. (1992). A decision support system for soil conservation planning. *Computers and Electronics in Agriculture, 7*(3), 187–202. https://doi.org/10.1016/S0168-1699(05)80019-5
26. Bilgili, M. (2011). The use of artificial neural network for forecasting the monthly mean soil temperature in Adana, Turkey. *Turkish Journal of Agriculture and Forestry, 35*(1), 83–93.
27. Levine, E. R., Kimes, D. S., & Sigillito, V. G. (1996). Classifying soil structure using neural networks. *Ecological Modelling, 92*(1), 101–108. https://doi.org/10.1016/0304-3800(95)00199-9
28. Tajik, S., Ayoubi, S., & Nourbakhsh, F. (2012). Prediction of soil enzymes activity by digital terrain analysis: Comparing artificial neural network and multiple linear regression models. *Environmental Engineering Science, 29*(8), 798–806. https://doi.org/10.1089/ees.2011.0313
29. Ghosh, P., Kumpatla, P., & S. (2022). GIS applications in agriculture. In *Geographic information systems and applications in coastal studies*. IntechOpen. https://doi.org/10.5772/intechopen.104786

30. Jasim, A. N., & Fourati, L. C. (2023). Agriculture 4.0 from IoT, artificial intelligence, drone, & blockchain perspectives. In *2023 15th International Conference on Developments in ESystems Engineering (DeSE)* (pp. 262–267). https://doi.org/10.1109/DeSE58274.2023.10099927
31. Vyas, S., Shabaz, M., Pandit, P., Parvathy, L. R., & Ofori, I. (2022). Integration of artificial intelligence and blockchain technology in healthcare and agriculture. *Journal of Food Quality, 2022*, 1–11. https://doi.org/10.1155/2022/4228448
32. Ji, L., & Peters, A. J. (2007). Performance evaluation of spectral vegetation indices using a statistical sensitivity function. *Remote Sensing of Environment, 106*(1), 59–65.
33. Song, H., & He, Y. (2005, July). Crop nutrition diagnosis expert system based on artificial neural networks. In *3rd International conference on information technology and applications.*
34. Yang, C.-C., Prasher, S. O., Landry, J.-A., & Ramaswamy, H. S. (2003). Development of a herbicide application map using artificial neural networks and fuzzy logic. *Agricultural Systems, 76*(2), 561–574. https://doi.org/10.1016/S0308-521X(01)00106-8
35. Henten, E. J. V., Hemming, J., Tuijl, B. A. J. V., Kornet, J. G., Meuleman, J., Bontsema, J., & Os, E. A. V. (2002). *An autonomous robot for harvesting cucumbers in greenhouses.* Springer.
36. Snehal, S. S., & Sandeep, S. V. (2014). Agricultural crop yield prediction using artificial neural network approach. *International Journal of Innovative Research in Electrical, Electronics, Instrumentation and Control Engineering, 2*(1), 683–686.
37. Lal, H., Jones, J. W., Peart, R. M., & Shoup, W. D. (1992). FARMSYS—A whole-farm machinery management decision support system. *Agricultural Systems, 38*(3), 257–273. https://doi.org/10.1016/0308-521X(92)90069-Z
38. Plant, R. E. (1989). An artificial intelligence based method for scheduling crop management actions. *Agricultural Systems, 31*(1), 127–155. https://doi.org/10.1016/0308-521X(89)90017-6
39. Pilarski, T., Happold, M., Pangels, H., Ollis, M., Fitzpatrick, K., & Stentz, A. (2002). *The demeter system for automated harvesting.* Springer.
40. Sharma, V., Tripathi, A. K., & Mittal, H. (2022). Technological revolutions in smart farming: Current trends, challenges & future directions. *Computers and Electronics in Agriculture, 201*, 107217. https://doi.org/10.1016/j.compag.2022.107217
41. Reddy, O. G. P., Dwivedi, B. S., & Ravindra, G. C. (2023). Applications of geospatial and big data technologies in smart farming. In *Smart agriculture for developing nations: Status, perspectives and challenges* (pp. 15–31). Springer Nature Singapore. https://doi.org/10.1007/978-981-19-8738-0_2
42. Foster, L., Szilagyi, K., Wairegi, A., Oguamanam, C., & de Beer, J. (2023). Smart farming and artificial intelligence in East Africa: Addressing indigeneity, plants, and gender. *Smart Agricultural Technology, 3*, 100132. https://doi.org/10.1016/j.atech.2022.100132
43. Siregar, R. R. A., Seminar, K. B., Wahjuni, S., & Santosa, E. (2022). Vertical farming perspectives in support of precision agriculture using artificial intelligence: A review. *Computers, 11*(9), 135. https://doi.org/10.3390/computers11090135
44. Liu, G., Yang, X., Ge, Y., & Miao, M. (2006, April). An artificial neural network–based expert system for fruit tree disease and insect pest diagnosis. In *International conference on networking, sensing and control.*
45. Munirah, M. Y., Rozlini, M., & Siti, Y. M. (2013). An expert system development: Its application on diagnosing oyster mushroom diseases. In *2013 13th International Conference on Control, Automation and Systems (ICCAS 2013)* (pp. 329–332). https://doi.org/10.1109/ICCAS.2013.6703917
46. Siraj, F., & Arbaiy, F. (2006, June). Integrated pest management system using fuzzy expert system. In *Knowledge management international conference & exhibition* (pp. 169–176).
47. Virparia, P. (2007). A web based fuzzy expert system for insect pest management in groundnut crop 'Prajna'. *Journal of Pure & Applied Sciences, 15*, 36–41.
48. Kolhe, S., Kamal, R., Saini, H. S., & Gupta, G. K. (2011a). A web-based intelligent disease-diagnosis system using a new fuzzy-logic based approach for drawing the inferences in crops. *Computers and Electronics in Agriculture, 76*(1), 16–27. https://doi.org/10.1016/j.compag.2011.01.002

49. Jesus, J., Panagopoulos, T., Vaz, T. N., Artunes, M., & Carlos, D. (2008). New aspects of energy, environment, ecosystems and sustainable development. In *Proceedings of the 4th IASME/WSEAS international conference on Energy, Environment, Ecosystems and Sustainable Development (EEESD '08)*. WSEAS Press.
50. Kolhe, S., Kamal, R., Saini, H. S., & Gupta, G. K. (2011b). An intelligent multimedia interface for fuzzy-logic based inference in crops. *Expert Systems with Applications, 38*(12), 14592–14601. https://doi.org/10.1016/j.eswa.2011.05.023
51. Balleda, K., Satyanvesh, D., Sampath, N. V. S. S. P., Varma, K. T. N., & Baruah, P. K. (2014). Agpest: An efficient rule-based expert system to prevent pest diseases of rice & wheat crops. In *2014 IEEE 8th international conference on Intelligent Systems and Control (ISCO)* (pp. 262–268). https://doi.org/10.1109/ISCO.2014.7103957
52. Charania, I., & Li, X. (2020). Smart farming: Agriculture's shift from a labor intensive to technology native industry. *Internet of Things, 9*, 100142. https://doi.org/10.1016/j.iot.2019.100142
53. Gani, A., Hussain, A., Pathak, S., & Omar, P. J. (2024). Analysing heavy metal contamination in groundwater in the vicinity of Mumbai's landfill sites: An in-depth study. *Topics in Catalysis*, 1–15.
54. Javaid, M., Haleem, A., Singh, R. P., & Suman, R. (2022). Enhancing smart farming through the applications of Agriculture 4.0 technologies. *International Journal of Intelligent Networks, 3*, 150–164. https://doi.org/10.1016/j.ijin.2022.09.004
55. Hati, A. J., & Singh, R. R. (2021). Smart indoor farms: Leveraging technological advancements to power a sustainable agricultural revolution. *AgriEngineering, 3*(4), 728–767. https://doi.org/10.3390/agriengineering3040047
56. Bacco, M., Barsocchi, P., Ferro, E., Gotta, A., & Ruggeri, M. (2019). The digitisation of agriculture: A survey of research activities on smart farming. *Array, 3–4*, 100009. https://doi.org/10.1016/j.array.2019.100009
57. Gani, A., Singh, M., Pathak, S., & Hussain, A. (2024). Recent advancements in chlorine applications for water quality control. In *Drinking water disinfection by-products: Sources, fate and remediation* (pp. 35–58). Cham.
58. López-Granados, F. (2011). Weed detection for site-specific weed management: Mapping and real-time approaches. *Weed Research, 51*(1), 1–11. https://doi.org/10.1111/j.1365-3180.2010.00829.x
59. Karimi, Y., Prasher, S. O., Patel, R. M., & Kim, S. H. (2006). Application of support vector machine technology for weed and nitrogen stress detection in corn. *Computers and Electronics in Agriculture, 51*(1–2), 99–109. https://doi.org/10.1016/j.compag.2005.12.001
60. Yang, C. C., Prasher, S. O., Landry, J., & Ramaswamy, H. S. (2002). Development of neural networks for weed recognition in corn fields. *Transactions of the ASAE, 45*(3). https://doi.org/10.13031/2013.8854
61. Gerhards, R., & Christensen, S. (2003). Real-time weed detection, decision making and patch spraying in maize, sugarbeet, winter wheat and winter barley. *Weed Research, 43*(6), 385–392. https://doi.org/10.1046/j.1365-3180.2003.00349.x
62. Brazeau, M. (2018). *Fighting weeds: Can we reduce, or even eliminate, herbicides by utilizing robotics and AI?* geneticliteracyproject.org/2018/12/12/fighting-weeds-can-we-reduce-or-even-eliminate-herbicide-use-through-robotics-and-ai.
63. Stigliani, L., & Resina, C. (1993). SELOMA: Expert system for weed management in herbicide-intensive crops. *Weed Technology, 7*(3), 550–559. https://doi.org/10.1017/S0890037X00037337
64. Tobal. (2014). Weeds identification using evolutionary artificial intelligence algorithm. *Journal of Computer Science, 10*(8), 1355–1361. https://doi.org/10.3844/jcssp.2014.1355.1361
65. Ortiz, M. P., Gutiérrez, P. A., Peña, J. M., Torres-Sánchez, J., López-Granados, F., & Hervás-Martínez, C. (2016). Machine learning paradigms for weed mapping via unmanned aerial vehicles. In *2016 IEEE symposium series on computational intelligence* (pp. 1–8). IEEE.
66. Moallem, P., & Razmjooy, N. (2012). A multi layer perceptron neural network trained by invasive weed optimization for potato color image segmentation. *Trends in Applied Sciences Research, 7*(6), 445–455. https://doi.org/10.3923/tasr.2012.445.455

67. Shaikh, T. A., Rasool, T., & Lone, F. R. (2022b). Towards leveraging the role of machine learning and artificial intelligence in precision agriculture and smart farming. *Computers and Electronics in Agriculture, 198*, 107119. https://doi.org/10.1016/j.compag.2022.107119
68. Gani, A., Pathak, S., & Hussain, A. (2024). Comprehensive analysis of modified heavy metal pollution index and health risk assessment in the Yamuna River of Delhi, India: Crucial study for environmental health management. *Journal of Hazardous, Toxic, and Radioactive Waste, 28*(4), 04024021.
69. Malhotra, K., & Firdaus, M. (2022). Application of artificial intelligence in IoT security for crop yield prediction. *ResearchBerg Review of Science and Technology, 2*(1), 136–157.
70. Gani, A., Pathak, S., Hussain, A., Shukla, A. K., & Chand, S. (2024). Emerging pollutant in surface water bodies: A review on monitoring, analysis, mitigation measures and removal technologies of micro-plastics. *Environmental Geochemistry and Health, 46*(7), 1–31.
71. Siva Barathi, A., Manapragada, N. V. S. K., Rai, P. K., Shukla, S., & Shukla, A. K. (2024). Artificial intelligence and machine learning-based building solutions: Pathways to ensure occupant comfort and energy efficiency with climate change. In G. Tripathi, A. Shakya, S. Kanga, S. K. Singh, & P. K. Rai (Eds.), *Big Data, artificial intelligence, and data analytics in climate change research. Advances in geographical and environmental sciences*. Springer. https://doi.org/10.1007/978-981-97-1685-2_4
72. Kwaghtyo, D. K., & Eke, C. I. (2023). Smart farming prediction models for precision agriculture: A comprehensive survey. *Artificial Intelligence Review, 56*(6), 5729–5772. https://doi.org/10.1007/s10462-022-10266-6
73. Chukkapalli, S. S. L., Gupta, M., & Joshi, A. (2023). Towards a sustainable era: Leveraging technology for positive impact. *International Research Journal of Modernization in Engineering Technology and Science*. https://doi.org/10.56726/IRJMETS41271
74. Ahmed, F. (2023). Towards a sustainable era: Leveraging technology for positive impact. *International Research Journal of Modernization in Engineering Technology and Science*. https://doi.org/10.56726/IRJMETS41271
75. Thilakarathne, N. N., Yassin, H., Bakar, M. S. A., & Abas, P. E. (2021). Internet of Things in smart agriculture: Challenges, opportunities and future directions. In *2021 IEEE Asia-Pacific Conference on Computer Science and Data Engineering (CSDE)* (pp. 1–9). https://doi.org/10.1109/CSDE53843.2021.9718402
76. Holzinger, A., Saranti, A., Angerschmid, A., Retzlaff, C. O., Gronauer, A., Pejakovic, V., Medel-Jimenez, F., Krexner, T., Gollob, C., & Stampfer, K. (2022). Digital transformation in smart farm and forest operations needs human-centered AI: Challenges and future directions. *Sensors, 22*(8), 3043. https://doi.org/10.3390/s22083043
77. Javaid, M., Haleem, A., Khan, I. H., & Suman, R. (2023). Understanding the potential applications of artificial intelligence in agriculture sector. *Advanced Agrochem, 2*(1), 15–30. https://doi.org/10.1016/j.aac.2022.10.001

Hydrodynamic Analysis of Raceway Pond Using Large Eddy Simulation Turbulence Model

Sweety Rajput, B. S. Das, Anil Kumar Sharma, Jnana Ranjan Kuntia, and Kamalini Devi

1 Introduction

The process of photosynthesis proves to be an effective means of reducing greenhouse gas emissions; therefore, microalgae have been used extensively for producing biofuel [1]. The utilization of CO_2 and sunlight for the growth of microalgae also proves the use of raceway ponds widely suitable [2]. The paddlewheel is a mixing mechanism used in raceway ponds for the proper distribution of nutrients, CO_2, and O_2 within the cell of microalgae. The mixing efficiency can be increased by increasing the rotation speed of the paddlewheel, but it may also increase the power consumption. To minimize the settlement of algae cells at the bottom of the pond, the mixing velocity maintained should be more than 0.1 m/s [3–6]. For experimental setups, large-scale models require more construction time and cost; therefore, numerical approaches have been suggested to minimize cost and time. The modification in the geometry of the raceway pond has been proposed by various authors to understand the effect of velocity, shear stress, and dead zones [7, 8]. Similar studies to understand the flow mechanism were proposed by Xu et al. [9]. Computational fluid dynamics (CFD) is an effective tool widely used to understand

S. Rajput (✉) · B. S. Das · A. K. Sharma
National Institute of Technology, Patna, Bihar, India
e-mail: sweetyr.phd20.ce@nitp.ac.in; bsd.ce@nitp.ac.in; aks.ce@nitp.ac.in

J. R. Kuntia
Chaitanya Bharathi Institute of Technology, Hyderabad, India
e-mail: jnanaranjan_civil@cbit.ac.in

K. Devi
Civil Engineering Department, National Institute of Technology Warangal, Warangal, India
e-mail: kamalinidevi@nitw.ac.in

S. Pathak et al. (eds.), *Intelligent Infrastructure and Smart Materials*,
https://doi.org/10.1007/978-3-031-92421-7_12

the hydrodynamic behavior in RP to enhance its efficiency. The simulation performed in CFD showed promising agreement with experimental results. The present study focuses on the simulation of various geometry of raceway ponds using the ANSYS Fluent workbench. The simulation of flow at varying positions of the paddlewheel (beginning of center wall, 25%, and 50%) using the turbulence LES model is performed to identify the optimal configuration and turbulence model that can minimize the maximum dead zone from the channel.

2 Research Gap

The simulation using various turbulence models like the Reynold stress model (RSM), SST model, and K-ω model needs to be performed as very few studies have been performed on these models. Most of the study has been performed using the k-ε model. To understand the phenomenon of sedimentation and erosion, further modification needs to be proposed. Various factors like the depth of the pond, flow velocity, and geometry of RP, which require experimental support need to be studied. The recent advancement is inclined toward its combination of RP with wastewater treatment plants, and photobioreactors providing further scope of research. The study on modification of bend geometry, location of the paddlewheel, and gas spargers is essential to optimize flow dynamics, enhance mixing efficiency, and improve overall system performance in raceway ponds. The geometry of the delta wing provides future potential for research.

3 Methodology

The simulation using ANSYS Fluent is performed using the following steps:

- Geometry creation
- Mesh generation
- Setting up of physics
- Post-processing solver

These four steps form the toolbox setup for the analysis of fluid flow using the ANSYS workbench.

3.1 Geometry Creation

The creation of geometry is the primary step in modeling using ANSYS fluent workbench. In the Ansys Design Modeler, fluid and solid domains are generated. In three-dimensional design coordinates, x depicts the direction of flow, direction Y

Table 1 Geometrical parameters used for simulation

L (m)	*W* (m)	*D* (m)	*V* (m/s)	AR = *L*/*B*	*Lc* (m)	*d* (m)	*r* (m)
20	2	0.4	0.3	10	15.85	0.15	2.075
40	2	0.4	0.3	20	35.85	0.15	2.075

Table 2 Meshing component of different components

AR	Position of paddlewheel	Element	Node	Skewness
10	Beginning of the center wall	326,592	375,246	0.9
10	25%	327,848	376,605	0.9
10	50%	327,176	375,858	0.9
20	Beginning of the center wall	588,232	676,674	0.9
20	25%	588,232	676,674	0.9
20	50%	588,232	676,674	0.9

depicts the width of the channel, and *Z* direction depicts the depth of flow. The open–closed loop channel is represented by the creation of an oval object and zone of momentum by a rectangular object in the design modeler. To inculcate boundary conditions, the geometry is assigned suitable domain names. The assigned names for created geometry are Inlet, Outlet, Wall, Bed, and Free Surface. The raceway pond of aspect ratios 10 and 20 are designed for simulation [10]. The paddlewheel of width 0.5 m is placed at the varying position in the raceway pond. The simulation was carried out along the varied depths of the raceway pond. The central wall is placed at the center of the raceway pond whose thickness (*d*) is 0.15 m in all cases of simulation. The length of the central wall is taken as ($L_{cw} = L - 2W - d$) where *L* is the length of the raceway pond, W is the width of the raceway pond, and *D* is the depth of the raceway pond. The radius (*r*) of the U-turn of the raceway pond is taken as $r = (W + d/2)$ [10]. Table 1 represents all the geometrical parameters taken for simulation using different aspect ratios by varying the position of the paddlewheel at the beginning of the center wall, and at 25% and 50% length of the raceway pond.

3.2 *Meshing*

The raceway pond is divided into two domains, namely, the center wall and the source of momentum generation. The mesh of two-aspect ratio geometry was created using the Meshing tool. The hexahedral mesh is created for both geometries using a Lagrangian reference frame. The element size of 0.05 m is provided for both geometries. The element and node contained in geometry having aspect ratios of 10 and 20 for different locations of the paddlewheel are represented in Table 2. The total skewness targeted was 0.9 as the default value for both aspect ratios.

3.3 Setting Up Physics

The setting up of physics in ANSYS Fluent is an important and deciding tool. After generating a mesh of high quality, choosing an appropriate turbulence model and assigning physical properties of the flow domain with appropriate boundary conditions is required to obtain an appropriate result. After choosing the physics module, the solver type such as pressure-based or time as transient condition is chosen, and thereafter gravitational component value is assigned. The resolved gravity component ($\rho g\sin\phi$, 0, $-\rho g\cos\phi$) in x, y, and z directions were chosen, where ϕ is the angle measured between the channel bed and horizontal level bed. Then after appropriate model needs to be selected. In the multiphase option, the volume of fluid (VOF) needs to be selected. The VOF in CFD helps to understand the free surface location and its interaction with a wide range of pressure discretization options. The analysis of the theory of the multiphase system can be done using the Euler–Lagrange method or the Euler–Euler method. The VOF is the Euler–Euler approach so for open channel flow Euler–Euler approach is selected. This approach helps to understand the interaction between the free surface and its interface. Then two Eulerian phases, i.e., air and water, are created. The primary phase is selected as air and the secondary phase is selected as water. The properties of the second phase, such as density, viscosity, specific heat, and thermal conductivity, are assigned for the respective flow domain. The finite volume method is used for discretization [11–13].

The simulation using the three-dimensional flow of fluid was performed using the software ANSYS Fluent. The simulation was performed considering the flow is transient, as the LES turbulence model is applied for transient flow. The LES equation is discretized in both time and space hence it is used for simulating the free surface [6, 13]. The generation of flow is placed in the longest direction of RP. The LES turbulence model is used to determine the performance of RP. The symmetry boundary condition is chosen at the top surface, whereas at the bottom, inner and outer walls, the wall boundary condition is adopted. Also, the no-slip condition was applied at the wall and free surface. The time-dependent Navier–Stokes equation is filtered in either Fourier space or physical space to obtain the governing LES equation. This filtering process significantly filters the eddies of smaller scale than the filter width or grid spacing used in computation and therefore governs the dynamics of large eddies. The RANS equation obtained is represented by Eqs. (1) and (2):

$$\frac{\partial \rho}{\partial t}+\frac{\partial}{\partial x_i}\left(\rho \underline{U_i}\right)=0 \tag{1}$$

$$\frac{\partial}{\partial t}\left(\rho \underline{U_i}\right)+\frac{\partial}{\partial x_j}\left(\rho \underline{U_i U_j}\right)=\frac{\partial}{\partial x_j}\left(\sigma_{ij}\right)-\frac{\partial \underline{p}}{\partial x_i}-\frac{\partial \tau_{ij}}{\partial x_j} \tag{2}$$

where u is the velocity of flow (ms^{-1}) in x, y, and z directions, ρ is the density of the fluid (kg m^{-3}), p is pressure(Pa), τ_{ij} is Reynolds stress tensor (m^2 s^{-2}), and σ_{ij} is tress tensor due to molecular viscosity. The stress tensor is represented by Eq. (3):

$$\tau_{ij} = \rho \underline{U}_i \underline{U}_j - \rho \underline{U}_i \underline{U}_j \tag{3}$$

The LES turbulence model used in the simulation is represented by Eq. (4).

$$\frac{\partial}{\partial t}(pU_i) + \frac{\partial}{\partial x_j}(pU_iU_j) = -\frac{\partial p}{\partial x_i} + \left[\mu\left(\frac{\partial U_i}{\partial x_j} + \frac{\partial U_j}{\partial x_i} - \frac{2}{3}\delta_{ij}\frac{\partial U_t}{\partial x_t}\right)\right] - \frac{\partial \tau_{ij}}{\partial x_j} \tag{4}$$

From Eq. (3), τ_{ij} is achieved by using a subgrid stress model. The present study uses the standard Smagorinsky–Lily model to get the solution.

3.4 *Post-process Solver*

Once the turbulence model is selected, the boundary condition for the assigned domain is incorporated. For the flow domain, the following boundary condition is adopted as mentioned:

- Pressure inlet
- Pressure outlet
- Symmetry conditions
- Wall conditions

In the present study consisting of two Eulerian phases, the open channel flow is chosen as the VOF model. Also, it is modeled as the implicit volume of the model. At the inlet flow domain, the boundary condition with absolute reference of the frame the velocity of flow assigned $v = 0.3$ m/s for various aspect ratios at different positions of the paddlewheel. For all cases, the direction specification method used is normal to the boundary condition. At the outlet, the pressure outlet boundary condition is adopted. The static pressure, which is the same as atmospheric pressure, is determined at the outlet. In the boundary condition near the wall of the channel, a no-slip condition is provided where the tangential component of velocity is mentioned as zero. At the free surface mentioned on the top surface of geometry, the pressure maintained is equal to atmospheric pressure. At the center wall also the no-slip condition is set near the wall. In the last step to run the calculation, the number of time steps set was 1000. This is required to converge the solution. The coupling of pressure–velocity in the transport equation and other equations also used are discretized by second-order upwind precision and the algorithm used is PISO for validation. After completion of the simulation, the post-process menu analyses the result obtained.

4 Result and Discussion

The raceway pond of varying lengths (L) of 20 m and 40 m and depth of flow (D) of 0.4 m was simulated by taking width (B) of 2 m and velocity of flow as 0.3 m/s. A total of 54 cases were studied. In an open channel with two aspect ratios 10 and 20, the effect of varying the position of the paddlewheel was studied. Figures 1, 2 and 3 show the velocity contour at different depths for different positions of the paddlewheel, i.e., at the beginning of the center wall, and at 25% and 50% of the length of the raceway pond, using the LES turbulence model. The medium depth considered for the study is 0.05 m, 0.2 m, and 0.35 m [14, 15]. The formation of dead zones is

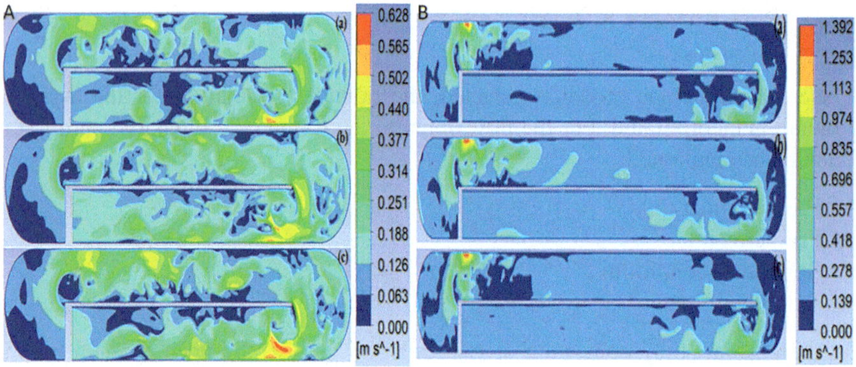

Fig. 1 Longitudinal velocity contour at a different depth from the bed used for the beginning position of paddlewheel from the center wall for (**A**) aspect ratio 10 and (**B**) aspect ratio 20 for inlet velocity $v = 0.3$ m/s. (**a**) 0.05 m; (**b**) 0.2 m; (**c**) 0.35 m

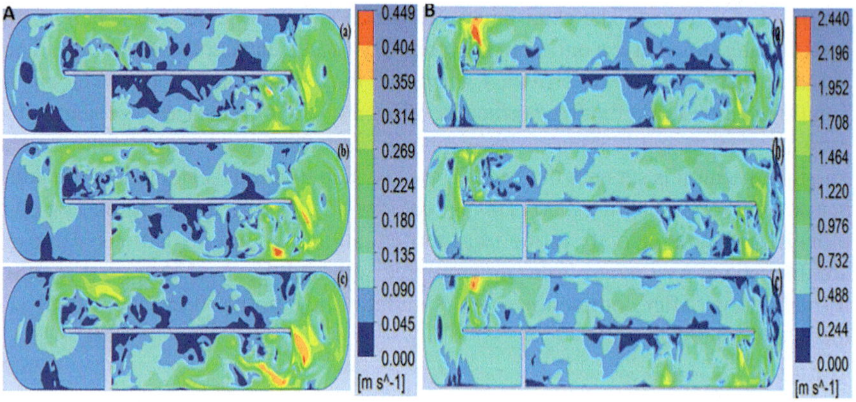

Fig. 2 Longitudinal velocity contour at a different depth from the bed using the LES model for 25% distance of paddlewheel for (**A**) aspect ratio 10 and (**B**) aspect ratio 20 for inlet velocity $v = 0.3$ m/s. (**a**) 0.05 m; (**b**) 0.2 m; (**c**) 0.35 m

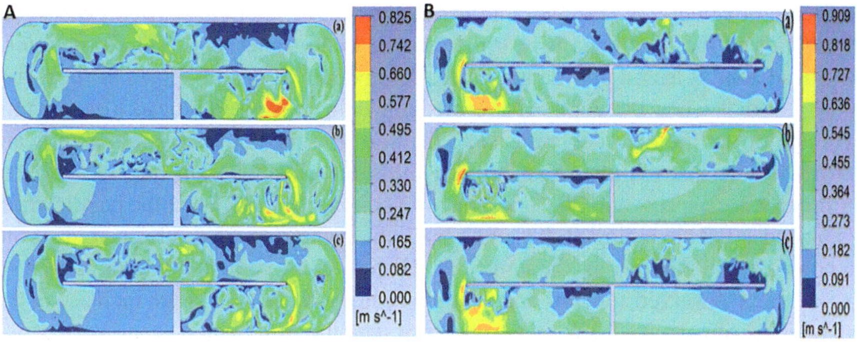

Fig. 3 Longitudinal velocity contour at a different depth from the bed using the LES model for 50% distance of paddlewheel for (**A**) aspect ratio 10 and (**B**) aspect ratio 20 for inlet velocity v = 0.3 m/s. (**a**) 0.05 m; (**b**) 0.2 m; (**c**) 0.35 m

usually observed at bends and near ends of central walls. It means this region is an area of no flow or no mixing. The effect of the formation of such a stagnation zone leads to the accumulation of some amount of nutrients with the reaction mixture, which is carried by mainstream inside the dead zone, and this is released only later after average storage time [16]. Therefore, uneven velocity distribution may lead to such formation. The formation of dead zone is mostly observed at the bend and near the central wall; this may also be due to the formation of vortex flow with high velocity at the smallest distance near the edge of the central wall and low velocity at a longer radius of curvature near the curved bend [6, 17]). Also, around the curved portion, the level of liquid may vary from higher hydrostatic pressure near the curved portion to lower hydrostatic pressure near the edge of the center wall. As per Bernoulli's principle, the outer wall having higher pressure is followed by lower flow velocity, whereas the low pressure near the inner curve is followed by large flow velocity, and this explains the phenomenon of dead zone [18].

4.1 Effect of Velocity at Different Positions of Paddlewheel

The raceway pond of aspect ratio 10 is designed to determine the effect of the varying magnitude of velocity. The paddlewheel is placed at the beginning of the raceway pond. For the inlet boundary condition, the inlet velocity is given as 0.3 m/s. At a medium depth of 0.05 m, as shown in Fig. 1A (a), a slight increase in velocity 0.502–0.565 m/s is observed at the side wall, which decreases as the flow progresses toward the edge of the center wall. At a depth of 0.2 m, the inlet velocity increases from 0.126 m/s to 0.502 m/s, and a very slight increase in velocity near the wall is observed, as shown in Fig. 1A (b). At the bottom depth, the maximum increase in velocity near the side wall is observed, as shown in Fig. 1A (c). This sudden increase in the velocity of flow is observed as the depth increases from top to bottom. This increase in velocity is not favorable for the growth of microalgae. In all cases, the formation of the dead zone is near the side wall and center wall. The formation of the dead zone is less at mid-depth, i.e., at 0.2 m depth of medium from the bed. It is reported that when the depth of the medium is increasing, the head loss per unit volume of the medium decreases. The depth of the medium has a less significant effect on the distribution of velocity at the same channel from the free surface. Similarly, for aspect ratio 20, for an inlet velocity of 0.3 m/s when the paddlewheel is placed at the beginning of the center wall, the volume of the dead zone reduces to a considerable amount, as shown in Fig. 1B (a)–(c). A very small amount of dead zone can be observed at bends and near the center wall. Also, a slight increase of the magnitude of velocity 1.392 m/s is observed near the wall, which keeps on decreasing as fluid flows toward the center of the raceway pond. The increase of velocity near the wall is considerably less at the top surface, as shown in Fig. 1B (a). The observation shows the increasing trend of velocity near the wall as depth increases to 0.2 m, then again decreases at the bottom surface. The maximum magnitude of the velocity of flow, i.e., 0.139 m/s, at the inlet is observed at the mid-depth of the pond, as shown in Fig. 1B (b). The region where the cell tends to settle down is characterized by where the velocity of flow is <0.10 m/s as reported in the literature [18, 19]. It is observed from Fig. 1 that the formation of a dead zone mostly occurs after a bend where there may be chances of occurrence of head loss [19]. The literature reported that an average velocity of 0.05 m/s is sufficient to keep the algal cell in suspension in a medium [11]. However, maintaining such low velocity at a bend is challenging because of frictional losses at the channel and corner. In all cases, the dead zone is observed near the bends where the velocity is <0.1 m/s, this may be due to the effect of the wall near the bend also.

The paddle wheel at the location of 25% distance along the length of the raceway pond is placed for aspect ratio 10 and 20, respectively, as shown in Fig. 2A, B. The dead zone is observed near the bend and the center wall. In Fig. 2A (a) slight increase in velocity near the wall from the anticlockwise direction of flow is observed, which decreases as the flow progresses toward the center of RP. The velocity decreases from the sidewall to the edge of the center wall, as shown in Fig. 2A (a)–(c). The less formation of the dead zone is observed in Fig. 2A (b). At the top surface, as shown in Fig. 2A (a), the sudden increase in velocity is less as compared to the bottom and mid surface. The maximum uneven distribution of velocity can be observed at the bottom surface, i.e., at a depth of 0.3 m, as shown in Fig. 2A (c). For an aspect ratio of 20, as shown in Fig. 2B (a)–(c), less formation of the dead zone can be observed. It can also be stated that the velocity of flow increases as the aspect ratio increases. The maximum magnitude of velocity at the inlet is 0.488 m/s in all depths of aspect ratio of 20. As shown in Fig. 2B (a), the increase of velocity of flow near the wall is more at the top surface, which reduces at the mid surface, i.e., at a depth of 0.2 m, as shown in Fig. 2B (b), then again slight increase is observed at a bottom depth of 0.35m, as shown in Fig. 2B (c). Among all cases, Fig. 2B (b) shows very little unevenness in velocity distribution, which keeps on decreasing as depth decreases. Along the free surface, the distribution of the magnitude of velocity was analyzed for aspect ratios 10 and 20, as shown in Fig. 3A, B (a)–(c). The paddlewheel is placed at a distance of 10 m and 20 m, respectively, for aspect ratios 10 and 20 along the length of the raceway pond. The inlet boundary condition of 0.3 m/s is adopted to understand the effect of flow in the raceway pond for both cases. The distribution of velocity significantly affects the performance of the raceway pond. It affects the distribution and assimilation of nutrients, light and dark cycle (*L*/*D*), and, therefore, affects the yield of microalgae. In all cases of Fig. 3, the zones of low flow generally termed dead zone (velocity < 0.1 m/s) are observed along the central wall and near the bends. For the narrow width of the raceway pond ($B = 2$ m), the formation of a dead zone adjacent to the central wall is more. This indicates that there is no development of flow along the straight channel. The flow development is progressive, and the distribution is evenly distributed along the cross-stream direction in the straight section. Such development of flow is more noticeable in a lower aspect ratio. The formation of the dead zone is observed at bends and near the central wall in all cases. In Fig. 3A (a), at a depth of 0.05 m, the velocity at the inlet is observed as 0.168 m/s. The slight increase in the magnitude of velocity 0.412 m/s at the inlet was observed at 0.35 m depth of the medium, i.e., at the bottom depth from the bed, as shown in Fig. 3A (c). Among all cases, Fig. 3A (b) shows less formation of the dead zone. Also, a sudden increase in the velocity of flow near the wall is observed at the top surface, i.e., at a depth of 0.05 m, as shown in Fig. 3A (a), which keeps on decreasing as depth increases, as shown in Fig. 3A (b)–(c). Similarly, for aspect ratio 20, placing the paddlewheel at 50% distance along the length of RP, the maximum magnitude of inlet velocity obtained is 0.182 m/s for all cases, as shown in Fig. 3B (a)–(c). As shown in Fig. 3B (a), at a depth of 0.05 m, the velocity increases from the edge of the center wall to the wall near the outlet and then decreases as the flow progresses toward the outlet. This observation decreases at a depth of 0.2 m, as shown in Fig. 3B (b), and then eventually decreases at the bottom depth, i.e., at a depth of 0.35 m, as shown in Fig. 3B (c). Hence from all observations, it can be stated that the increase in aspect ratio has a better distribution of the velocity of flow as compared to a smaller aspect ratio.

4.2 *Effect of Eddy Viscosity on Different Positions of Paddlewheel*

The paddlewheel is the only effective mechanism in the raceway pond to increase the mixing of nutrients and increase its productivity. Therefore, to avoid damage to microalgae cells, paddlewheel must generate micro eddies with a length larger than the size of the cell [20, 21]. Also, it is reported in the literature that increasing the mixing speed may increase the hydrodynamic stresses that may damage the algal cell [21, 22]. The formation of eddy reported in Figs. 4, 5, and 6 indicates the zone of damage to algal cells. This eddy formation takes place because of the high velocity of flow in a particular region. In Fig. 4A (a)–(c), the eddy formation is represented with varying medium depth for aspect ratio 10 at placing the paddlewheel at the beginning of the central wall. It is observed that eddy formation is less at a depth

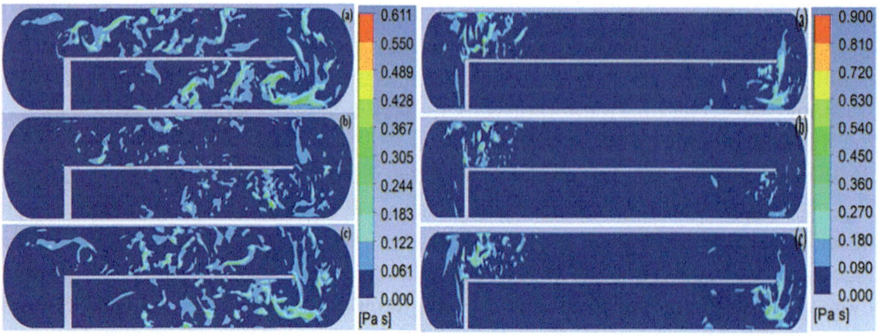

Fig. 4 Eddy viscosity contour at a different depth from the bed using the LES model for the beginning position of paddlewheel from the center wall for (**A**) aspect ratio 10 and (**B**) aspect ratio 20 for inlet velocity $v = 0.3$ m/s. (**a**) 0.05 m; (**b**) 0.2 m; (**c**) 0.35 m

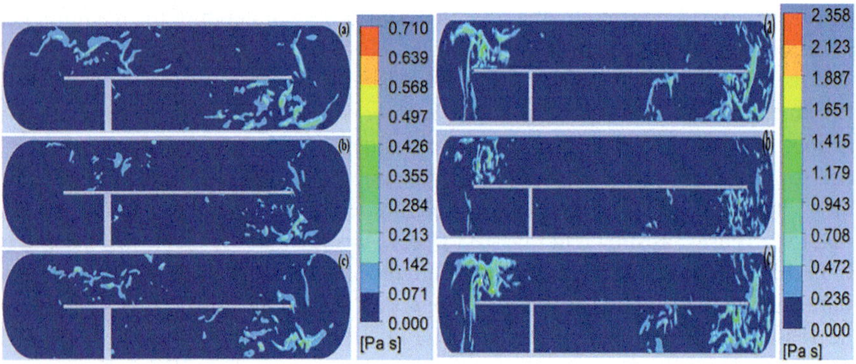

Fig. 5 Eddy viscosity contour at a different depth from the bed using the LES model for 25% distance of paddlewheel for (**A**) aspect ratio 10 and (**B**) aspect ratio 20 for inlet velocity $v = 0.3$ m/s. (**a**) 0.05 m; (**b**) 0.2 m; (**c**) 0.35 m

of 0.2 m, as shown in Fig. 4A (b), and increases eventually at a depth of 0.35 m, as shown in Fig. 4A (c). This formation is generally observed near the adjacent edge of the center wall and in between portions of RP in all cases of aspect ratio 10. The maximum magnitude of 0.122 Pa s eddy formation can be observed at the top surface of RP, which eventually decreases with the depth to 0.2 m and then eventually increases at a depth of 0.35 m, as shown in Fig. 4A (c). Similarly, Fig. 4B (a)–(c) represents the considerably less formation of eddies for aspect ratio 20 as compared to aspect ratio 10. In Fig. 4B (b), the formation of the eddy is more at the top surface of the raceway pond and considerably less formation is observed at the mid surface, as shown in Fig. 4B (b), which eventually rises at the bottom surface, as shown in Fig. 4B (c). Also, it can be stated that for a larger aspect ratio, the formation of eddy viscosity is less as compared to a smaller aspect ratio. Also, the formation of eddies is more at the top surface and bottom surface with a smaller aspect ratio. The paddlewheel positioned at a 25% distance along the length of the raceway pond is

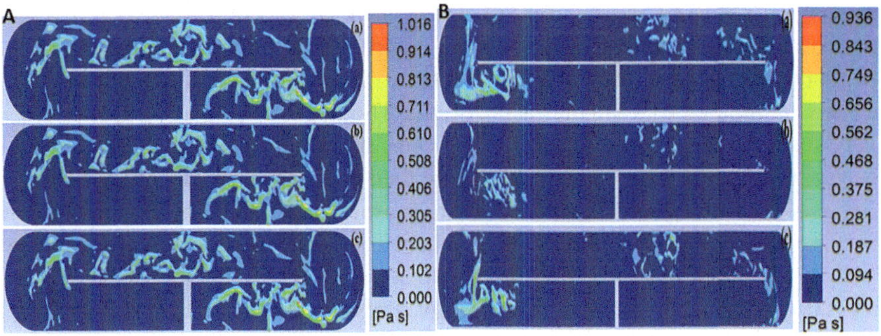

Fig. 6 Eddy viscosity contour at a different depth from the bed using the LES model for 50% distance of paddlewheel for (**A**) aspect ratio 10 and (**B**) aspect ratio 20 for inlet velocity $v = 0.3$ m/s. (**a**) 0.05 m; (**b**) 0.2 m; (**c**) 0.35 m

analyzed to determine the impact of eddy viscosity on the raceway pond, as shown in Fig. 5A, B (a)–(c) for different aspect ratios. The maximum magnitude of 0.142 Pa s is observed in Fig. 5A (a) for aspect ratio 10 at the top surface of the pond. At mid-depth, as shown in Fig. 5A (b), the formation is less. Similarly, for all cases (aspect ratio 20) of Fig. 5B (a)–(c), the formation of eddy viscosity is more as compared to aspect ratio 10. In Fig. 5B (b), at a mid-depth of the medium, there are fewer chances of damage to the algal cell.

The paddlewheel when placed at a 50% distance along the length of the raceway pond for different aspect ratios shows a significant change, as shown in Fig. 6A, B (a)–(c). The maximum formation of eddies takes place at the top surface while changing the position of the paddlewheel, which keeps on increasing as the depth increases, as shown in Fig. 6A (a)–(c). For the same position of paddlewheel for aspect ratio 20, the formation of eddies is considerably less, which indicates less damage to cells of algae. This formation of eddies is more near the edge of the center wall at the top surface, as shown in Fig. 6B (b), which decreases at the mid-depth of the pond, as shown in Fig. 6B (b). Hence it can be stated that an increase in aspect ratio and placing the paddlewheel at a 25% distance along the length of the raceway pond may increase the productivity of the pond.

4.3 *Effect of Pressure on Different Positions of Paddlewheel*

A significant change in the distribution of pressure can be observed in the raceway pond while changing the position of the paddlewheel. The paddlewheel is positioned at the beginning of the center wall for aspect ratios 10 and 20, as shown in Fig. 7A, B (a)–(c). For aspect ratio 10, it is observed that at the top surface, as shown in Fig. 7A (a), the negative pressure is experienced near the edge of the center wall and pressure significantly increases as it moves toward the bend. The negative pressure is experienced near the adjacent edge of the center wall and an increase of pressure is observed at a bend in all cases, as shown in Fig. 7A (b)–(c). It is also observed that there is no significant effect on pressure distribution by increasing the depth of the medium. Similarly, for aspect ratio 20, the pressure distribution is more even as compared to a smaller aspect ratio. The negative pressure experienced is very small near the center of the raceway pond up to a certain distance between the center wall and the side wall. The pressure then increases near the left side bend of RP, as shown in Fig. 7B (a)–(c). In a larger aspect ratio, there is also no effect of varying depth of medium on pressure distribution. The location of the paddlewheel is changed at 25% distance along the length of the raceway pond for aspect ratios 10 and 20, as shown in Fig. 8A, B. The negative pressure is observed at the mid portion of the raceway pond, which extends up to a certain portion toward the bend, and then starts increasing as the flow progresses toward the outlet. The maximum pressure is observed at the outlet, as shown in Fig. 8A (a)–(c). Similarly, for aspect ratio 20, the mid-section of the raceway pond along with the bend experiences negative pressure, as shown in Fig. 8B (a)–(c). The varying depth of the medium does not affect pressure distribution. Changing the position of the paddlewheel at 50% distance along the length of the raceway pond for aspect ratios 10 and 20 does not show

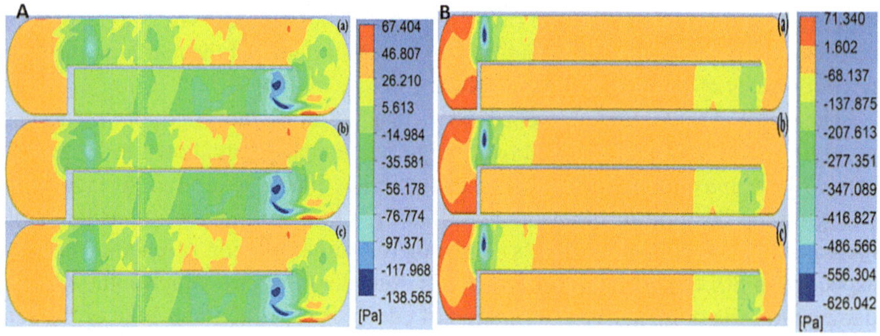

Fig. 7 Pressure contour at a different depth from the bed using the LES model for the beginning position of paddlewheel from the center wall (**A**) aspect ratio 10 and (**B**) aspect ratio 20 for inlet velocity $v = 0.3$ m/s. (**a**) 0.05 m; (**b**) 0.2 m; (**c**) 0.35m

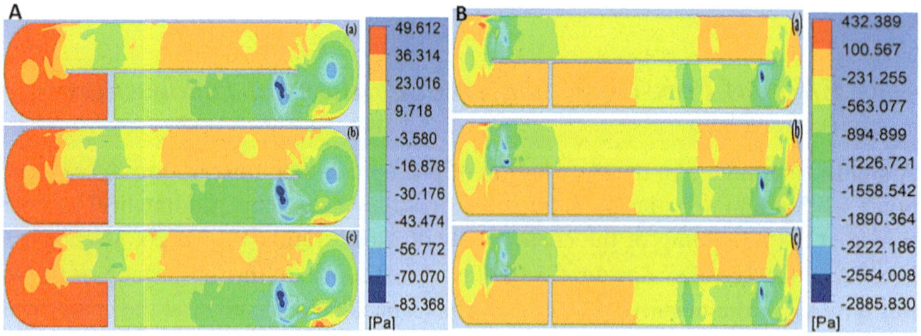

Fig. 8 Pressure contour at a different depth from the bed using the LES model for 25% distance of paddlewheel for (**A**) aspect ratio 10 and (**B**) aspect ratio 20 for inlet velocity $v = 0.3$ m/s. (**a**) 0.05 m; (**b**) 0.2 m; (**c**) 0.35 m

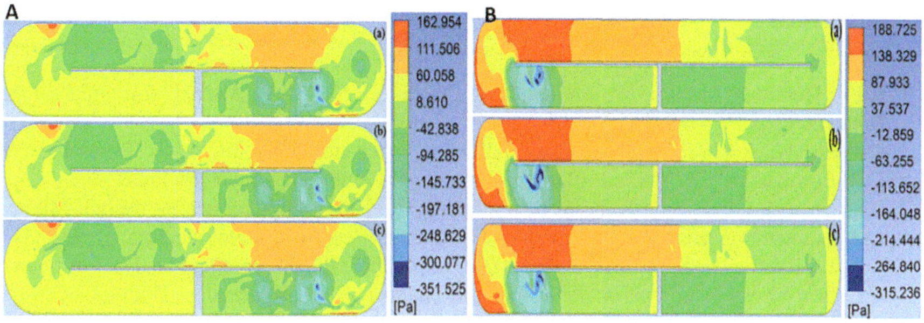

Fig. 9 Pressure contour at a different depth from the bed using the LES model for 50% distance of paddlewheel for (**A**) aspect ratio 10 and (**B**) aspect ratio 20 for inlet velocity v = 0.3 m/s. (**a**) 0.05 m; (**b**) 0.2 m; (**c**) 0.35 m

any significant effect on varying the medium depth. The negative pressure for aspect ratio 10 is observed near the inlet, which increases toward the bend. The adjacent side of the center wall also experienced negative pressure, as shown in Fig. 9A (a)–(c). Similarly, for aspect ratio 20, the negative pressure is experienced at the inlet, and pressure increases as the flow progresses toward the outlet of RP, as shown in Fig. 9B (a)–(c).

5 Conclusion

In the present study, the LES turbulence model has been applied to analyze the velocity, eddy viscosity, and pressure distribution in the raceway pond by varying the position of the paddlewheel and aspect ratio. The following conclusion has been drawn from the present study.

- The longitudinal velocity in the raceway pond is found to be increasing with an increase in depth of flow. The uneven distribution of velocity is observed in RP of a smaller aspect ratio when the paddlewheel is placed at a 50% distance along the length of the raceway pond.
- The formation of the dead zone decreases as the aspect ratio increases and maximum decrease is observed when the paddlewheel is placed at 50% along the length of RP.
- The formation of eddy viscosity decreases by increasing the aspect ratio. The maximum decrease of formation of eddy viscosity is observed at a depth of 0.2 m when the paddlewheel is placed at a 50% distance along the length of RP.
- There is no significant effect of pressure distribution on the varying depth of the medium.

References

1. Li, Y., Horsman, M., Wu, N., Lan, C. Q., & Dubois, C. N. (2008). Biofuels from microalgae. *Biotechnology Progress, 24*(4), 815–820.
2. Oswald, W. J. (1988). Micro-algae and waste-water treatment. In M. A. Borowitzka & L. J. Borowitzka (Eds.), Micro-algal biotechnology. Cambridge University Press.
3. Weissman, J. C., Goebel, R. P., & Benemann, J. R. (1988). Photobioreactor design: Mixing, carbon utilization, and oxygen accumulation. Biotechnology and Bioengineering, 31, 336–344.
4. Das, L., Khatua, K. K., & Das, B. S. (2022). Experimental and numerical analyses of boundary shear stress in non-prismatic compound channel. River Hydraulics: Hydraulics, Water Resources and Coastal Engineering, 2, 37–58.
5. Vonshak, A., Abeliovich, A., Boussiba, S., Arad, S., & Richmond, A. (1982). Production of Spirulina biomass: effects of environmental factors and population density. Biomass, 2, 175–185.
6. Kumari, R., Das, B. S., Devi, K., Khuntia, J. R., & Mohanty, M. P. (2024). Hydrodynamic performance of raceway pond using k-ω and LES turbulence models. Journal of Renewable and Sustainable Energy, 16(6).
7. Liffman, K., Paterson, D. A., Liovic, P., & Bandopadhayay, P. (2013). Comparing the energy efficiency of different high-rate algal raceway pond designs using computational fluid dynamics. Chemical Engineering Research and Design, 91(2), 221–226.
8. Hadiyanto, H., Elmore, S., Gerven, T. V., & Stankiewicz, A. (2013). Hydrodynamic evaluations in high rate algae pond (HRAP) design. Chemical Engineering Journal, 217, 231–239.
9. Xu, B., Li, P., & Waller, P. (2014). Study of the flow mixing in a novel ARID raceway for algae production. Renewable Energy, 62, 249–257.
10. Lima, A., Marinho, B., & Morais, T. (2020). Hydrodynamic analysis of flow in raceway ponds for algae cultivation under versatile conditions. Aquaculture International, 29(5), 1–17.
11. Rajput, S., Das, B. S., & Sharma, A. K. (2023). Different modeling approaches in increasing efficiency of raceway pond. In P. V. Timbadiya, P. L. Patel, V. P. Singh, & B. Barman (Eds.), Fluid mechanics and hydraulics (HYDRO 2021), lecture notes in civil engineering (Vol. 314, pp. 165–177). Springer.
12. Das, B. S., & Khatua, K. K. (2018). Numerical method to compute water surface profile for converging compound channel. Arabian Journal for Science and Engineering, 43(10), 5349–5364.
13. Das, B. S., Devi, K., Proust, S., & Khatua, K. K. (2018). Flow distribution in diverging compound channels using improved independent subsection method. In River flow 2018: 9th International conference on fluvial hydraulics, Issue no 05068 (Vol. 40, p. 8).
14. Amini, H., Hashemisohi, A., Wang, L., Shahbazi, A., Bikdash, M. K. C. D., & Yuan, W. (2016). Numerical and experimental investigation of hydrodynamics and light transfer in open raceway ponds at various algal cell concentrations and medium depths. Chemical Engineering Science, 156, 11–23.
15. Ali, H., Cheema, T. A., Yoon, H. S., Do, Y., & Park, C. (2015). Numerical prediction of algae cell mixing feature in raceway ponds using particle tracing methods. *Biotechnology and Bioengineering, 112*(2), 297–307.
16. Sawant, S. S., Khadamkar, H. P., Mathpati, C. S., Pandit, R., & Lali, A. M. (2017). Computational and experimental studies of high depth algal raceway pond photobioreactor. Renewable Energy, 118, 152–159.
17. Rajput, S., Das, B. S., & Sharma, A. K. (2025). Review on modelling approaches of raceway pond: experimental, numerical and machine learning approaches. ISH Journal of Hydraulic Engineering, 31(2), 382–397. https://doi.org/10.1080/09715010.2025.2452581
18. Sompech, K., Chisti, Y., & Srinophakun, T. (2012). Design of raceway ponds for producing microalgae. Biofuels, 3(4), 387–397.

19. Huang, J. K., Qu, X. X., Wan, M. X., Ying, J. G., Li, Y. G., Zhu, F. C., Wand, J., Shen, G., Chen, J., & Li, W. (2015). Investigation on the performance of raceway ponds with internal structures by the means of CFD simulations and experiments. Journal of Algal Research, 10, 64–71.
20. Thomas, W. H., & Gibson, C. H. (1990). Effects of small-scale turbulence on microalgae. Journal of Applied Phycology, 2, 71–77.
21. Berdalet, E., Peters, F., Koumandou, V. L., Roldan, C., Guadayol, O., & Estrada, M. (2007). Species-specific physiological response of dinoflagellates to quantified small-scale turbulence. Journal of Phycology, 43, 965–977.
22. Al-Homoud, A., & Hondzo, M. (2007). Energy dissipation estimates in oscillating grid setup: LDV and PIV measurements. Environmental Fluid Mechanics, 7, 143–158.

Water Scarcity and Climate Change in Intelligent Infrastructure and Smart Materials: Sustainable Technologies for a Greener Future

Biswadeep Pal

1 Introduction

Climate change presents an escalating global challenge, increasingly impacting various sectors, including water resources, infrastructure, and urban development. As the planet continues to warm due to anthropogenic activities, issues such as water scarcity are becoming more pronounced. Water scarcity, defined as the insufficient availability of water resources in a specific region due to smart infrastructure and human factors, is a growing concern exacerbated by unsustainable management practices, pollution, and wastage [1, 2].

The rapid pace of urbanization further complicates this issue. The global urban population has surged from 29.6% in 1950 to 56.2% in 2020 and is projected to reach 68.4% by 2050 [3]. This urbanization trend is a significant driver of water scarcity in many cities, affecting the daily lives of urban residents, the environment, and economic development. With population growth and urbanization, the demand for water in cities is expected to increase by 50–80% over the next three decades, compounded by the impacts of climate change on water availability [4].

Historically, developed cities have addressed water shortages through engineering solutions such as reservoirs, desalination plants, and inter-basin water transfers (IBTs). At the same time, these effective solutions are often costly, resource-intensive, and environmentally disadvantageous. Different studies highlighted the limitations, including a focus on specific cities without considering broader water wastage issues, leading to uncertainties in estimating affected populations. Potential solutions for water scarcity, such as groundwater exploitation and desalination, must account for factors like severity of water shortage, geography, socioeconomic conditions, and environmental capacity [5].

B. Pal (✉)
Department of Chemical Engineering, Parul University, Vadodara, Gujarat, India

S. Pathak et al. (eds.), *Intelligent Infrastructure and Smart Materials*,
https://doi.org/10.1007/978-3-031-92421-7_13

Despite the absence of a predominant global water scarcity crisis, numerous regions face persistent shortages. The pressure on water resources is particularly acute in areas such as Bangladesh, India, Pakistan, the Middle East, and various parts of Africa [6, 7]. In 2016, research indicated that 933 million urban residents lived in water-shortage-prone areas, with India and China leading in the number of city inhabitants facing water scarcity [8]. Additionally, 36.7% of large cities and 9 out of 30 megacities worldwide are experiencing significant water shortages due to rising temperatures and increased evaporation, resulting in deeper groundwater levels [9].

This review aims to provide insights into mitigating global water scarcity and enhancing the sustainability and livability of cities. It explores the role of intelligent infrastructure and smart materials in addressing climate change challenges, offering sustainable technologies for a greener future. A comprehensive understanding of water scarcity in urban areas and potential solutions is essential to create sustainable and livable urban environments, aligning with the goals outlined in the Sustainable Development Goals for 2030 (SDG-2030).

2 Consequences of Water Scarcity and Climate Change

Water scarcity and climate change have a profound impact on food production as they directly affect crops and livestock and ultimately observed malnutrition, starvation, and hunger. Additionally, water shortages limit the basic need for people to maintain basic hygiene, such as showering, laundering clothes, and cleaning their homes. In some cases, children are unable to attend school due to taking long journey to access water sources. Even health weakness also hampers the learning process in the childhood phase and families get distracted due to their children's illness [10]. Water scarcity compels people to consume unsafe water, leading to sanitation problems. Water shortage causes a lack of daily water requirement, increasing the risk of unhygienic contamination, which creates upbringing grounds for disease-carrying mosquitos. This situation can result in diseases such as dengue, malaria, trachoma (an eye infection causing blindness), plague, and typhus. In Fig. 1, changes in climate conditions and access to water have become a significant global economic concern and a potential driver of international tensions [11–13]. With the growing population in the world and increasing demands for water, such tensions are likely to proliferate in the future. Water scarcity negatively impacts canals, ponds, and other freshwater ecosystems by raising salt concentrations, causing nutrient pollution, and depleting natural resources and wetlands. Eco-diversity, which includes freshwater fish populations and marine habitats, is under threat due to the scarcity of water resources.

Precipitation in India varies significantly, from 11,000 mm to 90 mm annually [14, 15]. The average annual per capita water availability has plummeted from 5000 cubic meters in 1950 to 1545 cubic meters in 2011, with projections showing further declines to 1341 cubic meters by 2025 and 1140 cubic meters by 2050 [16, 17]. Agriculture, employing 54.6% of the population, uses over 90% of groundwater for

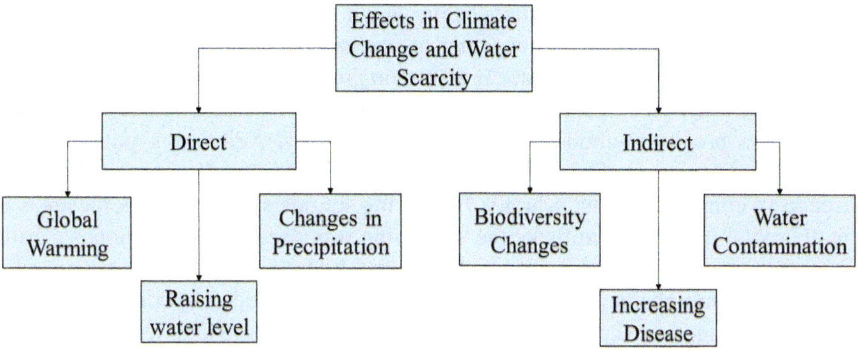

Fig. 1 Limitation of climate change

irrigation, leading to water table depletion in 64% of districts from TE2002 to TE2016. Triennium ending (TE) year is considered to decrease particular year extreme climatic variations [18, 19]. Government efforts have increased utilized irrigation potential to 87 million hectares (Mha), with a goal of 140 Mha [20]. Micro-irrigation, like drip and sprinkler systems, with efficiencies of 90% and 70%, respectively, are key to improving water use efficiency compared to surface irrigation's 40% [21]. The Pradhan Mantri Krishi Sinchayee Yojna (PMKSY) was established in 2015 to promote micro-irrigation. However, the efforts in rainwater harvesting in India often lack due to comprehensive planning and economic analysis. Six critical issues include mismatched supply and demand, complex economic evaluations, trade-offs between hydrological benefits and cost-effectiveness, disparities in catchment demand, redistribution of benefits rather than augmentation, and poor integration between surface and groundwater systems. Urban water scarcity, exacerbated by high population density and groundwater extraction, is responsible for mitigating graywater (GW) treatment and reuse. Despite challenges like social barriers and technical difficulties, constructed wetlands (CW) and biofilm reactors are promising solutions for GW treatment, especially for agriculture. Global warming and urban heat island (UHI) effects highlight the benefits of green roofs and walls, which can reduce energy consumption by up to 10.5% and district cooling demand by 13%, improving thermal comfort and air quality [22]. Constructed wetlands, using materials like gravel, vegetation, and biochar, are cost-effective for treating various wastes, offering sustainable sewage treatment alternatives for India. Green roofs also reduce energy demand and mitigate UHI effects, making them a key area for future research in India.

3 Preventing Water Scarcity and Climate Change

Prioritizing the enhancement of water infrastructure is essential. Water conservation and efficiency play a pivotal role in sustainable water management. Utilizing clean technologies like solar desalination and smart irrigation systems is crucial, with a particular focus on the agriculture sector, which is the largest consumer of water. Rainwater harvesting and the recycling of wastewater offer effective strategies to alleviate water scarcity and reduce pressure on groundwater and natural water bodies. Groundwater recharge, enabling water to move from surface sources to groundwater, is a proven method to combat water scarcity. Addressing pollution and maintaining water quality through measurement and monitoring are essential aspects of combating water scarcity. Enhancing sewage systems in specific areas is another vital measure to mitigate the worsening water scarcity. Education is a critical component of resolving the water crisis. Addressing future water scarcity requires a comprehensive overhaul of consumption practices, encompassing both individual usage and supply chains within large companies. Heightened awareness and education are necessary to facilitate this transformation. Establish collaborative, adaptive governance structures to manage water and drought risk effectively. While drought is a natural hazard, caused by large-scale climatic variability, water scarcity is caused by the long-term unsustainable use of water resources and inadequate infrastructure, as well as a lack of water availability. Governance arrangements that enable a greater understanding of impacts and vulnerabilities, as well as an effective response to changing situations, are vital in order to address these risks effectively. In particular, cities should first involve all relevant city departments, as well as private companies and civil society stakeholders. These agencies and stakeholders are involved in the assessment of risks and impacts associated with water scarcity, as well as make development strategies to address them. This should include authorities involved in managing the public water supply, water-dependent tourism and recreation, buildings and infrastructure (particularly where there is a risk of soil shrinkage), water-intensive industries, parks and environment, food supply, health and welfare and, where relevant, water-based transport and local power generation that relies on water (e.g., for cooling). Create a clear mandate for public agencies in the management of water issues. Second, establish adaptive processes to monitor and respond to evolving drought risks. Drought risk governance should be iterative, evidence-based, and supported by active situational monitoring. Strengthen collaboration with national, regional, and neighboring municipal governments to enhance resilience. Water scarcity is typically influenced by factors and decisions made outside the jurisdiction of the city, such as the upstream water concept. This makes it crucial to work with other authorities in tackling these issues. Corruption also severely impacts water-related city services, resilience, infrastructure, and revenues, with informal settlements most profoundly affected [23, 24]. In Mumbai, for example, residents in informal settlements pay between 52 and 206 times more for water than those in high-income areas [25–27]. Procurement corruption is particularly prominent. Solving these issues requires political engagement, participatory assessment of local needs, rights recognition, clarification of responsibilities, dedicated funding, and anticorruption regulation.

4 Financing Effective Green Infrastructure Management

Effective management of green infrastructure is crucial for enhancing water security in urban areas, particularly in the face of increasing drought risks and impacts. The sustainable water partnership advocates for local decision-makers to utilize various funding mechanisms aimed at implementing water security measures. These measures include assessing and mapping local drought risks and impacts, which are essential for understanding the full scope and costs associated with water scarcity. Collaborating with research partners such as universities and meteorological offices can provide critical data on drought risks, even in regions where comprehensive modeling is challenging. Ideally, drought risk assessments should be integrated into broader climate change risk assessments (CCRA) [28–30]. These assessments not only quantify direct impacts like water shortages and deteriorating water quality but also consider indirect effects such as food insecurity, price hikes due to agricultural disruptions, and stress on community services. Such assessments are essential for informed decision-making and prioritizing investments in green infrastructure. Water funds emerge as a promising financial and governance mechanism to unite stakeholders across public, private, and civil society sectors. These funds focus on nature-based solutions and sustainable watershed management practices, which are integral to improving water security in cities and their surrounding areas. By leveraging green infrastructure investments through water funds, communities can mitigate drought impacts and enhance resilience to water-related challenges. However, the effectiveness of these initiatives is often hindered by inadequate data collection, particularly in Global South nations where drought risks are most acute. Approximately 40% of World Meteorological Organization (WMO) country members lack basic hydrological data, crucial for effective water resource management. Improving data collection efforts at regional and national scales is therefore imperative for proactive monitoring and management of water resources, especially in regions vulnerable to drought [31, 32].

Financing effective green infrastructure management requires a strategic approach that integrates robust risk assessments, collaborative funding mechanisms like water funds, and improved data collection efforts. By investing in green infrastructure and leveraging nature-based solutions, cities can enhance their water security, mitigate drought impacts, and build resilience to future water challenges. This holistic approach not only safeguards water resources but also promotes sustainable development and community well-being.

5 Manage Demand for Green Infrastructure

Managing demand for green infrastructure in cities facing water stress is crucial not only during the summer season but also as a long-term strategy to alleviate pressure on water systems and promote sustainable water use. Implementing public information campaigns and behavioral interventions can significantly influence water use patterns. Strategies such as using smart meters, personalized notifications, and public dashboards can empower residents to monitor and manage their water consumption more effectively. Following the EAST framework (Easy, Attractive, Social, Timely), these campaigns can promote the use of recycled water, specifically graywater, for nonpotable purposes [32]. Successful examples from San Diego demonstrate that targeted campaigns can substantially increase public acceptance and support for water reuse initiatives [33, 34]. Adopting volumetric pricing mechanisms like block tariffs can incentivize water conservation among consumers [35]. By charging higher rates as water consumption increases, cities can send price signals that encourage efficient water use. Examples from Cape Town and Texas illustrate how such tariff structures can effectively reduce water demand during periods of scarcity [36–39]. Incorporating means-tested subsidies, as seen in Chile, ensures that water pricing remains equitable while encouraging conservation efforts [40]. Enhancing the accuracy and comprehensiveness of water availability and use estimates is essential. This includes better methodologies for estimating groundwater recharge and improving the collection of surface water data like dependable runoff. Addressing the methodological limitations and expanding the scope of water quality monitoring are critical steps toward sustainable water management. Moving away from standardized norms to more nuanced and data-driven approaches in estimating water demand across various sectors is vital. Utilizing real-time data and adjusting estimations based on actual usage patterns rather than theoretical norms can improve the precision of demand forecasts. Ensuring that water resource data collection is inclusive of social, economic, and environmental factors is imperative. Current biases toward hydrological aspects should be balanced with broader considerations to better reflect the diverse needs of communities and ecosystems. Benchmarking against successful water utilities worldwide provides valuable insights into best practices for managing water scarcity and drought resilience. Cities like Sydney, Singapore, and San Antonio are exemplary in their proactive approaches to water management, showcasing innovative strategies that integrate green infrastructure to enhance water security [41–43].

These approaches directly contribute to ensuring the availability and sustainable management of water and sanitation for all. Managing demand for green infrastructure supports the development of inclusive, safe, resilient, and sustainable cities. By promoting water conservation and efficient use through green infrastructure, cities contribute to climate resilience and adaptation efforts. Collaboration among stakeholders, government, private sector, and civil society is essential for implementing these strategies effectively and achieving water-related SDGs.

6 Agendas of SDG-2030

In 2015, the United Nations established the SDGs as a universal framework comprising 17 goals and 169 associated targets to guide global action by 2030 [44]. This ambitious agenda aims to address poverty eradication, improve sanitation, strengthen economies, and meet societal needs while promoting peace and prosperity. At its core is the goal to eliminate all forms of poverty, particularly extreme poverty, recognizing it as a critical barrier to sustainable development. The central agenda is the commitment to collaborative partnerships involving all countries and stakeholders, aiming to emancipate humanity from poverty and protect the planet. The SDGs encompass diverse areas such as human rights, gender equality, and environmental sustainability, emphasizing their interconnectedness across economic, social, and environmental dimensions. Over the next 15 years, the SDGs will drive transformative actions to ensure dignified and equitable living conditions for everyone while safeguarding natural resources and mitigating climate change impacts [45, 46]. Achieving these goals is essential for fostering resilient societies, promoting global peace, and revitalizing a global partnership focused on solidarity and inclusivity, especially for the most vulnerable populations.

7 Role of International Cooperation and Partnerships

The SDGs represent a continuation of the Millennium Development Goals (MDGs), which started in 2000 and concluded in 2015. The MDGs significantly contributed to lifting nearly one billion people out of extreme poverty, addressing hunger, and promoting the education of girl children. Unlike the MDGs, which relied primarily on government and nonprofit funding, the SDGs call on the private sector to play a role in achieving sustainable consumption and production patterns. Businesses, like Novozymes, have aligned their objectives with the SDGs, focusing on technologies that reduce water consumption for waste treatment [47]. However, stronger private sector engagement and more businesses addressing these goals are needed. Despite some progress, it has been observed that the SDGs are not advancing at the required pace to meet the 2030 deadline. Challenges include limited poverty reduction, a recent increase in world hunger, and persistent gender inequality.

India faces a pressing dual challenge concerning water resources. First, there is a scarcity of freshwater due to diminishing natural supplies and the growing demand for food production, drinking water, industrial use, and ecosystem management. Second, there are increasing conflicts related to water sharing. These needs involve maintaining a balance between supply and demand to tackle the growing scarcity and equitably allocating water across sectors to resolve the issues. However, addressing this challenge requires scientific data on water resources concerning various social, economic, and environmental objectives. This data is essential for water budgeting, allocation planning, and decision-making in water management. Reliable estimates of water supply and demand are crucial.

In 2012, it became clear that the Millennium Development Goals (MDGs) would fall far short of their targets. This realization led to discussions at the Rio+20 event in Rio de Janeiro, the birthplace of Agenda 21, where it was agreed that a more inclusive and thoughtful approach to sustainable development post-2015 was needed [48, 49]. Working groups were established to create a post-2015 Sustainable Development Declaration and Sustainable Development Goals (SDGs) in line with the spirit of Agenda 21; it was also recognized that progress on these goals required more rigorous monitoring than the self-reporting used for the MDGs [50, 51].

These events like MGD and SDG had a profound impact on the development community by implementing alternative approaches, adaptive management, emerging vision, acknowledging uncertainty, and seeking programmatic solutions. It also involves local stakeholders and collaboration, ensuring proper consideration of the project context. Solutions are considered provisional, recognizing the constantly changing global environment.

Adaptive management also addressed concerns raised by the OECD about local obstacles to effective water governance and the poor management of stakeholder interactions. Rather than focusing on specific methodologies, it advocates for a networked, agile approach to address the unique characteristics of each locality. This response is vital in the quest for a sustainable future, emphasizing the need to prevent and adapt to disruptions. It aims to equip everyone with the knowledge to navigate these challenges effectively [52–54].

8 Institutional, Innovation, and Systematic Progression

National Water Policy (2012) was implemented by the Ministry of Water Resources, Government of India, to promote efficient water usage and recycling. On the other hand, missions like Atal Bhujal Yojana (2018) and Jal Jeevan Mission (2019) emphasized sustainable water management. Additionally, programs like Namami Gange (2014) are dedicated to the conservation and rejuvenation of the Ganga River.

Overall, India faces formidable challenges in water resource management that require a multifaceted approach which is shown in Fig. 2. Enhance the generation of domestic resources, supported by international assistance in strengthening tax and revenue collection capabilities in developing nations. Ensure that developed countries fully honor their commitments to official development assistance, striving for a target of 0.7% of official development assistance (ODA)/gross national income (GNI) to be allocated to developing countries and 0.15–0.20% of ODA/GNI to least developed countries. Developed nations are encouraged to contemplate a 0.20% goal for least developed countries [55, 56]. Acquire additional financial funds for developing nations from a variety of funding origins. Aid developing countries in achieving sustainable debt levels through harmonized approaches to debt financing, relief, and restructuring.

Promotion of sustainable technologies helps in the advancement, transfer, sharing, and spread of knowledge to developing nations under various MOUs, laws, and

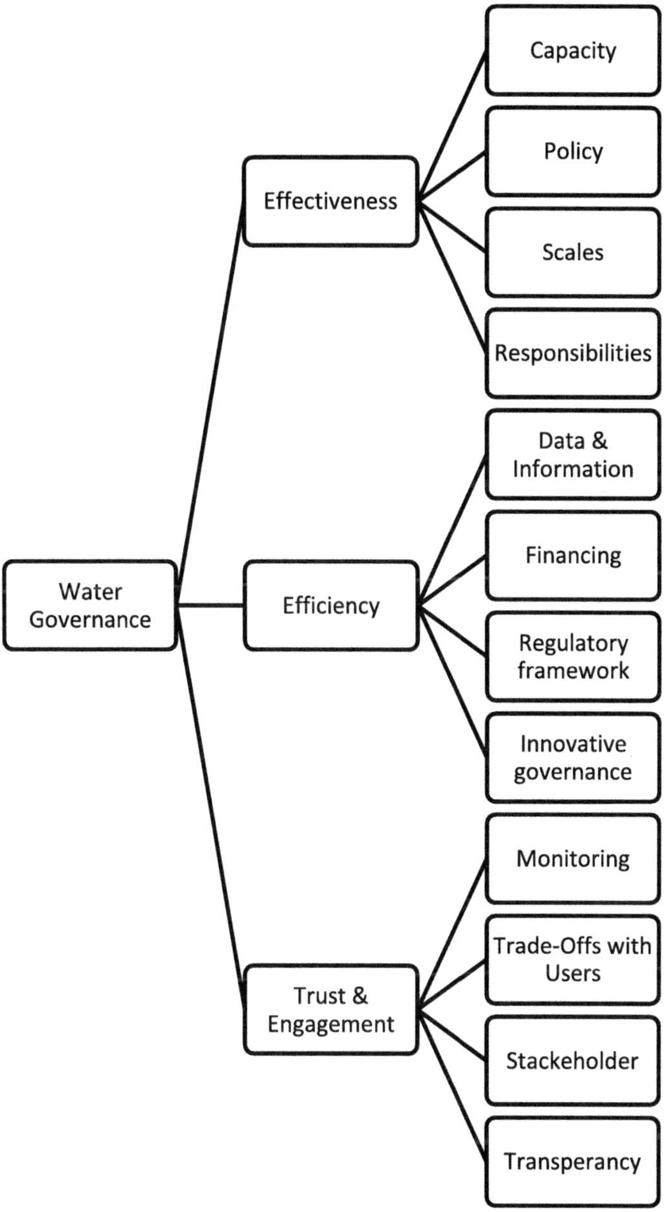

Fig. 2 Multilevel gap in water governance

treaties. Ensure that the technology bank and capacity-building mechanisms for the least developed countries are fully operational by 2017 [57]. Focus should be directed toward the use of empowering technologies, especially information and communication technology. Boost international aid to provide customized capacity-building support in developing nations, ensuring alignment with national sustainable development plans. Utilize North-South, South-South, and triangular cooperation. Support the establishment of a global, rules-based, open, impartial, and fair multilateral trading system under the World Trade Organization (WTO), and finalize the negotiations outlined in the Doha Development Agenda [58, 59]. Significantly, export capacities of developing nations enhanced their export by targeting a doubling of the least developed countries' share of global exports by 2020. Ensure continuous duty-free and quota-free market access for least developed countries, with clear and simple preferential rules of origin to ease market entry. Improve global macroeconomic stability through coordinated and coherent policies for sustainable development, respecting each nation's autonomy and leadership in shaping policies for water conservation and sustainable development [60, 61].

To strengthen the global partnership for sustainable development, it is essential to promote and support collaborations among governments, private sector entities, civil society organizations like Greenpeace, Oxfam, the Nature Conservancy, and international bodies like OECD, WHO, IMF, and World Bank. The promotion of knowledge sharing is achieved through collaborative research like Horizon 2020, Future Earth, and joint ventures like the Green Climate Fund (GCF) by UNICEF, the Carbon Trust, and EV100. The per capita water availability in corners without consideration of disparities in water allocation and access is required, particularly in a diverse country like India [62]. To address this issue, new indices are required that can account for social and economic inequalities in water usage. Short-term approaches to ensuring food security can jeopardize agricultural sustainability due to increased water usage. Water demand for agriculture is set to rise significantly, and efficient water use policies are needed. There is a need to design appropriate policies for industrial water use. Addressing conflicts arising from varying water use priorities and quantities is essential. Water quality problems in India have become a major concern. There is a lack of models and policies to manage wastewater from industrial and domestic sources, contributing to pollution and water scarcity. Understanding the intersection of gender, class, caste, and water is vital. Gender issues are interconnected with social hierarchies, impacting access to and control over water resources. To promote gender-sensitive policies, there is an urgent need for gender-disaggregated data and addressing gender inclusion within organizations managing water resources [63, 64]. Current engineers may lack the skills needed for effective water management. There is a need for education and training in new water management methods. Efforts should focus on training in-service engineers to adapt to evolving water management approaches. India faces water scarcity issues, which require efficient water utilization, changes in cropping patterns, rainwater harvesting, and development of a network of rivers and canals. The river-linking project should be expedited. Educational initiatives, awareness campaigns, sustainable development, and good governance are essential. Each drop of rainwater should be conserved [65, 66]. Industrial water usage should be closely monitored, and water-efficient technologies adopted. Water pricing may be considered to encourage responsible use. Water wastage in homes and communities should be minimized through individual efforts and raising awareness. Scale up existing initiatives aimed at creating measurements of progress related to sustainable development that complement the gross domestic product. Simultaneously, strengthen efforts for building statistical capacities in developing countries by 2030.

9 Recommendations for Sustainable Water Management

Only 0.5% of the total water on earth is available for use. The remaining 95% of the water is ocean water which is salty or on glaciers or too polluted or too far to be extracted [67, 68]. Water is one of the important resources available on the planet Earth. It is very essential for life on the earth and also keeps the balance within nature. However, the amount of available freshwater is decreasing due to consumption and not increasing as per the current situation. The increasing population has put a lot of stress on the water resources and some of the disturbance in the natural cycle has also contributed to the requirement to manage the water resources efficiently.

It has been known that the consumption of water has increased drastically. Within the last 50 years, consumption has increased four times against the increase in population up to two times [69, 70]. It has been estimated that 3.5 billion people worldwide lack access to safe water [71]. An estimation indicates that 47 countries have not gotten a sufficient amount of water and some of the countries are extracting more water than is replenished by the natural process. These countries cover one-third of the total world population [72]. This scenario indicates the scarcity of fresh and usable water on our water surface. It is a time to rethink our practices and generate ideas to save our available water resources for future generations to come to enjoy life as or better than what is available to us. It is the need of the hour to adopt the practices for efficient and effective water management to make it a sustainable resource. Per capita water supply in India is projected to decrease from 24 to 21 gallons per day (gpd) by 2025 and further to 18 gpd by 2050. It advocates a shift from traditional "gray" infrastructure to more eco-friendly "green" solutions for water management [73, 74]. This includes using natural processes like stormwater capture and green infrastructure to enhance water quality and supply.

Nature-based solutions (NBSs) are utilized in diverse sectors such as agriculture, sustainable urban development, disaster risk management, and water quality enhancement to tackle water scarcity problems in India. So, it is needed to adopt drip and sprinkler irrigation to reduce water wastage [75]. It emphasizes the importance of recharging aquifers, including through the restoration of watersheds and wetlands, to support sustainable groundwater resources. Government needs to create an infrastructure to manage stormwater runoff and reduce water pollution. This includes creating pervious areas for infiltration and promoting the use of green infrastructure. It also recommends improving water accounting and metering systems to enhance supply reliability [76, 77]. The potential of water reuse, particularly nonpotable water for car washing, irrigation, and industrial use, needs to be identified. It calls for curbing the unplanned expansion of urban areas and minimizing water distribution system leaks. The passage highlights the need for a holistic approach to water management in India, emphasizing the importance of international collaboration and the integration of social, environmental, and economic needs. It also mentioned India's "smart city" initiatives, which involve comprehensive development to address issues like congestion, air pollution, and resource depletion [78]. These initiatives also promote green infrastructure and ecological balance.

There are different strategies available, like public awareness campaigns, water-saving technologies, and regulatory measures which are very encouraging and necessary for water usage for effective sustainability. Conserve and manage water effectively by promoting awareness and implementing innovative technologies. By implementing different laws for responsible consumption in households or employing sustainable practices in agriculture, industries, and manufacturing, every effort counts toward securing this vital resource for future generations. Simple measures, such as fixing leaks, using low-flow appliances, and collecting rainwater, are somewhat mitigating water wastage. Aging and inadequate water infrastructure can result in significant water losses and inefficiencies. Investing in modernizing water supply systems is crucial for ensuring access to clean, safe water and for preserving the environment. Upgrading pipelines, treatment plants, and sewage systems not only reduces water leakage but also minimizes pollution, safeguarding both public health and ecosystems. By incorporating advanced technologies like smart meters and sensors, authorities can monitor water usage more efficiently, detect leaks promptly, and manage resources better. Moreover, improved treatment facilities like reserve osmosis, microfiltration, ion exchange, electrodialysis, and multi-effect evaporation help purify water effectively, reducing contaminants and ensuring higher-quality water for consumption and ecosystem health. Recycling and reusing water can significantly reduce the demand for freshwater sources. Implementing systems for treating and reusing wastewater in agricultural and industrial processes can help alleviate water scarcity issues. Additionally, the installation of graywater systems (a graywater system is a setup designed to collect, filter, and reuse wastewater generated from sources like sinks, showers, baths, and laundry machines within a household or building) for homes is currently encouraged to reuse for household purposes. Protecting natural water sources like rivers, lakes, aquifers, and watersheds is fundamental for maintaining a sustainable water supply. Implementing regulations and conservation efforts in these areas can help maintain the quality and quantity of available water. Agriculture is one of the largest consumers of water globally. Sustainable farming practices, such as drip irrigation, crop rotation, and reduced water-intensive crops, significantly decrease the water footprint of agriculture. Additionally, educating farmers through new technologies like IOT-based water spreading system for efficient water use makes a substantial difference between trading and modern techniques. Rainwater harvesting is a simple yet effective method of collecting rainwater for various uses. Promoting and incentivizing the installation of rainwater harvesting systems in residential and commercial buildings can help reduce the demand for treated freshwater. Ensuring clean and safe water is essential for human health and ecosystem sustainability. Regular monitoring of water quality, strict regulations, and prompt responses to pollution incidents are necessary. Collaborative efforts among government agencies, communities, and industries are key to maintaining water quality. Urban and land development should consider water sustainability in their planning. Implementing green infrastructure, such as permeable surfaces and green roofs, can help reduce stormwater runoff and improve water quality. Moreover, protecting natural wetlands and natural floodplains can mitigate flood risks and maintain water balance. Many water bodies, such as rivers and lakes, are shared by multiple countries. International cooperation is crucial for effective water management. Treaties and agreements should be established to ensure equitable access and sustainable use of transboundary water resources. Educating the public about the value of water and the importance of responsible water management is fundamental. This includes teaching water-saving habits in schools, conducting outreach programs, and raising awareness about the impact of climate change on water resources. Encourage research and innovation in water management technologies and practices. Funding for research and development can lead to the discovery of more efficient and sustainable solutions for water-related challenges. Recent advances in research and technology have led to the development of several techniques to purify water that was previously considered unfit for consumption. These techniques include reverse osmosis, electro-dialysis reversal (EDR), desalination, nanofiltration, and solar and UV filtration. This method uses semi-permeable membranes to remove impurities and larger particles from water, making it safe to drink. It can remove various contaminants, including bacteria, and is used in household filtration devices. EDR utilizes direct current to move ions through ion-selective membranes. This method is suitable for purifying water with low suspended solids or high silica content. Desalination is used to purify salty or brackish water by separating dissolved salts and minerals. It is crucial for producing fresh water from seawater and other sources. This method removes two-valent ions and heavy metals from water, often used to soften water and reduce salt content in brackish water. It uses membranes that are less fine than those in reverse osmosis systems, resulting in lower fouling rates. Solar and UV filtration combines solar energy and ultraviolet light to purify water contaminated with biological organisms, making it safe to drink.

Using these techniques is vital in combating water scarcity and providing access to clean and safe water for more people. To further address this issue, organizations take initiatives like Charity Water, Water.org, and others to play a significant role in providing clean water to those needy people [79]. Donations to these organizations can make a substantial difference. Funding goes toward building wells, constructing water treatment facilities, and implementing sustainable water solutions in underserved areas. Educating people about water scarcity, conservation, and importance of supporting water organizations is essential. Donations to these organizations play a crucial role in funding projects that build wells, implement sanitation systems, and offer education on water hygiene and conservation. Develop educational programs in schools and communities to teach about the importance of water conservation, sustainable usage, and the effects of water scarcity. Host community events focused on water conservation, such as water-saving competitions, community clean-ups near water sources, or educational fairs. Organize awareness campaigns through social media, traditional media outlets, and public events to reach a larger audience. Utilize powerful visuals, stories, and statistics to engage and inform people. Work with local governments, NGOs, and businesses to organize workshops, seminars, and collaborative initiatives focused on water conservation and management. Establish community gardens or projects that showcase water-efficient practices like drip irrigation, rainwater harvesting, or native plant gardening. Implementing technologies like rainwater harvesting, smart irrigation systems for recycling rainwater, and household water helps to prevent water scarcity and reduce costs. Collecting rainwater from rooftops or surfaces and storing it in tanks or reservoirs for later use is a simple yet effective technique. This water can be used for various purposes such as irrigation, toilet flushing, or even for drinking after proper treatment. Graywater refers to gently used water from sinks, showers, laundry, and other household sources. Implementing systems that treat and recycle graywater for non-potable purposes like landscape irrigation, flushing toilets, or cleaning can significantly reduce overall water consumption. Installing low-flow faucets, toilets, and water-efficient appliances significantly reduces household water usage, contributing to conservation efforts. Implementing smart irrigation systems that utilize sensors and data to optimize watering schedules based on weather conditions and plant needs helps prevent water wastage in agriculture and landscaping. Agriculture and irrigation are significant contributors to water scarcity. Improved farming techniques like drip irrigation and precision irrigation, soil moisture monitoring, mulching, crop selection and rotation, conservation tillage, and improved irrigation infrastructure are significantly implemented to reduce water usage while maintaining or increasing crop yields. In the drip irrigation system, water is directly injected into the roots of the plants, which helps to reduce water evaporation as well as decrease the water shortage. Applying mulch to the soil surface helps retain moisture, reduces evaporation, and suppresses weed growth, enabling farmers to use less water for irrigation [80–82]. Effective drainage and sewage systems are essential for maintaining clean drinking water. Upgrading these systems can prevent contamination of underground water. Climate change exacerbates water scarcity. Sustainable approaches to renewable energy and water management are essential for addressing this challenge. Effective population control measures can help alleviate this pressure. Transferring water conservation technologies to dry regions in developing countries can aid in combating water scarcity. Holistic approaches that consider economic, cultural, and ecological factors are essential for managing natural resources effectively. These strategies can collectively address water scarcity, ensuring that more people have access to clean and safe water.

Governments at all levels should develop comprehensive water management policies and legislation. These should include regulations, standards, and incentives to promote sustainable water practices and hold those who misuse water accountable. In the field of water management, a significant shift has occurred, especially in Adaptive Water Management (AWM) [83, 84]. Water resource management is increasingly recognized as a political issue, leading to a rebranding of the task as "water governance." AWM, which begins at the grassroots level, is gaining prominence over the dominant top-down, prescriptive approach of Integrated Water Resources Management (IWRM) [85]. Various AWM models have emerged, differing mainly in how power is distributed among stakeholders. While it is often believed that project owners should have the final say for the best results, many funding agencies are reluctant to relinquish control. AWM models have evolved to meet the needs of different donors or funding agencies. The World Bank initiated Adaptive Management experiments in 2012 as part of the "Doing Development Differently" (DDD) program [86, 87]. This was a response to the realization that many development initiatives fail due to oversimplified approaches and ineffective handling of political barriers. The World Development Report 2017 stressed the importance of focusing on how institutions function and challenged some of the Bank's core beliefs, such as the significance of international tendering and the concept of "Best Practices" [88]. John Dewey's theory of social change, which emphasizes the role of those experiencing problems in initiating change, aligns with these ideas [89–91]. He proposed that cultural evolution results from ongoing testing of gradual modifications. Adaptive planning, although time-consuming, is still quicker than the integration of new approaches within institutional layers, which can take 20 years or more, according to Dutch scholars. The COVID-19 pandemic has shown that under pressing circumstances, transformations can happen more rapidly.

10 Conclusion

In conclusion, the sustainable management of water resources is a global imperative. By adopting these recommendations, we can work toward a future where water is abundant, clean, and accessible for all. It is not only our responsibility but also our necessity to preserve this precious resource for generations to come. The conclusion summarizes the key findings of the review and highlights the urgent need for concerted efforts to address water scarcity in the face of climate change. It emphasizes the importance of adopting sustainable water management practices to achieve SDG-2030 targets and ensure a water-secure future.

References

1. Rijsberman, F. R. (2006). Water scarcity: Fact or fiction? *Agricultural Water Management, 80*(1–3), 5–22.
2. Tzanakakis, V. A., Paranychianakis, N. V., & Angelakis, A. N. (2020). Water supply and water scarcity. *Water, 12*(9), 2347.
3. Kalenyuk, I., & Uninets, I. (2020). Factors and features of the smart-economy development. *Economics & Education, 5*(2), 42–47.
4. Su, Q., & Karthikeyan, R. (2023). Regional water stress forecasting: Effects of climate change, socioeconomic development, and irrigated agriculture—A Texas case study. *Sustainability, 15*(12), 9290.

5. Gude, V. G. (2016). Desalination and sustainability–an appraisal and current perspective. *Water Research, 89*, 87–106.
6. Hassan, M. S., Hed, N. M., & Kamilan, I. H. (2022). Parliamentary reforms and sustainable development goals (SDG): The way forward for an inclusive and sustainable parliament. *The Journal of Legislative Studies, 28*(4), 578–605.
7. Maftouh, A., El Fatni, O., Fayiah, M., Liew, R. K., Lam, S. S., Bahaj, T., & Butt, M. H. (2022). The application of water–energy nexus in the Middle East and North Africa (MENA) region: A structured review. *Applied Water Science, 12*(5), 83.
8. Sahu, P., & Debsarma, C. (2023). Urban water scarcity: A global challenge and impending solutions. In *Impacts of urbanization on hydrological Systems in India* (pp. 129–142). Springer International Publishing.
9. Mallanna, C., & Kolhapure, D. A. (2020). Drinking water supply status in habitations with population coverage—A geographical study. *Research Journey, 153*.
10. Cordes, C., & Miller, E. (2000). *Fool's gold: A critical look at computers in childhood.*
11. Fazeli-Dinan, M., Azarnoosh, M., Özgökçe, M. S., Chi, H., Hosseini-Vasoukolaei, N., Haghi, F. M., Zazouli, M. A., et al. (2022). Global water quality changes posing threat of increasing infectious diseases, a case study on malaria vector Anopheles stephensi coping with the water pollutants using age-stage, two-sex life table method. *Malaria Journal, 21*(1), 178.
12. Malavige, G. N., Sjö, P., Singh, K., Piedagnel, J.-M., Mowbray, C., Estani, S., Lim, S. C. L., et al. (2023). "facing the escalating burden of dengue: Challenges and perspectives." PLOS. *Global Public Health, 3*(12), e0002598.
13. Huzortey, A. A., Kudom, A. A., Mensah, B. A., Sefa-Ntiri, B., Anderson, B., & Akyea, A. (2022). Water quality assessment in mosquito breeding habitats based on dissolved organic matter and chlorophyll measurements by laser-induced fluorescence spectroscopy. *PLoS One, 17*(7), e0252248.
14. Dikshit, K. R., Dikshit, J. K., Dikshit, K. R., & Dikshit, J. K. (2014). Weather and climate of north-East India. In *North-East India: Land, people and economy* (pp. 149–173).
15. Jain, S. K., Agarwal, P. K., Singh, V. P., Jain, S. K., Agarwal, P. K., & Singh, V. P. (2007). Physical environment of India. In *Hydrology and water resources of India* (pp. 3–62).
16. Kumar, Shiv, Indira Khurana, and P. R. K. Sobhanbabu. "Scoping study: India's global resource footprint in food, energy and water (FEW)." (2014).
17. Rode, S., Saraf, R., Veigas, J., Shetty, N., & Shardul, S. (2023). Design and fabrication of IoT based agricultural automation system. In *2023 5th biennial international conference on nascent Technologies in Engineering (ICNTE)* (pp. 1–6). IEEE.
18. Kishore, P., Singh, D. R., Chand, P., & Prakash, P. (2020). What determines groundwater depletion in India? A meso level panel analysis. *Journal of Soil and Water Conservation, 19*(4), 388–397.
19. Jain, R., Kishore, P., & Singh, D. K. (2019). Irrigation in India: Status, challenges and options. *Journal of Soil and Water Conservation, 18*(4), 354–363.
20. Lal, R. (2006). Land area for establishing biofuel plantations. *Energy for Sustainable Development, 10*(2), 67–79.
21. Micklin, P. P. (1991). The Water Management Crisis in Soviet Central Asia. *Technology, 4*, 6.
22. Mutani, G., & Todeschi, V. (2020). The effects of green roofs on outdoor thermal comfort, urban heat Island mitigation and energy savings. *Atmosphere, 11*(2), 123.
23. Vlachos, E., & Braga, B. (2001). The challenge of urban water management. In *Frontiers in urban water management: Deadlock or hope* (pp. 1–36). IWA Publishing.
24. Greve, P., Kahil, T., Mochizuki, J., Schinko, T., Yusuke Satoh, P., Burek, G. F., et al. (2018). Global assessment of water challenges under uncertainty in water scarcity projections. *Nature Sustainability, 1*(9), 486–494.
25. Gimelli, F. (2021). *Envisioning a capability approach to water service development in urban informal settlements*. PhD diss., Monash University.
26. Button, C. (2014). *Domesticating infrastructure: Mumbai's middle-class housing and rainwater harvesting*. PhD diss., Durham University.

27. Satterthwaite, D., Archer, D., Colenbrander, S., Dodman, D., Hardoy, J., & Patel, S. (2018). Responding to climate change in cities and in their informal settlements and economies. *International Institute for Environment and Development, Edmonton, Canada, 61*.
28. Aalst, V., Maarten, K., Cannon, T., & Burton, I. (2008). Community level adaptation to climate change: The potential role of participatory community risk assessment. *Global Environmental Change, 18*(1), 165–179.
29. Mechler, R., S. Schindler, N. Hanke, M. Högl, and M. Siebert. "Assessment of climate-related risks. A 6-step methodology." (2021).
30. Wilby, R. L., Xianfu, L., Watkiss, P., & Rodgers, C. A. (2021). Towards pragmatism in climate risk analysis and adaptation. *Water Policy, 23*(S1), 10–30.
31. Dixon, H., Sandström, S., Cudennec, C., Lins, H. F., Abrate, T., Bérod, D., Chernov, I., Ravalitera, N., Sighomnou, D., & Teichert, F. (2022). Intergovernmental cooperation for hydrometry–what, why and how? *Hydrological Sciences Journal, 67*(16), 2552–2566.
32. Yihdego, Y., & Eslamian, S. (2017). Drought management: Initiatives and objectives. In *Handbook of drought and water scarcity* (pp. 1–25). CRC Press.
33. Anderson, J., Baggett, S., Jeffrey, P., McPherson, L., Marks, J., & Rosenblum, E. (2008). Public acceptance of water reuse. In *Water reuse: An international survey of current practice, issues and needs* (pp. 332–350).
34. Jato-Espino, D., & Pathak, S. (2021). Geographic location system for identifying urban road sections sensitive to runoff accumulation. *Hydrology, 8*(2), 72.
35. Ma, X., Wu, D., & Zhang, S. (2018). Multiple goals dilemma of residential water pricing policy reform: Increasing block tariffs or a uniform tariff with rebate? *Sustainability, 10*(10), 3526.
36. Calverley, C. M., & Walther, S. C. (2022). Drought, water management, and social equity: Analyzing Cape Town, South Africa's water crisis. *Frontiers in Water, 4*, 910149.
37. Schulz, C. E., & Jansen, A. (2006). Water demand and the urban poor. In *A study of the factors influencing water consumption among households in Cape Town, South Africa*.
38. Scott, C. A., & Pasqualetti, M. J. (2010). Energy and water resources scarcity: Critical infrastructure for growth and economic development in Arizona and Sonora. *Natural Resources Journal, 50*, 645.
39. Ward, F. A., Booker, J. F., & Michelsen, A. M. (2006). Integrated economic, hydrologic, and institutional analysis of policy responses to mitigate drought impacts in Rio Grande Basin. *Journal of Water Resources Planning and Management, 132*(6), 488–502.
40. Donoso, G. (2017). Urban water pricing in Chile: Cost recovery, affordability, and water conservation. *Wiley Interdisciplinary Reviews: Water, 4*(2), e1194.
41. Kim, H. M., Sabri, S., & Kent, A. (Eds.). (2020). *Smart cities for technological and social innovation: Case studies, current trends, and future steps*. Academic Press.
42. Barber, B. R. (2017). *Cool cities: Urban sovereignty and the fix for global warming*. Yale University Press.
43. Powell, M. (Ed.). (2022). *The climate city*. Wiley.
44. Park, J., Cuervo-Cazurra, A., & Montiel, I. (2023). The evolution of sustainability concerns over business activities: From local to cross-national to global. In *Research handbook on international corporate social responsibility* (pp. 7–24). Edward Elgar Publishing.
45. Powers, M. (2024). *A livable planet: Human rights in the global economy*. Oxford University Press.
46. Candy, D. (2022). The fragility of our planet. *Alternative Law Journal, 47*(3), 166.
47. Pedersen, C. S. (2018). The UN sustainable development goals (SDGs) are a great gift to business! *Procedia CIRP, 69*, 21–24.
48. Wilkinson, R., & Hulme, D. (Eds.). (2012). *The millennium development goals and beyond: Global development after 2015* (Vol. 65). Routledge.
49. van der Straaten, J., Balagopal, G., & UNICEF. (2013). A passport to protection: A guide to birth registration programming. In *United Nations children's fund, A passport to protection: A guide to birth registration programming*. UNICEF.

50. Gore, C. (2015). The post-2015 moment: Towards sustainable development goals and a new global development paradigm. *Journal of International Development, 27*(6), 717–732.
51. Filho, L., Walter, U. A., Alves, F., Pace, P., Mifsud, M., Brandli, L., Caeiro, S. S., & Disterheft, A. (2018). Reinvigorating the sustainable development research agenda: The role of the sustainable development goals (SDG). *International Journal of Sustainable Development & World Ecology, 25*(2), 131–142.
52. Akhmouch, A., & Clavreul, D. (2017). Stakeholder engagement for inclusive water governance. In *Routledge handbook of water law and policy* (pp. 162–175). Routledge.
53. Widmer, A. M. (2014). *The governance of climate change adaption in Switzerland: Issues, actors, and processes at the national and cantonal level and in land-use relevant policies*. PhD diss., ETH Zurich.
54. Özerol, Gül. "Aligning the multiplicities in natural resource governance: A study on the governance of water and land resources in irrigated agriculture." (2013).
55. Muniz, B. N. B. (2020). *Is the 0.7% goal of ODA/GNI still adequate?* Master's thesis, Universidade de Lisboa (Portugal).
56. Stokke, O., & Stokke, O. (2019). The follow-up on the 0.7% target: A bird's-eye perspective. *International Development Assistance: Policy Drivers and Performance*, 41–52.
57. Mishra, R. K., & Janaki Krishna, P. S. (2021). Capacity building and technical cooperation for achieving sustainable development goals. In *Peace, justice and strong institutions* (pp. 33–43). Springer International Publishing.
58. Fitzgerald, Oonagh E., Mira Burri, Chios Carmody, Dan Ciuriak, Aaron Cosbey, Thomas Cottier, Henry Gao et al. "Modernizing the World Trade Organization." (2020).
59. Desai, S., Manapragada, N. V. S. K., Shukla, A. K., & Pignatta, G. (2022). Mould-growth study in building materials exposed to warm and humid climate using heat and mass transfer (HAMT). *EnergyPlus Simulation Method. Sustainability, 14*(14), 8292-1-28. https://doi.org/10.3390/su14148292
60. Chakraborty, L. (2020). *Macroeconomic policy coherence for SDG 2030: Evidence from Asia Pacific*. No. 20/292.
61. Lencucha, R., Kulenova, A., & Thow, A. M. (2023). Framing policy objectives in the sustainable development goals: Hierarchy, balance, or transformation? *Globalization and Health, 19*(1), 5.
62. Shi, L., Ahmad, S., Shukla, P., & Yupho, S. (2021). Shared injustice, splintered solidarity: Water governance across urban-rural divides. *Global Environmental Change, 70*, 102354.
63. Fletcher, A. J., & Schonewille, R. (2015). Overview of resources on gender-sensitive data related to water. WWAP Gender Series 4. Paris: UNESCO.
64. Pangare, V., & UNESCO. (2015). *Guidelines on how to collect sex-disaggregated water data*. UNESCO.
65. Shinde, V. R., Mishra, R. R., Bhonde, U., & Vaidya, H. (Eds.). (2023). *Managing urban rivers: From planning to practice*. Elsevier.
66. Islam, S. N. (2020). *Rivers and sustainable development: Alternative approaches and their implications*. Oxford University Press.
67. Pathak, S., Sharma, S., Banerjee, A., & Kumar, S. (2024). A methodology to assess and evaluate sites with high potential for stormwater harvesting in Dehradun, India. *Big Data Research, 35*, 100415.
68. Döös, K. (1995). Interocean exchange of water masses. *Journal of Geophysical Research: Oceans, 100*(C7), 13499–13514.
69. Cleland, J. (2013). World population growth; past, present and future. *Environmental and Resource Economics, 55*, 543–554.
70. Lam, D. (2011). How the world survived the population bomb: Lessons from 50 years of extraordinary demographic history. *Demography, 48*(4), 1231–1262.
71. du Plessis, A. (2023). Water resources from a global perspective. In *South Africa's water predicament: Freshwater's unceasing decline* (pp. 1–25). Springer International Publishing.
72. Giampietro, M., & Pimentel, D. (1993). *The tightening conflict: Population, energy use, and the ecology of agriculture*. Negative Population Growth, Incorporated.

73. Chakraborti, R. K., Kaur, J., & Kaur, H. (2019). Water shortage challenges and a way forward in India. *Journal-American Water Works Association, 111*(5), 42–49.
74. Gupta, A., Mishra, S., Bokde, N., & Kulat, K. (2016). Need of smart water systems in India. *International Journal of Applied Engineering Research, 11*(4), 2216–2223.
75. Gupta, A., & Shukla, A. K. (2024). Optimal approaches in global warming mitigation and adaptation strategies at city scale. *Discover Sustainability, 5*, 272. https://doi.org/10.1007/s43621-024-00497-8
76. Gogate, N. G., & Rawal, P. M. (2012). Sustainable stormwater management in developing and developed countries: A review. In *International conference on advances in design and construction of structures (ADCS 2012), Bangalore*, India.
77. Shukla, S., Deeksha, Chand, S., Rai, P. K., Kumar, S. S., & Shukla, A. K. (2024). Measuring sustainable development of cities using remote sensing and geospatial technologies: A review. In P. K. Rai, A. Singh, B. Beingachhi, & S. K. Singh (Eds.), *Natural resources management and sustainable livelihoods in the mountainous region* (Advances in geographical and environmental sciences). Springer. https://doi.org/10.1007/978-981-97-2100-9_3
78. Rani, S., Mishra, R. K., Usman, M., Kataria, A., Kumar, P., Bhambri, P., & Mishra, A. K. (2021). Amalgamation of advanced technologies for sustainable development of smart city environment: A review. *IEEE Access, 9*, 150060–150087.
79. Waterhouse, K. B. (2018). Acknowledging the good work of water-aid organizations. *Journal-American Water Works Association, 110*(9), 68–72.
80. El-Beltagi, H. S., Basit, A., Mohamed, H. I., Ali, I., Ullah, S., Kamel, E. A. R., Shalaby, T. A., Ramadan, K. M. A., Alkhateeb, A. A., & Ghazzawy, H. S. (2022). Mulching as a sustainable water and soil saving practice in agriculture: A review. *Agronomy, 12*(8), 1881.
81. Prem, M., Ranjan, P., Seth, N., & Patle, G. T. (2020). Mulching techniques to conserve the soil water and advance the crop production—A review. *Current World Environment: An International Research Journal of Environmental Sciences, 15*, 10–30.
82. Demo, A. H., & Bogale, G. A. (2024). Enhancing crop yield and conserving soil moisture through mulching practices in dryland agriculture. *Frontiers in Agronomy, 6*, 1361697.
83. Hjorth, P., & Madani, K. (2023). Adaptive water management: On the need for using the post-WWII science in water governance. *Water Resources Management, 37*(6), 2247–2270.
84. der Brugge, V., & Rutger, and Roel Van Raak. (2007). Facing the adaptive management challenge: Insights from transition management. *Ecology and Society, 12*(2).
85. Sharma, S., Pathak, S., & Kumar, S. (2024). A planning-support tool for spatial suitability assessment of harvesting sites for stormwater infrastructure. *Journal of Hazardous, Toxic, and Radioactive Waste ASCE, 28*(1), 04023044.
86. Pellini, A., Karetji, P. C., & Soekadis, A. (2018). Doing development differently at scale. In *Knowledge, politics and policymaking in Indonesia* (pp. 131–146).
87. Rogers, P., & Macfarlan, A. (2020). What is adaptive management and how does it work? In *Better evaluation: Monitoring and evaluation for adaptive management, working paper 2.*
88. Auld, E., Rappleye, J., & Morris, P. (2019). PISA for development: How the OECD and World Bank shaped education governance post-2015. *Comparative Education, 55*(2), 197–219.
89. VanWynsberghe, R., & Herman, A. C. (2015). Education for social change and pragmatist theory: Five features of educative environments designed for social change. *International Journal of Lifelong Education, 34*(3), 268–283.
90. Pedwell, C. (2017). Transforming habit: Revolution, routine and social change. *Cultural Studies, 31*(1), 93–120.
91. Boyles, D. (2020). *John Dewey's imaginative vision of teaching: Combining theory and practice* (pp. 1–183).

Advanced Geotechnical Analysis and Mitigation of Slope Stability in Dharchula, Pithoragarh, Uttarakhand, Using GeoStudio Software

Gaurav Bharti, Parikshit Hurukadli, Bishnu Kant Shukla, Amit Tripathi, Pushpendra Kumar Sharma, Ashish Singh, Gyanendra Singh, Harsh Patel, and Kunal Nigam

1 Introduction

Dharchula, situated in the Pithoragarh district of Uttarakhand, is a region renowned for its cultural significance and stunning natural surroundings. Beneath the attractive appearance, however, are major geological issues such as landslides and slope instability. The rugged topography of Dharchula, which is characterised by steep slopes and intricate geographical formations, naturally threatens the region's stability. Extreme rainfall, seismic activity, geological faults, and human activities all contribute to the region's increased susceptibility to landslides. In addition to causing property damage and casualties, landslides in Dharchula have also endangered local residents' means of subsistence and damaged vital infrastructure. The frequency and intensity of these occurrences highlight how urgently a thorough understanding of slope stability dynamics in the area is needed. The purpose of this research is to examine the complex interactions between anthropogenic, geomorphological, and geological elements that lead to landslides and slope instability in Dharchula.

The primary aim of the present work is to identify the fundamental mechanisms causing slope failure in this susceptible location by looking at the geological

G. Bharti · P. Hurukadli · A. Tripathi · A. Singh · G. Singh · H. Patel · K. Nigam
Department of Civil Engineering, JSS Academy of Technical Education, Noida, Uttar Pradesh, India

B. K. Shukla (✉)
Department of Civil Engineering, JSS Academy of Technical Education, Noida, Uttar Pradesh, India

School of Civil Engineering, Lovely Professional University, Phagwara, Punjab, India

P. K. Sharma
School of Civil Engineering, Lovely Professional University, Phagwara, Punjab, India

S. Pathak et al. (eds.), *Intelligent Infrastructure and Smart Materials*,
https://doi.org/10.1007/978-3-031-92421-7_14

environment, rainfall patterns, land use practices, and past landslide occurrences. Additionally, this study aims to provide methods for reducing the dangers related to landslides in Dharchula. The main purpose is to promote sustainability and resilience in the face of natural hazards by combining hazard mapping, policy interventions, community engagement, and geotechnical evaluations. By doing this, the research hopes to add to the body of knowledge on landslide mitigation and slope stability, providing insights that could guide future planning and development projects in Dharchula and other mountainous areas across the world. Particularly in hilly areas like the Himalayas, stability caused by landslides has a substantial impact on ground subsidence, which can seriously threaten nearby residents and infrastructure.

It is a complicated phenomenon that is impacted by the previously listed geological, geomorphological, and hydrological elements. Important factors include the hydrological conditions, slope features, weathering, and the differential strength of the rock mass. In this instance, the research is located close to Dharchula Village and the Tawaghat Bridge on Kailash Mansarovar Road. For the purpose of creating efficient mitigation methods, it is vital to comprehend these elements. Several actions can be taken to reduce the dangers related to ground sinking in the research area: Risk Evaluation and Tracking: Conduct a thorough risk assessment to ascertain the issue's extent and any potential dangers. Install monitoring devices to detect ground movement and changes in the landscape.

In order to reduce the risks of ground subsidence in the study area, it is crucial to carry out thorough geotechnical studies [1–3]. These studies will assess the stability of slopes and identify areas that are prone to sinkholes. This information will then be used to make informed decisions about land use planning and infrastructure development [4]. Additionally, it is important to manage hydrological systems in order to minimise water infiltration and decrease slope instability. Raising awareness and preparedness within the community is also essential. This can be achieved by educating residents about early warning systems and the associated risks, as well as promoting disaster preparedness and response initiatives. Lastly, retrofitting existing infrastructure is necessary to improve its resilience against ground movement, ensuring the long-term sustainability and safety of the region.

1.1 Objective of the Research

The purpose of this research is to use various techniques to provide a thorough assessment of the landslide near the Tawaghat Bridge on Kailash Mansarovar Road, close to Dharchula Village in Pithoragarh District, Uttarakhand. The goal of this research is to make a significant contribution to the field of geotechnical engineering by offering insights that are relevant to the safety and sustainable development of Uttarakhand's mountainous terrain. Through the pursuit of these goals, the study hopes to improve resistant people's quality of life and enable them to make well-informed decisions.

- To ascertain chemical and physical characteristics, such as bulk density and specific gravity, as well as texture, colour, and particle size.
- To find out shear parameter of soil.
- To integrate it with different geotechnical software such as GeoStudio, QGIS.
- Compare the factor of safety (FS) for different sites in the region.

1.2 Types of Slope Stability Methods

For the purpose of determining the safety and stability of slopes and for the purpose of implementing appropriate mitigation measures, slope stability analysis is necessary. Evaluation of slope stability can be accomplished using a variety of approaches, each of which has its own set of guiding principles, benefits, and drawbacks. The following are the major methods: limit equilibrium method (LEM), numerical modelling, empirical methods, and probabilistic methods. Field monitoring and geotechnical investigations are also included.

1.2.1 Method of Limit Equilibrium

The LEM is predicated on the idea that a slope would collapse when the driving forces (caused by gravity) are greater than the resisting forces (caused by the strength of the soil). An evaluation of the FS is carried out using this technique by contrasting the driving forces with the resisting forces. Stability is indicated by an FS that is more than 1, whereas the possibility of failure is indicated by an FS that is less than 1. When using the slice methods, the slope is cut into vertical slices, and the equilibrium of forces and moments is determined for each individual slice. Methods such as the Bishop Simplified method, the Janbu method, and the Morgenstern–Price method are examples of traditional slice methods. In the Wedge Analysis, it is utilised for rock slopes in which it is presumed that the failure mechanism would take place along flat surfaces that are arranged in the shape of a wedge.

1.2.2 Modelling Using Numerical Data

The process of solving the governing equations of equilibrium, compatibility, and constitutive behaviour of soil through the use of computational tools is what is known as numerical modelling. The behaviour of soil and rock masses under a variety of stress circumstances is modelled and simulated by these instruments. When using the finite element method (FEM), the slope is first divided into a mesh of elements, and then the equations of equilibrium for each element are solved simultaneously. This takes into consideration intricate geometries as well as a wide range of material characteristics. Discrete equations are discretised using a different method in the finite difference method (FDM), which is comparable to the FEM. The discrete element method (DEM) is a technique that simulates the behaviour of particular particles or blocks, as well as the relationships between them.

1.2.3 Empirical Method

It is made in the field and previous case studies are used to create correlations and associations, which are the foundation of empirical methodology. In most cases, these techniques are utilised for providing preliminary evaluations and for places that have access to a substantial amount of historical data. According to factors such as uniaxial compressive strength, rock grade designation, and joint conditions, Rock Mass Rating (RMR) is a classification system that is used to categorise rock masses. Slope Mass Rating (SMR) is an extension of RMR that was developed expressly for slopes and takes into consideration additional aspects such as slope orientation and weathering.

1.2.4 The Probabilistic Methods

In order to take into consideration, the inherent uncertainties that are present in soil parameters, geometry, and loading conditions, probabilistic approaches are utilised. Rather than presenting a single FS, these techniques examine the likelihood of failure occurring. Calculating the chance of failure is accomplished by the use of Monte Carlo Simulation, which generates different scenarios of soil parameters and loading situations through the use of random sampling. The first-order second-moment (FOSM) method is used to compute the mean and variance of the FS by utilising the statistical features of the parameters that are supplied.

1.2.5 An Investigation of the Geotechnical Situation and Field Monitoring

Investigations and monitoring in the field provide data that is vital for comprehending the circumstances that exist on a slope in its natural state. When doing an appropriate examination of slope stability, this information is absolutely necessary. Information on the stratigraphy, strength, and other qualities of the soil may be obtained through the process of borehole drilling and sampling [2]. For the purpose of determining the characteristics of the soil, in situ testing consists of shear vane tests, cone penetration tests (CPT), and standard penetration tests (SPT) [3]. Real-time monitoring of slope movements, pore water pressures, and other data is accomplished through the use of inclinometers, piezometers, and other sensors by the Instrumentation and Monitoring department.

It is appropriate for preliminary designs and a wide range of soil conditions because of its simplicity and extensive application. On the other hand, the LEM assumes preset failure surfaces, which may not always be true, and it does not take into consideration progressive failure mechanisms. Numerical modelling, such as the FEM, offers precise insights into the behaviour and geometry of complex soils; nevertheless, it is computationally demanding, takes a substantial amount of knowledge, and requires a large amount of data. In spite of the fact that empirical approaches are less precise and have limited application outside the settings under which correlations were generated, they are suitable for early evaluations since they are quick and easy to apply and require just a small amount of data. Probabilistic approaches take into account variability and uncertainty, providing a measure of failure likelihood that is useful for risk assessment. However, when compared with deterministic methods, probabilistic methods require more precise statistical data, are more complicated, and require more people and resources to implement. Despite the fact that field monitoring and geotechnical investigations provide reliable, site-specific data that is essential for validating models and devising mitigation measures, these methods are frequently time-consuming, expensive, and need specialised staff for appropriate data collection and interpretation.

In order to determine which method of slope stability analysis is the most suitable choice, several factors are taken into consideration [5]. These factors include the specific features of the slope, the level of detail that is required, and the resources that are at one's disposal. Every approach comes with its own mix of benefits and drawbacks. It is usual practice to apply a combination of diverse techniques when it comes to acquiring a comprehensive understanding of slope stability and devising suitable mitigation measures. This is because it is the most effective way to achieve these goals.

The future scope will be developing Real-Time Monitoring which helps us to establish real-time monitoring systems using sensors to measure soil moisture, pore water pressure, and ground movement. These can provide early warning signals for potential slope failures. These systems can be integrated with IoT (Internet of Things) technologies to transmit data for immediate analysis. The development of predictive models using machine learning and artificial intelligence can help forecast landslides by analysing historical data, weather patterns, and real-time monitoring data. The future scope of slope stability analysis in the Dharchula region of Pithoragarh, Uttarakhand, is vast and multifaceted. By leveraging advanced technologies, engaging communities, and promoting sustainable practices, it is possible to mitigate landslide risks and enhance the safety and resilience of the region. Continued research and collaboration will be key to addressing the complex challenges associated with slope stability in this geologically sensitive area.

2 Study Area

A little distance away from the Tawaghat Bridge, in the Dharchula, Pithoragarh district of Uttarakhand State, India, is where the study area may be found. This encompasses a section of the Kailash Mansarovar Road that is 25 km in length, specifically from Malpa to Budhi. Pithoragarh is situated on a ridge on the eastern side of the Nepal highlands, and it is situated at a distance of 600 km (by NH-9) from the capital of the nation, New Delhi. Lamaree and Pangla are the villages that are found along this road. They are the settlements that are seldom inhabited and are located on the left bank of the Kali River. Dharchula is a town and municipality that can be found in the Pithoragarh district of the state of Uttarakhand in India. This location is situated on the banks of the Kali River, which serves as a border between India and Nepal. The significance of Dharchula's position as a commercial centre and a gateway to the trans-Himalayan area is closely linked to the time period in which it was established. The purpose was to determine the chemical and physical characteristics, such as bulk density and specific gravity, in addition to the texture, colour, and particle size of the substance. When it comes to the likelihood of landslides occurring in the region, the terrain has a significant role. The untamed Himalayan area, which is marked by the severity of precipitation, brittle geological formations, and steep inclines, is where Dharchula is located.

Figure 1 comprises a series of maps depicting the geographical context and specific study sites in the Dharchula region of the Pithoragarh district, Uttarakhand, India. The maps are arranged in a hierarchical manner, zooming in from the national level to the specific locations of interest within Dharchula. The top left map shows India, highlighting Uttarakhand state in red. This provides a broad geographical

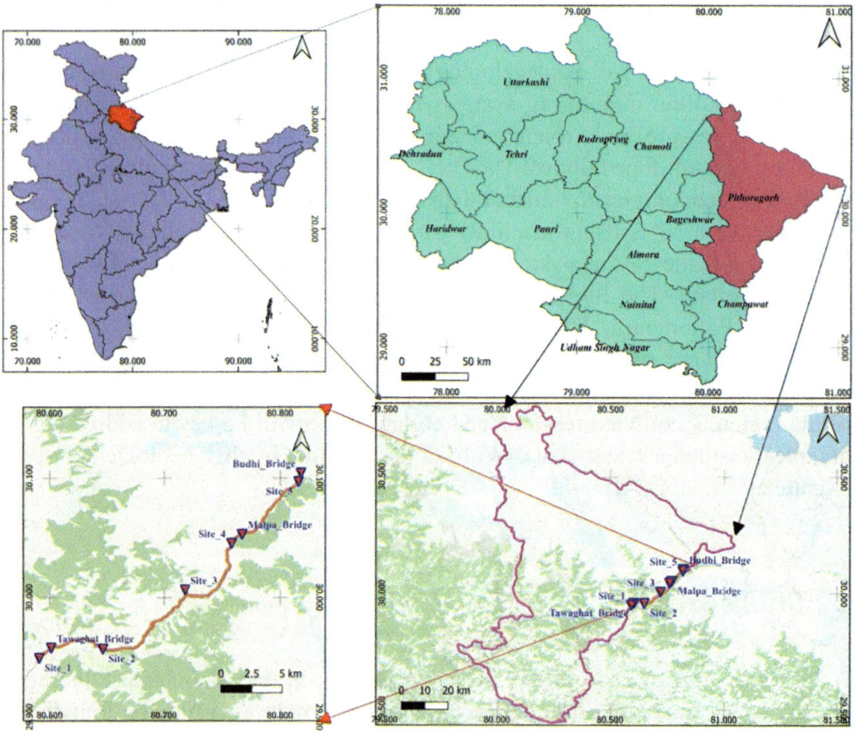

Fig. 1 Google Earth map of Dharchula with locations 1–5. (Source: USGS)

context, indicating where Uttarakhand is located within India. The top centre map zooms in on Uttarakhand, showing its administrative divisions. Pithoragarh district is highlighted, indicating its position within the state. This map provides a clearer view of the district's location relative to other districts in Uttarakhand. The top right map focuses on the Pithoragarh district, specifically highlighting the Dharchula region. This map helps to further narrow down the area of interest within the district, showcasing its position and the administrative boundaries. The bottom right map shows a more detailed view of the Dharchula region, with a focus on the road network and key geographical features. It highlights the study sites (sites 1–5) along with major landmarks such as bridges (Tawaghat Bridge, Malpa Bridge, and Budhi Bridge). The bottom left map provides a detailed view of the specific study sites (sites 1–5) within Dharchula. The sites are marked along a road, showing their relative positions and proximity to each other. The map also indicates nearby landmarks such as the Tawaghat, Malpa, and Budhi bridges, providing reference points for the locations of the study sites.

Sites 1–5 are specific locations within Dharchula where detailed studies or measurements are conducted. They are marked with blue triangles on the map. Tawaghat Bridge, Malpa Bridge, and Budhi Bridge are significant landmarks that help in identifying the exact locations of the study sites. The image is sourced from the US Geological Survey (USGS), indicating that the geographical data and maps are derived from reputable and reliable sources. This hierarchical series of maps provides a comprehensive geographical context, starting from the national level and zooming into specific study sites within Dharchula. It is useful for understanding the location, spatial distribution, and geographical context of the study sites, which is essential for field investigations and analysis related to slope stability, geological studies, or environmental assessments in the region. Table 1 shows a list of sites along with their corresponding geographical coordinates. (latitude and longitude), labelled from S1 to S5. This column lists the names or identifiers of the sites. In this case, the sites are labelled from S1 to S5. This column lists the geographical coordinates (latitude and longitude) of each site.

The study area is an active landslide-prone area. It has previous history of life-taking landslides. On 8 October 2023, at one of these sites, a massive landslide occurred in which a Bolero car was crushed, as shown in Fig. 2. At the end of November, we were fortunate to visit the site and collect soil samples. Some articles about the incident are added below. It is known for its steep hills and receives a significant amount of rainfall, especially during the monsoon season. The climate of the region has a considerable impact on the development and characteristics of the

Table 1 Sites' coordinates

Site No.	Coordinates
S1	29.959120, 80.607103
S2	29.956460, 80.648224
S3	30.049511, 80.755127
S4	30.055365, 80.776051
S5	30.090445, 80.805731

Fig. 2 Car damaged by landslide in the region

soil, since abundant rainfall leads to soil erosion and leaching. Figure 3 shows the critical site in the region.

2.1 Geology of Study Locations

Situated within the Kumaon Himalayas, Dharchula, is a component of the greater Himalayan Mountain range. The South Tibetan Detachment System (STDS) and the Main Central Thrust (MCT), two prominent tectonic features that are close to the area and have greatly impacted its geology, are what define it. Rock formations including metamorphic, sedimentary, and igneous rocks are among the many types of rocks found in the area. Among the most notable are the following:

Metamorphic rocks are mostly from the Lesser Himalayan Sequence, schists, gneisses, and quartzites. Dolomites and limestones from the Tethyan Himalayan Sequence are examples of sedimentary rocks. Igneous rocks are mostly from areas that have intrusive gabbros and granites. Dharchula's geological history is characterised by a number of noteworthy occurrences. Sequences of sediments were deposited in a marine environment during the Paleozoic to Mesozoic Eras. The Cenozoic Era is defined by the tectonic activity that resulted from the collision of Asia and India, which raised and formed the Himalayas. The Seismic activity in the Dharchula region is prone to seismic activity because of its geological location. The region's earthquakes are caused by the continuous tectonic activity along the Himalayan thrusts. Figure 4 shows the elevation map of the Dharchula region.

Fig. 3 Critical site locations for landslides

Geomorphology: The area is characterised by rough topography with river valleys, deep gorges, and steep hills. The Kali River, which passes through Dharchula, has a significant influence on the local environment. Comprehending the geology of Dharchula is imperative for the advancement of infrastructure, evaluation of natural hazards, and sustainable resource management within the area [5].

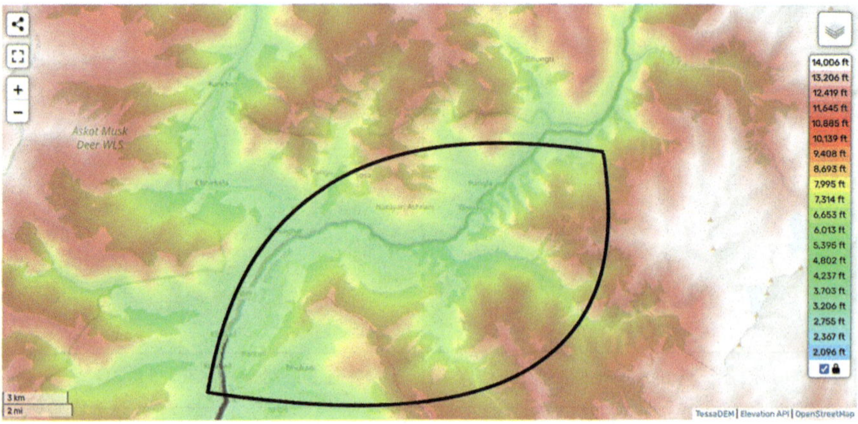

Fig. 4 Elevation map of Dharchula. (Source: USGS)

3 Literature Survey

The stability of slopes is a crucial concern in the Himalayan areas, since landslides and slope collapses can cause catastrophic consequences for populations and infrastructure. The Pithoragarh area in Uttarakhand, India, is very susceptible to such occurrences because of its intricate geology, steep inclines, and substantial human activities. This study consolidates studies on slope stability in the Pithoragarh region, with a specific emphasis on the geological and geotechnical properties, evaluation techniques, and strategies for reducing risks. The intricate geological characteristics of the Pithoragarh area are a major contributing factor to its slope stability problems. Bansal and Sarkar [5] introduced an innovative technique that integrates limit equilibrium techniques with machine learning ensembles to forecast slope stability in Kalimpong. This approach may be modified for application in the Pithoragarh region. Their research showcased the capacity of hybrid models to enhance the precision of slope stability forecasts. Remote sensing and Geographic Information System (GIS) techniques have been extensively employed for evaluating large-scale areas. The region's vulnerability to ground subsidence has also been acknowledged in their study. Ram et al. [6] examined the phenomenon of ground sinking at Dar village, located in the Darma Valley, and identified it as a potential impending tragedy. Their discoveries emphasise the necessity for ongoing surveillance and evaluation of ground stability in Pithoragarh. Solanki et al. [7] thoroughly examined the geological and geotechnical properties of the Khotila landslip in the Dharchula region. They emphasised the significance of heavily worn and fractured rock formations in causing slope instability. The study highlighted the need for comprehending the specific geological conditions in a given area in order to accurately forecast and efficiently control the hazards associated with landslides.

Various procedures have been used to evaluate slope stability in the area, including both conventional LEMs and sophisticated machine learning techniques. Gupta et al. [8, 9], in their research, performed a landslide susceptibility evaluation for Uttarakhand, employing GIS technologies to delineate and forecast locations at high risk of landslides. These methods allow for the monitoring of a vast area and can assist in the creation of successful plans to reduce the impact of landslides. Haigh et al. [10] highlighted the significance of environmental indicators in monitoring landslip activity. They recommended ongoing environmental evaluation and the implementation of bioengineering approaches to naturally improve slope stability. Additionally, numerical modelling approaches, such as the FEM have been utilised. Hearn and Shakya [11] conducted a study on the challenges of creating durable road access in the Himalayas. They highlighted the importance of integrating community-based approaches to effectively manage the risks posed by natural catastrophes. Verma et al. [12] utilised a combination of FEM and artificial neural networks (ANN) to forecast slope instability. Their study demonstrated favourable outcomes in terms of precision and dependability.

Reliability analysis of slope stability has utilised advanced simulation approaches, such as Monte Carlo simulations. Wang et al. [13] showcased the utilisation of these methodologies in real-world dependability research, which can be customised to suit the specific circumstances in Pithoragarh. Wang and Lin [14] utilised fuzzy statistics within the FEM to examine the stability of slopes. Wei et al. [15, 16] evaluated the efficacy of hybrid models, specifically WOA-SVR and HHO-SVR, in predicting the FS for circular failure slopes. Both structural and non-structural solutions are employed to mitigate slope stability challenges in the Pithoragarh region. Xie et al. [17] constructed a predictive model for the stability of rocky slopes by employing the random forest method, so showcasing the capacity of machine learning in geotechnical contexts. Yan et al. [18] gave a detailed evaluation of these methods. Through the binding of the soil and the reduction of surface runoff, vegetation is an essential component in the process of stabilising slopes. Similarly, Yang et al. [19] proposed a novel approach combining a hybrid grey wolf optimiser with feature weighted-multiple kernel support vector regression. Yue and Kang [20] examined the impact of shear strength factors on the stability of soil slopes using LEM-based slicing approaches. They emphasised the significance of precise measurement of these parameters. Zhang et al. [21] utilised a comprehensive artificial intelligence model to calculate the friction angle (FA) of clays, a critical factor in assessing slope stability in the area. Zhao et al. [22] enhanced the Levenberg–Marquardt backpropagation neural network by including particle swarm and whale optimisation methods. This improvement enables the prediction of the deflection of RC beams, which in turn can provide valuable insights for the development of efficient stabilisation structures. Zhou et al. [23–27], through a series of studies, created a hybrid model by combining the Jaya algorithm with the extreme gradient boosting machine to accurately predict blast-induced ground vibrations. Hybrid models enhance the precision of stability forecasts by amalgamating the advantages of several approaches. Zhu et al. [28] introduced a comprehensive framework for slope stability analysis using LEM. This approach has gained significant popularity due to its straightforwardness and efficiency. Azarafza et al. [29] conducted an analysis on the stability of jointed rock slopes using block theory. This analysis is crucial for comprehending the behaviour of rock masses in the Pithoragarh region. These investigations establish a basis for conducting more studies on stability and developing plans to reduce or prevent negative impacts. Conventional geotechnical studies continue to be a fundamental aspect of evaluating slope stability. Begum et al. [30] have suggested a framework that incorporates the use of machine learning algorithms for the purpose of interpreting spectrum data. This framework provides a way that is both quick and non-destructive for evaluating the qualities of materials. This strategy has the potential to be especially helpful in the Dharchula region, where conventional geotechnical studies may be difficult to conduct due to the topography. Recent breakthroughs in machine learning have greatly improved the capacity to forecast slope stability. Bhagat et al. [31] utilised logistic regression, CART, and random forest algorithms to forecast slope failures caused by blasts, demonstrating the effectiveness of ensemble learning methods in intricate stability evaluations. In their study, Bui et al. [32] proposed a novel hybrid artificial intelligence model that combines a decision tree with an evolution algorithm. This model was specifically designed to accurately anticipate slope collapse in open-pit mines. These models may be customised for the Pithoragarh region, resulting in more precise and dependable forecasts. Research by Chawla et al. [33] on landslide susceptibility mapping in the Darjeeling Himalayas, which shares similar geological and environmental conditions with Pithoragarh, highlights the influence of geological structures and rock types on landslide occurrence. This study emphasises the need for detailed geological mapping and characterisation to understand the spatial distribution of landslide-prone areas effectively. Chen et al. [34] introduced a machine learning-based classification method for rock discontinuity trace, integrating Synthetic Minority Over-sampling Technique (SMOTE) with Gradient Boosted Tree (GBT) ensemble learning. This approach addresses the challenges of imbalanced datasets and enhances the predictive performance of slope stability models.

When it comes to determining the stability of slopes, geotechnical and geophysical measuring techniques are absolutely necessary. Despite the fact that indirect optical measures of leaf area index (LAI) are predominantly utilised in agriculture and forestry contexts, they also have uses in monitoring plant cover and its influence on slope stability. These developments include the use of remote sensing technologies and ground-based optical sensors. It is possible to modify these approaches in order to conduct slope stability evaluations in areas such as Dharchula, with the purpose of monitoring changes in vegetation that may have an impact on slope stability. When it comes to providing early warning indications of possible slope instability, the ability to evaluate vegetation cover in a quick and precise manner can be quite useful. Innovative methods of geotechnical analysis can be obtained through the combination of machine learning and spectroscopy. Despite the fact that the estimation of coal rank was the major focus of this work, the approaches and technologies that were utilised can be extended to the investigation of soil and rock for slope stability evaluations. Analysing the qualities of soil and rock may be accomplished through the use of diffuse reflectance spectroscopy, which can then provide data that can be included in machine learning models in order to forecast slope stability. The combination of machine learning with advanced simulation approaches, such as the FEM, has demonstrated encouraging outcomes. Engineering interventions, such as the construction of retaining walls, the implementation of rock bolting techniques, and the regrading of slopes, continue to be crucial in ensuring slope stability. Various case studies offer valuable insights into certain slope stability concerns and methods for reducing their impact. Scientists have examined the application of fuzzy statistics and hybrid models to address the uncertainty related to slope stability studies. Engineering methods, such as the implementation of retaining walls, rock bolting, and slope regrading, are commonly employed to stabilise slopes. Biotechnical solutions, such as reforestation and the implementation of plant mats, have been proposed as organic approaches to enhance slope stability. These strategies help strengthen the soil and improve its ability to withstand shear stresses by using root networks. Community-based disaster risk management activities are crucial in enhancing awareness and preparedness among local populations. The Pithoragarh region in Uttarakhand has substantial difficulties with slope stability as a result of its intricate geology, weather circumstances, and human activity. Diverse evaluation approaches, including both conventional geotechnical investigations and sophisticated machine learning algorithms, have been utilised to comprehend and forecast slope stability. Implementing both structural and non-structural options is crucial for minimising landslip hazards and improving the region's resilience. For efficient management of slope stability in Pithoragarh, it is essential to have continuous monitoring, connect with the community, and integrate new technology.

3.1 History and Reasons for Landslides in the Area

In the Indian state of Uttarakhand's Pithoragarh district sits the town and municipality of Dharchula. It is located along the Kali River's banks, which separates India and Nepal. Dharchula's important location as a commerce hub and a gateway to the trans-Himalayan region is strongly associated with its history. To evaluate the texture, color, and particle size, as well as the chemical and physical properties such as bulk density and specific gravity, a comprehensive analysis was conducted in the region. The topography of the area greatly influences how prone it is to landslides. Dharchula is located in the untamed Himalayan region, which is distinguished by precipitation intensity, brittle geological formations, and steep inclines. These elements play a part in the area's regular landslide incidents [35–38].

1. **Geological Make-Up**: The Himalayan region is experiencing geological activity, with tectonic forces constantly altering the landscape. The intersection of the Indian and Eurasian plates, where Dharchula is situated, is a seismically active region that produces high mountain ranges and deep valleys. Because of the geological instability of the area, landslides can occur, especially during earthquakes and periods of heavy rainfall [35].
2. **Weather and Climate Patterns**: Dharchula experiences a broad range of weather, including heavy downpours during the monsoon season. Because of the

area's steep slopes and loose soil, it is prone to erosion and soil saturation, which increases the risk of landslides. In addition, the melting of snow and glaciers throughout the summer months exacerbates the terrain's fragility [36].

3. **Tree Planting and Land Utilisation**: Like many other Himalayan regions, Dharchula has endured years of deforestation and unsustainable land use practices. Because the removal of vegetation weakens the structure of the soil and reduces its ability to absorb water, it raises the risk of erosion and landslides [37].
4. **Development of Infrastructure and Human Settlements**: Both the development of human populations and the construction of infrastructure, such as roads, dams, and hydropower plants, can raise the risk of landslides. Poorly planned construction can undermine slopes and alter how water naturally drains the earth, increasing the risk of landslides in the area [38].

4 Methodology

In this study, we examine key findings and methodologies from geological, geomorphological, geotechnical, hydrological, and remote sensing studies to elucidate the diverse factors contributing to slope instability. A thorough stability analysis of the slope hosting Dharchula village was conducted, with a longitudinal profile plotted from the ridge top to the valley bottom. Utilising software and the LEM, the study aimed to determine critical parameters such as the safety factor, maximum shear strain, and total displacement.

4.1 Morgenstern–Price Method

A universal slice technique called Morgenstern–Price was created using limit equilibrium. Satisfactory balance between the forces and moments acting on the individual blocks is necessary. By splitting the soil above the slip surface into separating planes, the blocks are formed. In Fig. 5, forces acting on specific blocks are shown.

The Morgenstern–Price method for calculating the limit equilibrium of forces and moments on individual blocks is based on the following assumptions:

- The planes dividing the blocks are always vertical.
- The line of action of the weight of block W_i passes through the midpoint of the ith segment of the slip surface, denoted as point M.
- The normal force N_i acts at the centre of the ith segment of the slip surface, also at point M.
- The inclination of the forces E_i acting between blocks varies for each block delta, and at the endpoints of the slip surface, the inclination is delta = 0.

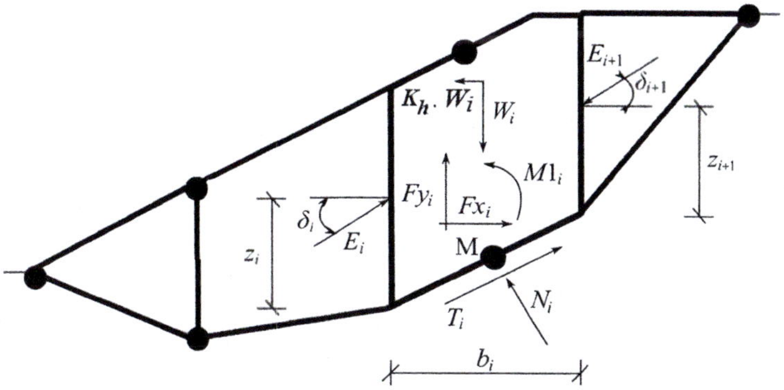

Fig. 5 Diagram for Morgenstern–Price method

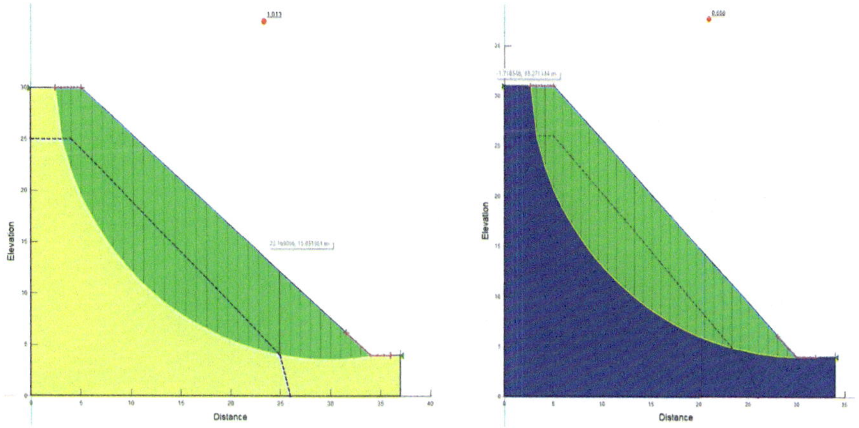

Fig. 6 Site 1 factor of safety for dry and saturated conditions

5 SLOPE/W Results

The current study uses the Morgenstern–Price (M–P) method to measure the FS for a range of critical cut slopes in Dharchula with varied soil parameters. SLOPE/W software is used to assist in GeoStudio 2021.4, and the results are verified through a field survey. The entire methodology, including field validation and mathematical models, is explained in this section. The SLOPE/W findings for several samples at S1 to S5 in dry (bulk) and saturated conditions are shown in Figs. 6, 7, 8, 9, and 10, and Tables 2 and 3 show the soil parameters and computed FS.

The FS values indicate the stability of the slopes. An FS value greater than 1 indicates a stable slope, while an FS value less than 1 indicates potential instability. In dry conditions, all sites have FS values greater than 1, indicating stable slopes. The highest FS is observed at S2 (1.349). In saturated conditions, the FS values are significantly lower, with some sites (e.g., S1 and S4) having FS values

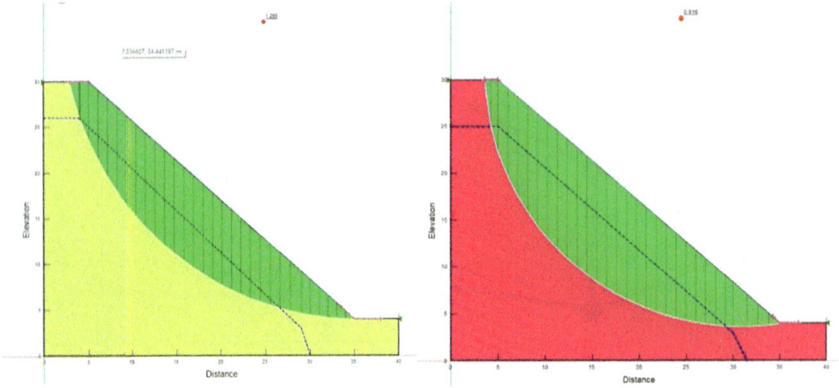

Fig. 7 Site 2 factor of safety for dry and saturated conditions

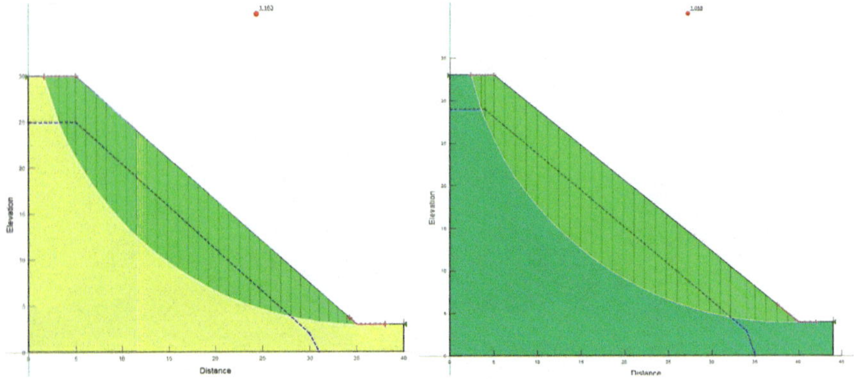

Fig. 8 Site 3 factor of safety for dry and saturated conditions

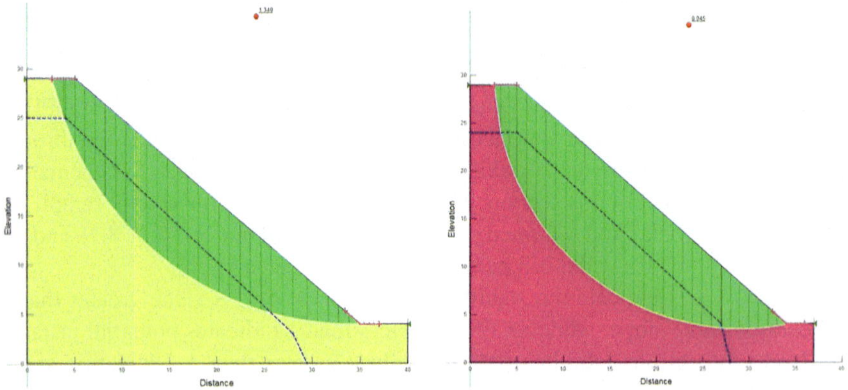

Fig. 9 Site 4 factor of safety for dry and saturated conditions

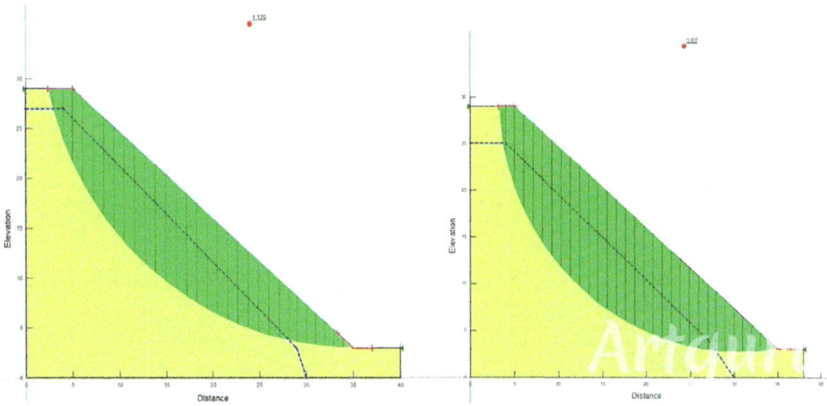

Fig. 10 Site 5 factor of safety for dry and saturated conditions

less than 1, indicating unstable slopes. This highlights the impact of water saturation on slope stability. Saturation increases the unit weight of the soil, reducing the FS. This is evident as the FS values decrease for all sites when saturated. The increase in unit weight and reduction in shear strength due to saturation make slopes more prone to failure. Cohesion values range from 14.8 to 18 kN/m^2. Higher cohesion contributes to slope stability by providing additional shear strength. FA values range from 29° to 31°, indicating moderate to high internal friction, which helps in resisting shear forces. Slope height varies significantly across sites, with the highest slope height observed at Site S1 (38.6 m) and the lowest at S3 (24.28 m). Slope angles (SA) range from 37° to 42°, with steeper slopes contributing to higher risk of instability, especially under saturated conditions. The stability of slopes in the Dharchula region is highly sensitive to water saturation. Under dry conditions, the slopes generally remain stable, but saturation significantly reduces the FS, indicating potential instability. Sites with higher cohesion and friction angles tend to have better stability. However, even these sites become vulnerable under saturated conditions.

Effective slope management in the Dharchula region should focus on drainage improvements to prevent saturation and reduce the risk of landslides. Measures such as retaining walls, vegetation cover, and proper water management can help maintain slope stability. Continuous monitoring and assessment are essential, particularly during the monsoon season when saturation levels increase, to ensure timely interventions and prevent slope failures. This analysis highlights the importance of understanding soil properties and the impact of environmental conditions on slope stability to develop effective mitigation strategies in the Dharchula region.

Table 2 Geotechnical analysis of collected soil samples

S.No.	Site/ stage		Classification of soil	OMC %	MDD gm/cc	Unit wt kN/ m^3	FA 0°	SA 0°	Slope height m	FS	Cohesion kN/m^3
1	S1	Dry	Silty Sand	14.3	1.81	18	29	40	38.6	1.013	18
2	S1	Sat	Silty Sand	14.3	1.81	20	29	40	38.6	0.658	18
3	S2	Dry	Silty Sand	13.8	1.82	16	29.7	37	28.63	1.349	16
4	S2	Sat	Silty Sand	13.8	1.82	18	29.7	37	28.63	0.945	16
5	S3	Dry	Silty Sand	13.4	1.84	15	31	40	24.28 2	1.163	16
6	S3	Sat	Silty Sand	13.4	1.84	17	31	40	24.28 2	1.088	16
7	S4	Dry	Silty Sand	14.8	1.79	13	30	41	30.6	1.265	15
8	S4	Sat	Silty Sand	14.8	1.79	15	30	41	30.6	0.916	15
9	S5	Dry	Silty Sand	14.5	1.8	12.7	30	42	37	1.12	14.8
10	S5	Sat	Silty Sand	14.5	1.8	14.6	30	42	37	0.927	14.8

Table 3 Factor of safety of collected soil samples

Site	FOS	Condition
S1 dry	1.013	Stable
S1 saturated	0.658	Unstable
S2 dry	1.349	Stable
S2 saturated	0.945	Unstable
S3 dry	1.163	Stable
S3 saturated	1.088	Stable
S4 dry	1.265	Stable
S4 saturated	0.916	Unstable
S5 dry	1.120	Stable
S5 saturated	0.927	Unstable

6 Conclusion

To examine the impact of topography, weather, and other climatic conditions on slopes, a comprehensive numerical analysis was carried out under static conditions for both dry and saturated scenarios using the LEM. The key findings of the study are summarised as follows:

- The modelling and evaluation of stability grades at significant locations in Dharchula Pithoragarh, which were discovered through site inspections and landslide inventory maps, are the focus of the current work. The future perspective of stability in these sites is determined by calculating the FS and slip surfaces using limit equilibrium software (SLOPE/W). The comparison of FS produced from simulation studies demonstrates that all slopes are stable in dry conditions, but four slopes are moderately unstable in saturated conditions and one slope is stable. This indicates that rainfall in the research area is fragile in both dry and saturated conditions.

- The output is the stability condition, and the five parameters—cohesion, internal FA, unit weight, SA, and slope height—are selected at random. The Inter-Criteria–based Correlation (CRITIC) approach yields results that show slope stability is most affected by slope height, with cohesion coming in second.
- A thorough analysis and experimental data, which are bolstered by the use of retaining walls and other anchoring techniques, demonstrate that slopes usually fail in saturated conditions or are only marginally stable. This method will be very helpful in minimising, foreseeing, and lessening the effects of these kinds of catastrophic events, which are among the main barriers to the country's socio-economic development.
- Preventing the erosion of finer particles is vital, as the settlement has hot springs and multiple mountain streams. Water percolation into the slope must be kept to a minimum. To prevent seepage into the slope, it is advised to direct water from the hill's upper elevations. Every precaution should be taken to guarantee that there is little to no water seeping into the slope.
- In addition, erosion control measures such as building retaining walls alongside road cuts should be used to stabilise the lower slope beneath the Kailash Mansarovar Road. This will lessen the likelihood of erosion caused by seepage of surface and subsurface water or toe erosion by the Sharda River (Kali Nadi) near the slope's base.
- However, these are only short-term fixes. The occurrence of more frequent extreme weather events could still lead to the deposits sliding in the context of climate change. To reduce the hazards to life and property, preventive actions including moving people to safer regions should be taken.
- To safeguard people's lives, property, and the environment from the damaging impacts of slope collapses, it is important to ensure safe and sustainable development in areas that are vulnerable to slope instability. While the main cause of slope failure is earthquakes, the main causes of landslide initiation are rainfall and the monsoon season. It is impossible to prevent landslides in many situations, particularly when there are earthquakes.

Therefore the development of the stability measures took into account two important factors: earthquake and monsoon activity. Given that the monsoon is the main cause of landslides, appropriate slope stabilisation measures were suggested to stop them from happening again. The proposed stabilisation measures involve the following:

1. Slope reinforcement: Implementing techniques like retaining walls, soil nailing, rock bolting, and geosynthetics to stabilise slopes.

2. Erosion control: Applying surface protection measures such as vegetation, erosion control blankets, and riprap to prevent surface erosion.
3. Ground improvement: Using methods like grouting, compaction, and soil stabilization to enhance the strength and stability of the slope materials.

References

1. Arya, P., Patel, S. B., Bharti, G., Shukla, B. K., & Hurukadli, P. (2023). Impact of using a blend of bagasse ash and polyester fiber in black cotton soil for improvement of mechanical and geotechnical properties of soil. *Materials Today: Proceedings, 78*, 738–743.
2. Gowda, S., Kunjar, V., Gupta, A., Kavitha, G., Shukla, B. K., & Sihag, P. (2024). Prediction of the subgrade soil California bearing ratio using machine learning and neuro-fuzzy inference system techniques: A sustainable approach in urban infrastructure development. *Urban Science, 8*(1), 4.
3. Bharti, G., Shukla, B. K., Parashar, B., Kumar, A., & Varadharajan, S. (2023). Experimental study on physico-mechanical properties of expansive soil by using wheat straw fiber. In *AIP conference proceedings* (Vol. 2800). AIP Publishing.
4. Maurya, N., Srivastav, Y., Rawat, S., Ranjan, Y., Srivastava, R., Shukla, B. K., & Varadharajan, S. (2023). Reinforcing civil infrastructure with waste glass-enhanced concrete: A comprehensive review of properties, performance and applications. *Materials Today: Proceedings, 93*(3), 239–248.
5. Bansal, V., & Sarkar, R. (2024). Prophetical modeling using limit equilibrium method and novel machine learning ensemble for slope stability gauging in Kalimpong. *Iranian Journal of Science and Technology, Transactions of Civil Engineering, 48*(1), 411–430.
6. Ram, B. K., Shawez, M., Gupta, V., & Rawat, G. (2024). Ground subsidence in Dar village (Darma valley), Pithoragarh district, Kumaun Himalaya, India: A Himalayan disaster in waiting. *Journal of Earth System Science, 133*(1), 34.
7. Solanki, A., Gupta, V., Bhakuni, S. S., Ram, P., & Joshi, M. (2019). Geological and geotechnical characterisation of the Khotila landslide in the Dharchula region, NE Kumaun Himalaya. *Journal of Earth System Science, 128*, 1–14.
8. Gupta, V., Kumar, S., Kaur, R., & Tandon, R. S. (2022). Regional-scale landslide susceptibility assessment for the hilly state of Uttarakhand, NW Himalaya, India. *Journal of Earth System Science, 131*(1), 2.
9. Gupta, V., Ram, B. K., Kumar, S., & Sain, K. (2022). A case study of the 12 July 2021 Bhagsunath (McLeod Ganj) flash flood in Dharamshala, Himachal Pradesh: A warning against constricting natural drainage. *Journal of the Geological Society of India, 98*(5), 607–610.
10. Haigh, M. J., Rawat, J. S., & Bartarya, S. K. (1989). Environmental indicators of landslide activity along the Kilbury road, Nainital, Kumaun lesser Himalaya. *Mountain Research and Development, 9*, 25–33.
11. Hearn, G. J., & Shakya, N. M. (2017). Engineering challenges for sustainable road access in the Himalayas. *Quarterly Journal of Engineering Geology and Hydrogeology, 50*(1), 69–80.
12. Verma, A. K., Singh, T. N., Chauhan, N. K., & Sarkar, K. (2016). A hybrid FEM–ANN approach for slope instability prediction. *Journal of The Institution of Engineers (India): Series A, 97*, 171–180.
13. Wang, Y., Cao, Z., & Au, S. K. (2011). Practical reliability analysis of slope stability by advanced Monte Carlo simulations in a spreadsheet. *Canadian Geotechnical Journal, 48*(1), 162–172.
14. Wang, Z., & Lin, M. (2021). Finite element analysis method of slope stability based on fuzzy statistics. *Earth Sciences Research Journal, 25*(1), 123–130.

15. Wei, W., Li, X., Liu, J., Zhou, Y., Li, L., & Zhou, J. (2021). Performance evaluation of hybrid WOA-SVR and HHO-SVR models with various kernels to predict factor of safety for circular failure slope. *Applied Sciences, 11*(4), 1922.
16. Wei, X., Zhang, L., Yang, H. Q., Zhang, L., & Yao, Y. P. (2021). Machine learning for pore-water pressure time-series prediction: Application of recurrent neural networks. *Geoscience Frontiers, 12*(1), 453–467.
17. Xie, H., Dong, J., Deng, Y., & Dai, Y. (2022). Prediction model of the slope angle of rocky slope stability based on random forest algorithm. *Mathematical Problems in Engineering, 2022*(1), 9441411.
18. Yan, G., Hu, R., Luo, J., Weiss, M., Jiang, H., Mu, X., Xie, D., & Zhang, W. (2019). Review of indirect optical measurements of leaf area index: Recent advances, challenges, and perspectives. *Agricultural and Forest Meteorology, 265*, 390–411.
19. Yang, H., Wang, Z., & Song, K. (2022). A new hybrid grey wolf optimizer-feature weighted-multiple kernel-support vector regression technique to predict TBM performance. *Engineering with Computers, 38*, 1–17.
20. Yue, Z. Q., & Kang, X. Y. (2021). Different contributions of two shear strength parameters to soil slope stability with limit equilibrium method based slice techniques. In *ISRM international symposium-Asian rock mechanics symposium* (Vol. 861, pp. ISRM–ARMS11). ISRM.
21. Zhang, H., Nguyen, H., Bui, X. N., Pradhan, B., Asteris, P. G., Costache, R., & Aryal, J. (2022). A generalized artificial intelligence model for estimating the friction angle of clays in evaluating slope stability using a deep neural network and Harris Hawks optimization algorithm. *Engineering with Computers, 38*, 1–14.
22. Zhao, J., Nguyen, H., Nguyen-Thoi, T., Asteris, P. G., & Zhou, J. (2022). Improved Levenberg–Marquardt backpropagation neural network by particle swarm and whale optimization algorithms to predict the deflection of RC beams. *Engineering with Computers, 38*(Suppl 5), 3847–3869.
23. Zhou, J., Li, E., Yang, S., Wang, M., Shi, X., Yao, S., & Mitri, H. S. (2019). Slope stability prediction for circular mode failure using gradient boosting machine approach based on an updated database of case histories. *Safety Science, 118*, 505–518.
24. Zhou, J., Li, X., & Mitri, H. S. (2016). Classification of rockburst in underground projects: Comparison of ten supervised learning methods. *Journal of Computing in Civil Engineering, 30*(5), 04016003.
25. Zhou, J., Qiu, Y., Khandelwal, M., Zhu, S., & Zhang, X. (2021). Developing a hybrid model of Jaya algorithm-based extreme gradient boosting machine to estimate blast-induced ground vibrations. *International Journal of Rock Mechanics and Mining Sciences, 145*, 104856.
26. Zhou, J., Shen, X., Qiu, Y., Li, E., Rao, D., & Shi, X. (2021). Improving the efficiency of microseismic source locating using a heuristic algorithm-based virtual field optimization method. *Geomechanics and Geophysics for Geo-Energy and Geo-Resources, 7*(3), 89.
27. Zhou, K. P., & Chen, Z. Q. (2009). Stability prediction of tailing dam slope based on neural network pattern recognition. In *2009 second international conference on environmental and computer science* (pp. 380–383). IEEE.
28. Zhu, D. Y., Lee, C. F., & Jiang, H. D. (2003). Generalised framework of limit equilibrium methods for slope stability analysis. *Geotechnique, 53*(4), 377–395.
29. Azarafza, M., Yarahmadi Bafghi, A. R., Asghari-Kaljahi, E., Bahmannia, G., & Moshrefy-far, M. (2013). Stability analysis of jointed rock slopes using key block method (Case study: Gas flare site in 6, 7 and 8 phases of south pars gas complex). *Geotechnical Geology, 9*(3), 169–185.
30. Begum, N., Maiti, A., Chakravarty, D., & Das, B. S. (2021). Diffuse reflectance spectroscopy based rapid coal rank estimation: A machine learning enabled framework. *Spectrochimica Acta Part A: Molecular and Biomolecular Spectroscopy, 263*, 120150.
31. Bhagat, N. K., Mishra, A. K., Singh, R. K., Sawmliana, C., & Singh, P. K. (2022). Application of logistic regression, CART and random forest techniques in prediction of blast-induced slope failure during reconstruction of railway rock-cut slopes. *Engineering Failure Analysis, 137*, 106230.

32. Bui, X. N., Nguyen, H., Choi, Y., Nguyen-Thoi, T., Zhou, J., & Dou, J. (2020). Prediction of slope failure in open-pit mines using a novel hybrid artificial intelligence model based on decision tree and evolution algorithm. *Scientific Reports, 10*(1), 9939.
33. Chawla, A., Chawla, S., Pasupuleti, S., Rao, A. C. S., Sarkar, K., & Dwivedi, R. (2018). Landslide susceptibility mapping in Darjeeling Himalayas, India. *Advances in Civil Engineering, 2018*(1), 6416492.
34. Chen, J., Huang, H., Cohn, A. G., Zhang, D., & Zhou, M. (2022). Machine learning-based classification of rock discontinuity trace: SMOTE oversampling integrated with GBT ensemble learning. *International Journal of Mining Science and Technology, 32*(2), 309–322.
35. Khanduri, S. (2017). Disaster hit Pithoragarh District of Uttarakhand Himalaya: Causes and implications. *Journal of Geography and Natural Disasters, 7*(3), 2–5.
36. Khanduri, S. (2022). Rain-induced slope instability: Case study of monsoon 2020 affected villages in Pithoragarh District of Uttarakhand, India. *International Journal of Earth Sciences Knowledge and Applications, 4*(1), 1–18.
37. Pande, R. K. (2006). Landslide problems in Uttaranchal, India: Issues and challenges. *Disaster Prevention and Management: An International Journal, 15*(2), 247–255.
38. Bhardwaj, D., & Sarkar, R. (2024). Landslide susceptibility mapping using probabilistic frequency ratio and Shannon entropy for Chamoli, Uttarakhand Himalayas. *Iranian Journal of Science and Technology, Transactions of Civil Engineering, 48*(1), 377–395.

Innovative Geopolymer Mortar: Enhancing Strength and Durability Through Fly Ash and Pond Ash Utilization

Aruna Lalandevanahalli Ashoka, B. Bharath, Prasad Bishetti, Pushpendra Kumar Sharma, and Bishnu Kant Shukla

1 Introduction

For the production of mortar, conventionally Ordinary Portland Cement (OPC) is used as primary binder. It is generally recognized that the manufacture of OPC has environmental consequences. For every ton of OPC produced, approximately 1 ton of carbon dioxide (CO_2) is emitted into the atmosphere as a result of the calcination of limestone and the combustion of fossil fuels. However, as fly ash is widely accessible and cost-free globally, there is a possibility to use this coal-burnt by-product in place of OPC to make mortar [1, 2]. The use of OPC as the main binder for mortar manufacturing has been crucial to the continuously changing built environment. OPC's utility and effectiveness are undeniable, but its manufacturing method creates serious environmental issues. In fact, the manufacturing of OPC makes a major contribution to the global emissions of greenhouse gases, mainly CO_2. According to studies, the calcination of limestone and the burning of fossil fuels release about one ton of CO_2 for every ton of OPC produced [3]. The manufacture of OPC uses a lot of energy, second only to the production of steel and aluminium, which has a

A. L. Ashoka · B. Bharath
Department of Civil Engineering, JSS Academy of Technical Education, Noida, Uttar Pradesh, India

P. Bishetti
Department of Civil and Infrastructure Engineering, Indian Institute of Technology, Dharwad, Karnataka, India

P. K. Sharma
School of Civil Engineering, Lovely Professional University, Phagwara, Punjab, India

B. K. Shukla (✉)
Department of Civil Engineering, JSS Academy of Technical Education, Noida, Uttar Pradesh, India

School of Civil Engineering, Lovely Professional University, Phagwara, Punjab, India

S. Pathak et al. (eds.), *Intelligent Infrastructure and Smart Materials*,
https://doi.org/10.1007/978-3-031-92421-7_15

negative impact on the environment [4]. Fly ash, a by-product of burning coal, is widely available and easy to get, which makes it a more affordable and environmentally friendly option for building mortar. Fly ash, when used as a partial replacement for OPC at room temperature and in the presence of water, reacts with calcium hydroxide to form calcium silicate hydrate (CSH) gel [5]. In the building industry, geopolymer binders have been proposed as OPC substitutes in the past. These binders can be made by polymerizing silicon (Si) and aluminium (Al) found in materials or byproducts that come from geological sources, like fly ash and rice husk ash, with alkaline liquids. In 1978, Davidovits was the first to refer to these binders as "geopolymers" [3]. Palomo et al. [4] suggest that pozzolanas such as blast furnace slag might be used to create binders by activating them with alkaline liquids further supports this. This method, which predominantly activates silicon and calcium, largely or entirely substitutes OPC in concrete and results in a C-S-H gel as the primary binder after the hydration procedure. The environmental effects of this dependence are worrisome, especially in India where the demand is high. The heavy extraction of river sand has had a number of negative repercussions, including saline water intrusion into land and coastal areas, lower riverbeds, bridge foundation deterioration, and failure of riverbanks [6]. Governmental bans on river sand mining have been implemented in some locations in reaction to this worsening environmental catastrophe, leading to severe natural sand scarcity in the building sector [7]. In order to reduce the amount of river sand extracted, these circumstances call for the investigation of alternate sources. Dune sand, offshore sand, synthetic sand (crushed rock sand), quarry dust, and pond ash are a few alternatives that could be used [8]. The waste product from burning coal or lignite in a thermal power plant boiler is formally known as pulverized fuel ash (PFA); however, according to IS:3812:2003, it is referred to as pond ash. Depending on the source of the ash and the manner of collection, the term PFA includes fly ash, bottom ash, and pond or mound ash [9]. The potential substitution of pond ash for river sand could help to control waste in thermal power plants as well as lessen the environmental effects of river sand mining. Numerous studies have been undertaken on the basis of this framework, adding to our knowledge of geopolymer materials and prospective applications. An increase in molarity can lead to improved compressive strength, according to research on the impact of different molarities on the compressive strength of geopolymer mortar [10]. Similar to this, the effects of pond ash replacement for sand have been investigated, with encouraging results in terms of compressive strength [11]. Although a growing corpus of research demonstrates the potential of geopolymer materials, little is known about their long-term behaviour. So, additional studies are necessary to close this knowledge gap and provide a thorough understanding of the impacts of molarity fluctuations and sand replacement on the durability of geopolymer mortar (GPM). The availability of experimental work on Geopolymer concrete is somewhat extensive, but the literature on Geopolymer mortar (GPM) is limited, especially in the context of ambient curing, where the available survey is very scarce. Sand is being created in an unsustainable way to fulfil the growing demand by washing surface soils (termed filter sands) and crushing rocks into the required sizes (termed manufactured sand). Given the enormous amounts of fly ash and pond ash that are sitting unused in India and the scarcity of river sand in the construction industry, bottom ash would meet the nation's dual needs. In order to make GPM, this study investigates the viability of employing fly ash and pond ash geopolymerization processes.

The current study mainly focuses on the creation of geopolymer mortar employing various molarities because there is a significant demand for ecologically acceptable alternatives to conventional construction materials and procedures. The work investigates the effects of various molarities, 8, 10, and 12 M on the compressive strength and durability of geopolymer mortar while taking into account the scientific developments made by pioneers like Davidovits [3] and Palomo et al. [4]. Additionally, we investigate the impact on compressive strength and durability of GPM by substituting 20, 40, and 60% of the sand.

2 Literature Review

A novel class of building materials known as geopolymer mortar is produced by alkaline activation of industrial wastes rich in aluminosilicates. Materials including fly ash, slag, and metakaolin are used in geopolymer mortar, as opposed to traditional OPC mortar. These materials undergo polymerization in the presence of alkaline activators such as sodium hydroxide (NaOH) and sodium silicate (Na_2SiO_3) solutions. By creating a three-dimensional network of aluminosilicate, this technique produces a binder with exceptional strength, durability, and environmental sustainability [3]. The potential of geopolymer mortar to tackle two major issues facing the building sector—the mitigation of CO_2 emissions and the environmentally friendly handling of industrial waste—makes it significant. The manufacturing of OPCs by traditional methods contributes significantly to global greenhouse gas levels and is a major source of CO_2 emissions. In contrast, the utilization of fly ash and other industrial leftovers in geopolymer mortar promotes a circular economy by reducing the need for OPC and diverting waste materials from landfills. In addition, geopolymer mortar has superior mechanical qualities, such as high early strength and resilience to chemical degradation, which makes it a competitive substitute for OPC in a range of structural uses [4].

2.1 Historical Development of Geopolymers

Joseph Davidovits, a materials scientist from France, first proposed the idea of geopolymers in the late 1970s [3]. According to Davidovits, it is possible to activate specific aluminosilicate materials with alkaline solutions to create a manmade substance that resembles stone and has characteristics similar to those of natural geological formations. His groundbreaking research established the groundwork for the advancement of geopolymer technology, highlighting its promise as a long-term substitute for conventional cementitious materials. Prior studies examined the basic chemistry and microstructure of geopolymers, as well as the relationships between different sources of aluminosilicate and alkaline activators. Preliminary research showed that it is possible to create geopolymer binders that are more durable and have better mechanical qualities than OPC. These results motivated additional research to optimize the performance properties, curing conditions, and mix design of geopolymer materials [3]. Significant progress was achieved in our knowledge and use of geopolymers in the 1990s and early 2000s. Researchers looked into using a variety of industrial wastes as raw materials for the synthesis of geopolymers, including slag from the production of steel and fly ash from coal-fired power plants. Standardized testing procedures for evaluating the mechanical and durability characteristics of geopolymers were also developed during this time, making it easier to compare them to traditional OPC-based materials. The emphasis has now shifted to the commercialization and widespread application of geopolymer technology. Geopolymer mortar and concrete have proven to be practical in real-world construction scenarios through pilot projects and large-scale applications, underscoring their potential to lessen the environmental impact of the construction sector. The goal of ongoing research is to overcome obstacles pertaining to geopolymers' scalability, affordability, and long-term performance, opening the door to wider use of these materials in environmentally friendly building techniques. The evolution of geopolymers over time demonstrates a steady transition towards more ecologically friendly and sustainable building materials. Geopolymer technology has come a long way from Davidovits' original theoretical ideas to the current range of actual applications being investigated. It presents a promising answer to some of the most critical problems facing the building sector [3, 6, 12].

2.2 *Environmental Impact of Ordinary Portland Cement*

2.2.1 CO_2 Emissions During OPC Manufacturing

Because fossil fuels are burned and limestone must be calcined, the manufacturing of OPC contributes significantly to global CO_2 emissions. Calcium carbonate ($CaCO_3$) is heated during the calcination process to produce CO_2 and calcium oxide (CaO). Much of the emissions linked to the manufacture of OPC is explained by this reaction alone. Furthermore, the production process's energy-intensive nature increases CO_2 emissions because it depends on fossil fuels for kiln heating.

Table 1 illustrates that 87% of the CO_2 emissions from the manufacturing of OPCs come from the combination of burning fossil fuels and limestone calcination. This significant carbon footprint highlights the need for substitute materials and binders that can lessen these negative effects on the environment [13–15].

Table 1 CO_2 emissions from OPC production processes

Process	CO_2 emissions (kg CO_2 per ton of OPC)	Percentage contribution to total emissions	Ref.
Calcination of limestone	540	50%	[13]
Combustion of fossil fuels	400	37%	
Electricity consumption	100	9%	[14]
Transportation and logistics	60	4%	[15]
Total	1100	100%	

Table 2 Energy consumption in OPC manufacturing

Stage of production	Energy consumption (MJ per ton of OPC)	Percentage of total energy use	Ref.
Quarrying and crushing	50	5%	[16]
Raw material grinding	150	15%	[17]
Clinker production (kiln)	700	70%	
Finish grinding and packing	100	10%	[18]

2.2.2 Energy Usage in the Production of OPCs

The production of steel and aluminium is more energy-intensive than the OPC manufacturing process, as depicted in Table 2. The main cause of the energy requirements is the high temperatures that must be reached in the kiln to speed up the calcination and clinker production processes. The kiln's efficiency, the fuel type, and the particular manufacturing techniques utilized all have an impact on the amount of energy consumed in the production of OPC.

OPC manufacturing and usage have a large negative environmental impact, resulting in high energy consumption, huge CO_2 emissions, and a host of other negative environmental effects. These elements highlight how urgent it is to investigate environmentally friendly substitutes like geopolymer mortar, which has the ability to lessen the building sector's environmental impact.

2.3 *Fly Ash as a Sustainable Binder*

2.3.1 Composition and Characteristics of Fly Ash

Fly ash is a fine particulate residue left over from burning pulverized coal in thermal power plants. As shown in Table 3, its primary ingredients are silica (SiO_2), alumina (Al_2O_3), iron oxide (Fe_2O_3), and calcium oxide (CaO), along with lesser components such as magnesium oxide (MgO), sulphur trioxide (SO_3), and trace elements. Fly ash may have a distinct specific chemical composition depending on the type of coal utilized and the combustion conditions.

Table 3 Typical chemical composition of fly ash

Component	Percentage by weight (%)
Silica (SiO_2)	45–65
Alumina (Al_2O_3)	20–30
Iron oxide (Fe_2O_3)	5–10
Calcium oxide (Cao)	1–5
Magnesium oxide (MgO)	0.5–3
Sulphur trioxide (SO_3)	0.5–2
Loss on ignition (LOI)	0.5–5

Because fly ash's main ingredients are Al_2O_3 and SiO_2, it can be used to create geopolymers. Fly ash's pozzolanic qualities are attributed to its high SiO_2 and Al_2O_3 concentration, which allows it to react with alkaline activators to generate stable, cementitious compounds.

2.3.2 Availability and Benefits of Fly Ash

Fly ash is a waste by-product of coal-fired power plants that is widely available. It is especially readily available in areas where there is a substantial generation of energy from coal. There are a number of financial and environmental advantages to using fly ash in building materials, such as:

Waste Utilization: By utilizing fly ash in buildings, industrial waste may be better managed, which lessens the demand for landfills and the related environmental risks [18].

Decrease in CO_2 Emissions: By partially avoiding the energy-intensive clinker generation process, replacing some of the OPC with fly ash can greatly lower the carbon footprint of producing concrete [18].

Increased Durability: Fly ash makes concrete stronger and more resilient over time by enhancing its capacity to withstand sulphate assault, the alkali-silica reaction, and thermal cracking [19].

Financial Savings: By partially substituting fly ash for cement, material costs can be reduced, which benefits the building sector financially [19].

2.4 *Geopolymerization Process*

2.4.1 Mechanism of Geopolymerization

Aluminosilicate minerals are converted into a three-dimensional network of silico-aluminate by the intricate chemical process known as geopolymerization. Three primary steps can be distinguished in the process: polycondensation, transportation or orientation, and dissolution.

Dissolution: The aluminosilicate materials (fly ash, metakaolin, or slag) dissolve when alkaline activators are present. Silicate (SiO_4^{4-}) and aluminate (AlO_4^{5-})

species are released into the solution as a result of the breaking of Si-O-Si and Al-O-Si bonds, which is facilitated by the high pH environment created by the alkaline activators (such as potassium or sodium hydroxide). Rearrangement and orientation of the dissolved silicate and aluminate species occur during their transportation in the solution. The ions rearrange themselves at this point to create the precursor species required for the ensuing polycondensation process [20, 21].

Polycondensation: Using the polymerization of the precursor species, a three-dimensional network is created during this stage. Silicate and aluminate oligomer condensation, which results in the creation of Si-O-Al and Si-O-Si bonds, is what defines this step. The ensuing network, referred to as a geopolymer matrix, demonstrates notable mechanical robustness and longevity. The following can be used to describe the general geopolymer network formula [22]:

$$\mathrm{Mn}\left[-\left(\mathrm{SiO_2}\right)_z-\mathrm{AIO_2}\right]_n\cdot\mathrm{wH_2O}$$

where z is the Si/Al ratio, n is the degree of polymerization, w is the amount of water integrated into the structure, and M stands for an alkali metal cation (e.g., Na^+, K^+).

The kind and concentration of the alkaline activator, the curing temperature, and the composition of the aluminosilicate source material are some of the variables that have a significant impact on the geopolymerization process. Comprehending these variables is essential to maximizing the geopolymerization procedure to attain the intended material characteristics [22].

2.4.2 Role of Alkaline Activators

Alkaline activators are an essential component of the geopolymerization process. They offer the required alkaline conditions for the aluminosilicate precursors to dissolve and the polymerization processes to start. The most widely used alkaline activators are potassium hydroxide (KOH) and NaOH. To improve reactivity and strength development, these alkaline activators are frequently combined with potassium silicate (K_2SiO_3) or sodium silicate (Na_2SiO_3) [18].

NaOH is a common solvent because it dissolves aluminosilicate compounds well and is highly soluble in it. The rate of dissolution and the total geopolymerization process are greatly influenced by the concentration of the NaOH solution. Although larger concentrations often result in faster dissolution and stronger compressive strengths, they can also make a material more brittle. Another powerful activator that is frequently used instead of NaOH is KOH, which can create geopolymers with a higher degree of thermal stability. By using KOH, the geopolymer network may become more open, improving the material's ability to withstand expansion and contraction caused by heat [18].

Adding Na_2SiO_3 to the alkaline activator solution (AAS) gives the process of polycondensation an extra source of SiO_2. Na_2SiO_3 functions as a binder,

Table 4 Influence of alkaline activators on geopolymer properties

Activator	Concentration	Compressive strength (MPa)	Setting time (h)	Thermal stability (°C)	Ref.
NaOH	8 M	30	24	600	[19]
NaOH	12 M	45	16	600	[20]
KOH	8 M	28	24	800	[21]
KOH	12 M	40	16	800	[22]
Na_2SiO_3 + NaOH	8 M + 10% Na_2SiO_3	50	8	600	[20]

encouraging the creation of a more cohesive and denser geopolymer matrix. A critical factor influencing the geopolymer's ultimate strength and setting time is the SiO_2/Na_2O ratio in the Na_2SiO_3 solution. Table 4 provides a summary of how various alkaline activators affect the attributes of geopolymers, such as compressive strength, setting time, and thermal stability [19].

2.4.3 Development of Compressive Strength in Geopolymers

The type and quantity of alkaline activators, the curing temperature, and the composition of the aluminosilicate source material all affect how compressive strength develops in geopolymers [20]. Increased curing temperatures and larger concentrations of alkaline activators often result in increased compressive strength for geopolymers.

Concentration of Alkaline Activator: Elevated concentrations of alkaline activators promote a more thorough dissolution of the aluminosilicate substance, resulting in a more widespread polymerization and a more compact geopolymer matrix. Compressive strength increases significantly when the concentration of NaOH is increased from 8 to 12 M, as Table 4 illustrates [20].

Curing Temperature: By supplying the thermal energy needed for the dissolution and polycondensation reactions, elevated curing temperatures quicken the geopolymerization process. The typical range of curing temperatures is 90 °C to ambient. Greater ultimate compressive strengths and quicker strength growth are typically the outcomes of higher curing temperatures. For instance, compressive strengths of up to 40 MPa can be attained in a day by geopolymers that are cured at 60 °C [19].

Character of the Source Material Aluminosilicate: The type of aluminosilicate material and its reactivity have a big impact on the geopolymerization process. Slag, fly ash, and metakaolin are often utilized source materials that give unique characteristics to the geopolymer. While metakaolin-based geopolymers are renowned for their quick setting times and strong early strengths, fly ash-based geopolymers usually have high compressive strength and low permeability [19].

Development of Compressive Strength in Fly Ash-Based Geopolymers: Fly ash-based geopolymers' compressive strength rises quickly in the first 7 days of curing and then more gradually in the next few weeks. The quick dissolution and polymerization of the aluminosilicate species are responsible for the first fast strength rise, but the continuous densification and maturation of the geopolymer matrix is the cause of the sustained strength growth [20].

Compressive strength development in geopolymers is influenced by a number of factors, including the curing temperature, the alkaline activator content, and the characteristics of the aluminosilicate source material. By adjusting these factors, geopolymers with unique mechanical properties suitable for a range of construction applications can be produced.

2.4.4 Alternatives to Natural River Sand

Natural river sand mining has major negative environmental effects that have sparked concerns around the world. Mining for river sand upsets ecosystems, resulting in a decline in biodiversity and changes to natural habitats. Important environmental concerns consist of the following:

Erosion and Sedimentation: Sand removal from riverbeds causes erosion to occur more quickly, which erodes riverbanks and increases sedimentation downstream. River courses may change as a result, affecting aquatic habitats and agriculture [23].

Loss of Aquatic Habitats: Many aquatic species suffer from the habitat destruction caused by sand mining. Water quality can be further deteriorated by the release of pollutants from the riverbed caused by the disruption of sediment layers [24].

Water Table Decline: Groundwater recharge may be impacted by extensive sand extraction, which lowers the water table. This may have detrimental effects on nearby water supplies, affecting both agricultural and human consumption [25].

Impact on Infrastructure: Roads, buildings, and bridges that are close to rivers may be in danger as a result of sand mining's instability of the riverbanks. There are serious risks to safety and finances from this [26]. It is imperative to find and use sustainable alternatives to natural river sand in light of these environmental effects.

2.4.5 Possible Substitutes: Quarry Dust, Synthetic Sand, Offshore Sand, and Dunes

To lessen the negative environmental effects of sand mining, a number of alternatives to naturally occurring river sand have been investigated. Every option has unique characteristics and environmental factors to take into account.

Dune Sand—Availability and Composition: Found in large quantities in arid areas, dune sand is mostly composed of quartz. Because of its tiny grain size, it can be used in a variety of building applications [27].

Impact on the Environment: Using dune sand eases the strain on river ecosystems. Before use, though, it might need to be cleaned to get rid of salt and organic contaminants [28].

Table 5 Comparison of sand alternatives

Alternative	Source	Primary composition	Environmental impact	Challenges	Ref.
Dune sand	Desert regions	Quartz	Reduces river sand mining; may impact desert ecosystems	Requires washing to remove impurities	[7, 8]
Offshore sand	Seabed	Silicate minerals	Alleviates pressure on rivers; potential marine habitat disruption	Must be managed to prevent coastal erosion	[9, 10]
Synthetic sand	Crushed rocks/ recycled concrete	Tailored composition	Promotes recycling; reduces natural sand demand	Resource-intensive production	[11, 12]
Quarry dust	Stone crushing by-product	Fine particles	Utilizes waste material; reduces natural sand demand	Needs treatment for suitable particle size	[13, 14]

Offshore Sand: The composition and availability of offshore sand. Because it comes from the seabed, offshore sand presents a viable substitute because of its abundance. Like river sand, it usually contains silicate minerals [29].

Environmental Impact: Offshore sand mining reduces the strain on river ecosystems, but it must be handled carefully to prevent damage to marine habitats and coastal erosion [30].

Synthetic Sand: Production, composition, and availability: Crushed rocks or recycled concrete are used to make synthetic sand. It can be modified to satisfy certain building specifications, guaranteeing uniformity in grading and quality [31].

Impact on the Environment: Although producing synthetic sand can require a lot of resources, it provides a sustainable substitute by encouraging the recycling of building materials and lowering dependency on natural sand [32].

Quarry Dust: Ingredients and Availability: In the process of making concrete, natural sand can be partially replaced with quarry dust, a fine substance produced as a by-product of crushing stones [33].

Impact on the Environment: Using quarry dust lowers the need for natural sand and aids in trash management. To guarantee a particle size distribution appropriate for construction and to get rid of extra fines, it may need to be treated [34].

A comprehensive analysis of these sand substitutes, encompassing their origins, main constituents, ecological ramifications, and obstacles, is presented in Table 5.

2.4.6 Pond Ash as a Sustainable Alternative

A sustainable substitute for natural river sand in building is pond ash, a by-product of burning coal in thermal power plants.

Composition and Properties: The main constituents of pond ash are iron oxide (Fe_2O_3), silicon dioxide (SiO_2), and aluminium oxide (Al_2O_3). It is appropriate for use in concrete due to its pozzolanic qualities and particle size dispersion [35].

Availability and Use: The growing amount of pond ash produced by coal-fired power plants makes it an easily accessible building material. By using pond ash in concrete, the environmental impact of its disposal is lessened [36].

Environmental Benefits: Waste management: By utilizing pond ash, thermal power stations can manage their enormous amounts of ash production, which lessens the demand for landfills and reduces pollution to the environment [37].

Resource Conservation: By substituting pond ash for natural sand, sand extraction has a smaller negative environmental impact and preserves natural resources [38].

Technical Performance: Research indicates that adding pond ash to concrete can make it more durable and easier to deal with. The geopolymer mortar's long-term strength growth is facilitated by its pozzolanic reaction [20].

The environmental problems related to sand mining can be resolved by using substitutes for natural river sand, such as pond ash, offshore sand, artificial sand, quarry dust, and dune sand. It is important to thoroughly analyse the advantages and disadvantages of each solution to ensure sustainable and effective construction procedures.

2.5 *Molarity and Its Impact on Geopolymer Mortar*

2.5.1 Influence of Molarity on Strength and Durability

One important factor that greatly affects the strength and longevity of geopolymer mortar is the molarity of the alkaline solution used in it. The concentration of the KOH or NaOH solution, or molarity, influences the subsequent geopolymerization process as well as the dissolution of the precursor aluminosilicate.

Developing Strength in Low Molarity (8 M): Because aluminosilicate materials dissolve more slowly at lower molarities, there are fewer reactive species available for polycondensation. The geopolymer mortar has a reduced compressive strength as a result. Research has indicated that after 28 days of curing, geopolymer mortar made with an 8 M NaOH solution has compressive strengths between 15 and 20 MPa [19].

Medium Molarity (10 M): Raising the molarity to 10 M facilitates the better dissolving of aluminosilicate components, resulting in a geopolymer network that is more widely distributed. After 28 days, this leads to a greater compressive strength, usually between 20 and 25 MPa [39].

High Molarity (12 M): The dissolution rate reaches its maximum at 12 M, resulting in a cohesive and dense geopolymer matrix with a much greater compressive strength that frequently reaches 25 MPa after 28 days. However, a mixture that is too viscous and may be difficult to deal with could have a detrimental influence on workability if the molarity is too high (above 12 M) [22].

Durability: Resistance to Chemical Attack: By generating a more impermeable matrix, solutions with higher molarity extend the life of geopolymer mortar. By reducing the entry of hostile ions, this enhanced impermeability strengthens resistance to chemical assaults, such as exposure to acids and sulphates [40].

Water Absorption and Sorptivity: The water absorption and sorptivity of geopolymer mortar are also influenced by the molarity of the alkaline solution. A denser microstructure brought about by higher molarity lowers sorptivity and water absorption. This increases the mortar's durability and qualifies it for use in difficult areas [41].

2.5.2 Optimal Molarity for Geopolymer Mortar

The ideal molarity for geopolymer mortar is determined by weighing the requirements for durability, workability, and mechanical qualities unique to the particular use.

Strength and Durability Optimization: A molarity of 12 M is frequently advised for applications needing strong compressive strength and endurance, such as structural components and infrastructure exposed to harsh climatic conditions. A strong and impermeable geopolymer matrix with better strength and resilience to chemical assaults is ensured by this concentration [21].

Workability Considerations: Although greater molarities increase strength, workability may suffer as a result. A balance needs to be established for projects where it is important to have ease of handling and placing the mortar, like intricate formworks or ornamental applications. A molarity of 10 M might be the ideal middle ground, delivering respectable workability and strong performance [42].

Cost and Environmental Impact: The geopolymer mortar's cost and environmental impact are also influenced by the molarity selection. More alkali activators are needed for higher molarities, which raises the possibility of handling risks and expenses. Thus, a thorough evaluation takes into account material costs and safety; thus, when choosing the ideal molarity for a certain application, consideration for the surroundings is essential [43].

Optimization Based on Case Specifics

Residential Construction: A molarity of 8–10 M may be enough for non-structural applications like paving and masonry units, balancing cost and performance [44].

Industrial and Infrastructure Projects: A molarity of 10–12 M is more suitable for heavy-duty applications needing improved mechanical qualities and durability, like bridges, highways, and maritime structures [44].

The precise criteria for strength, durability, workability, cost, and environmental impact determine the ideal molarity for geopolymer mortar. While a molarity of 8–10 M may be appropriate for situations where cost and workability are important, a molarity range of 10–12 M is often ideal for attaining good performance in strength and durability. It is advised to conduct more study and field testing to better customize the molarity to the particular requirements of every project [44].

2.6 Pond Ash as a Replacement for Sand

2.6.1 Characteristics of Pond Ash

Burning coal or lignite in thermal power plants produces pond ash, commonly referred to as PFA. It is made up of fine particles that have burned and gathered in ash ponds close to these power stations. Pond ash has different properties depending on where it comes from and how it is burned. Pond ash typically demonstrates the following characteristics:

- **Particle Size**: The majority of pond ash particles fall into the fine to medium particle size range, making them generally finer than sand [45].
- **Chemical Composition**: Depending on the coal source, the chemical composition of pond ash can include SiO_2, Al_2O_3, Fe_2O_3, CaO, and other minor components [45]. Pond ash has pozzolanic activity, which means that when it reacts with calcium hydroxide in the presence of water, it forms new cementitious compounds that strengthen and extend the life of mortar and concrete.
- **Availability**: Pond ash is widely accessible as a by-product of a large number of coal-fired power plants across the world, which makes it a desirable and sustainable substitute for natural sand in construction applications [45].

2.6.2 Benefits and Drawbacks of Pond Ash Replacement

Benefits:

1. **Sustainability**: Using pond ash in place of sand lessens the environmental damage caused by sand mining, which can result in the depletion of natural resources and the destruction of ecosystems [46].
2. **Waste Utilization**: The use of pond ash aids in the control of coal combustion by-product disposal, promoting sustainable waste management techniques and waste reduction [47].
3. **Strengthening**: The pozzolanic activity of pond ash can strengthen mortar or concrete, resulting in durable and long-lasting constructions [48].
4. **Cost-Effectiveness**: Using pond ash instead of typical sand in construction projects can save money because it is frequently available for less or for free [46].

Drawbacks:

1. **Variability**: When pond ash is utilized as a substitute material, its performance may vary depending on the coal supply, the burning procedure, and the collection techniques [49].
2. **Quality Control**: Strict testing procedures and quality control measures may be necessary to guarantee the uniform quality and characteristics of batches of pond ash [49].

Table 6 Previous research findings on pond ash in mortar

Study	Pond ash replacement (%)	Compressive strength (MPa) in %	Durability aspects	Ref.
Kumar et al. (2015)	15%	3.19%	Enhanced resistance to chloride ion penetration	[42]
Lakhiar et al. (2022)	10%	20%	Improved sulphate resistance and reduced alkali-silica reaction	[43]
Lal et al. (2019)	10%	40%	Reduced water absorption and increased flexural strength	[44]

3. **Chemical Compatibility**: Although pond ash can increase strength, mix designs and curing techniques may need to be adjusted to account for its chemical composition and reactivity in order to get the required results [50].
4. **Regulatory Compliance**: Adherence to relevant legislation and standards is necessary because regulatory norms and guidelines regarding the use of pond ash in construction may differ by location [50].

2.6.3 Previous Research on Pond Ash in Mortar

The performance and viability of replacing sand in mortar with pond ash have been examined in a number of studies. These investigations, presented in Table 6, have shed important light on the durability, mechanical characteristics, and environmental effects of these substitutes.

Pond Ash Replacement (%): Shows how much of the mortar mix's sand has been swapped out for pond ash. The range of compressive strength values attained in the relevant experiments is represented by the term compressive strength (MPa).

The main benefits in durability, such as resistance to sulphate attack, resistance to chloride penetration, the alkali-silica reaction, the reduction of water absorption, and the improvement of flexural strength, are summarized in the durability aspects section. Based on the replacement % and mix design considerations, these research results demonstrate the possibility of pond ash as a sustainable substitute for sand in mortar, with variable degrees of improvement in strength and durability attributes [43].

2.7 *Durability Characteristics of Geopolymer Mortar*

2.7.1 Water Absorption and Sorptivity Tests

Tests for water absorption and sorptivity are essential for determining the longevity of geopolymer mortar, especially in conditions where the material will be in contact with moisture or water. The porosity and capillary action of the mortar, which are markers of its resistance to water infiltration and associated deterioration, are ascertained with the aid of these tests:

- **Test for Water Absorption**: This test determines how much water the mortar absorbs over a predetermined time frame. It is a crucial sign of the porosity and possible longevity of the material. A denser, less porous material that is less prone to water-related damage is indicated by a lower water absorption rate [20].
- **Sorptivity Test**: This test measures how quickly water is absorbed through capillary action. It entails partially submerging a specimen in water and tracking the specimen's mass gain over time. This test sheds light on the initial rate of water intrusion, which is important to know when assessing the mortar's long-term resilience to cyclic wetting and drying conditions [20].

2.7.2 Long-term Performance of Geopolymer Mortar

One important component of geopolymer mortar's longevity is its long-term performance. The material's total lifetime is influenced by elements like mechanical wear, freeze–thaw cycles, and chemical attack resistance.

Chemical Resistance: The exceptional resilience of geopolymer mortars against chemical assaults, such as those involving acids, sulphates, and chlorides, is well documented. This is because, in contrast to the CSH structure of conventional Portland cement, the stable aluminosilicate structure generated during geopolymerization is less reactive [20].

Freeze–Thaw Resistance: In frigid areas, geopolymer mortar's resistance to freeze–thaw cycles is essential. The reduced water intrusion caused by the dense microstructure and low permeability minimizes the risk of freeze–thaw damage [51].

Mechanical Wear: Another measure of durability is long-term mechanical performance, which includes compressive strength retention and abrasion resistance. Geopolymer mortars are appropriate for a variety of structural and non-structural applications since they usually retain their mechanical qualities for long periods of time [52].

Environmental Impact: Because geopolymer mortar uses industrial wastes like fly ash and slag, it uses less energy than Portland cement, which has a higher carbon impact [53, 54].

Chemical Durability: Because geopolymer mortars are more resistant to chemical attacks, they are more suited for harsh environments like those found in maritime applications and the containment of industrial waste [55].

Mechanical Properties: Both kinds of mortar are capable of achieving high compressive strengths; however, because of their denser microstructure and decreased permeability, geopolymer mortars hold up better over time [56].

Sustainability: Because geopolymer mortar uses waste resources, it reduces landfill disposal and supports the circular economy, both of which are aspects of sustainability [47, 57].

Over the past 20 years, there has been a notable surge in geopolymer research due to the growing demand for eco-friendly building materials and the negative environmental effects of conventional cement production. Important patterns found in the literature consist of:

- *Waste Utilization:* An increasingly common practice in the manufacturing of geopolymers is the addition of industrial by-products such as fly ash, slag, and pond ash. By using fewer landfills, these materials not only offer a sustainable substitute for natural resources but they also aid in waste management [58].
- *Optimization of Alkaline Activators:* A lot of research has gone into determining the best kind and number of alkaline activators (NaOH, Na_2SiO_3, etc.) to improve the mechanical and long-term qualities of geopolymer mortars [59].
- *Durability and Performance Studies:* Research on the long-term durability of geopolymer mortars, including their ability to withstand chemical attack, freeze–thaw cycles, and mechanical wear, has grown. Studies on durability are essential for verifying the use of geopolymers in diverse environmental settings.
- *Development of Composite Materials:* Another developing field of study is the creation of geopolymer composites with additional fibres (such as glass, carbon, or natural fibres) to increase tensile strength and crack resistance.
- *Gaps in Current Knowledge:* Even with great progress, there are still a number of gaps in our understanding of geopolymer mortars that must be filled for a wider range of applications and comprehension.

The extant literature is deficient in crucial information regarding the preparation of geopolymer mortar. The purpose of this study is to determine the variables that affect the geopolymer mortar mixture proportions. In contrast to the custom of curing geopolymer mortar at increased temperatures, the specimens in this study are cured in ambient settings. The effects of various alkaline concentrations are also investigated in this study. As is customary, compressive strength is used as the primary development criteria for geopolymer mortar. A methodical trial-and-error approach backed by judgment and expertise was used in the development process.

3 Material Collection and Characterization

Fly ash: Fly ash for this study was obtained from the Raichur Thermal Power Plant, which is located in the Raichur District of Karnataka State, India. The physical parameters of fly ash were determined in a laboratory, while the chemical values were obtained from the plant manager of the Raichur thermal power station. Raichur fly ash is categorized as Class F based on the concentrations of SiO_2 (57%), Al_2O_3 (26%), and CaO (0.97%). The physical properties of the fly ash used in this investigation are listed in Table 7.

Alkaline activator solution: The AAS, which is composed of Na_2SiO_3 and NaOH, reacts chemically with the silicon and aluminium components in fly ash to

Table 7 Physical characteristics of the fly ash

Colour	Light brown
Specific gravity (SSD)	2.18
Bulk density (kg/m^3)	800
Specific surface area (Blane), (m^2/kg)	390
Moisture content, %	0.09
Autoclave expansion, %	0.04
Category	Class F

Table 8 Composition analysis of sodium silicate (Na_2SiO_3) and sodium hydroxide (NaOH)

Constituent	NaOH	Na_2SiO_3
Sodium oxide (Na_2O)	–	15.9%
Silicon dioxide (SiO_2)	–	31.4%
Water (H_2O)	–	52.7%
Potassium(K)	0.1%	–
Zinc(Zn)	0.02%	–
Sulphate (SO_2)	0.05%	–
Carbonate (Na_2CO_3)	2.0%	–
Chloride (Cl)	0.01%	–
Lead (Pb)	0.001%	–
Iron (Fe)	0.001%	–

make a binding paste that binds aggregates together and produces geopolymer mortar. The AAS helps activate the silicon and aluminium in fly ash. One essential ingredient is NaOH, which can be found in commercial or technical-grade pellets or flakes. In order to prevent excessive exothermic reactions, the NaOH solution for the experiment was carefully made by dissolving pellets in water 1 day before use. During casting, the hydroxide solution and Na_2SiO_3 solutions are mixed as part of the amalgamation process, producing the necessary AAS. Na_2SiO_3, also known as water glass or liquid glass, is usually supplied in liquid (gel) form; in this study, the ratio of Na_2O to SiO_2 is fixed at 2. In-depth information about the chemical and physical characteristics of NaOH and Na_2SiO_3 was acquired from SS Enterprises in Kamakshipalya, Bangalore.

Sodium silicate (Na_2SiO_3): Generally, Na_2SiO_3, also known as water glass or liquid glass, is available in liquid (gel) form. In the present investigation, the ratio of Na_2O to SiO_2 = 2 is used. The physical and chemical properties of Na_2SiO_3 and NaOH were provided by SS Enterprises Kamakshipalya Bangalore, which are depicted in Table 8.

Fine Aggregates: It has been shown that the fine aggregates typically used in regular mortar are sufficient for the creation of geopolymer mortar. These tiny aggregates underwent characterization analyses in compliance with IS 2386 (Parts 1–3)—1963 requirements. From the sieve analysis, shown in Fig. 1 and Table 9, the river sand and pond ash conform to zone 2.

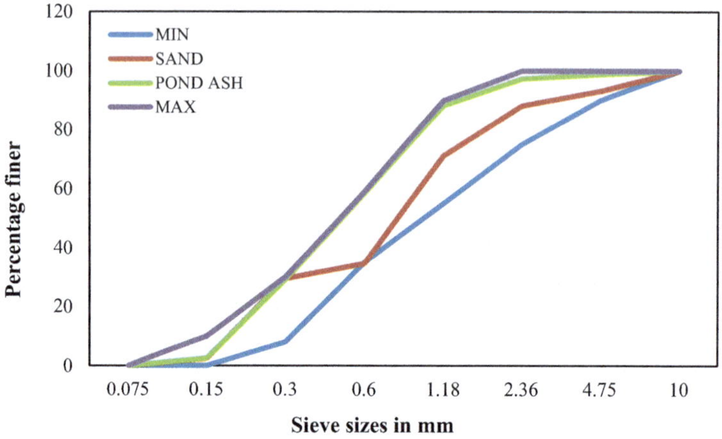

Fig. 1 Percentage finer using sieves for sand and pond ash

Table 9 Physical properties of fine aggregate

Property	River sand	Bottom ash
Specific gravity	2.52	2.2
Bulk density	1400 kg/m^3	812 kg/m^3
Fineness modulus	3.1	2.9

4 Results and Discussion

The objective of this research is to create geopolymer mortar (GPM) mixtures with a goal strength of roughly 21.2 MPa for a solution with a molecular weight of 12 under ambient curing circumstances. Different test mixtures were created while keeping the NaOH to Na_2SiO_3 ratio constant at 1. The experiment involved varying the concentration of NaOH to 8, 10, and 12 M, resulting in 12 distinct mixes being examined. In particular, four mixes were made with an 8 M concentration of NaOH, four more at a 10 M concentration, and four more at a 12 M concentration. The water-to-binder (W/B) ratio was adjusted by adding more water to change the workability. Table 10 contains comprehensive information about the trial mixes.

08FB0—8 Molarity mix with 0% alternating fine aggregate to pond ash.
08FB20—8 Molarity mix with 20% alternating fine aggregate to pond ash.
08FB40—8 Molarity mix with 40% alternating fine aggregate to pond ash.
08FB60—8 Molarity mix with 60% alternating fine aggregate to pond ash.
10FB0—10 Molarity mix with 0% alternating fine aggregate to pond ash.
10FB20—10 Molarity mix with 20% alternating fine aggregate to pond ash.
10FB40—10 Molarity mix with 40% alternating fine aggregate to pond ash.
10FB60—10 Molarity mix with 60% alternating fine aggregate to pond ash.
12FB0—12 Molarity mix with 0% alternating fine aggregate to pond ash.
12FB20—12 Molarity mix with 20% alternating fine aggregate to pond ash.
12FB40—12 Molarity mix with 40% alternating fine aggregate to pond ash.
12FB60—12 Molarity mix with 60% alternating fine aggregate to pond ash.

Table 10 Composition details for the GPM trial mixes

Mix ID	Fly ash (kg/m³)	Fine aggregate				NaOH	Na_2Sio_3	Molarity	Alkali/binder ratio
		Sand %	Pond ash %	Fine aggregate (kg/m³)	Pond ash (kg/m³)	(kg/m³)	(kg/m³)	(M)	
08FB0	400	100	0	700	0	32	160	8	0.5
08FB20	400	80	20	560	81.2	32	160	8	0.5
08FB40	400	60	40	420	162.4	32	160	8	0.5
08FB60	400	40	60	280	243.6	32	160	8	0.5
10FB0	400	100	0	700	0	40	160	10	0.5
10FB20	400	80	20	560	81.2	40	160	10	0.5
10FB40	400	60	40	420	162.4	40	160	10	0.5
10FB60	400	40	60	280	243.6	40	160	10	0.5
12FB0	400	100	0	700	0	48	160	12	0.5
12FB20	400	80	20	560	81.2	48	160	12	0.5
12FB40	400	60	40	420	162.4	48	160	12	0.5
12FB60	400	40	60	280	243.6	48	160	12	0.5

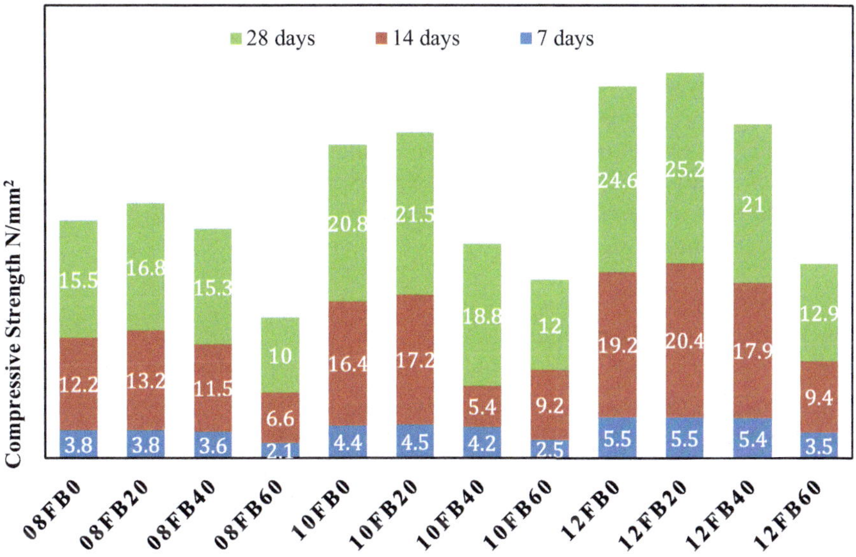

Fig. 2 Strength parameter of geopolymer mortar of 8, 10, and 12 M

Using compression testing equipment, the hardened geopolymer mortar was put through its compression test in compliance with IS 516 criteria. Compression testing was performed on cube specimens that were 100 mm × 100 mm × 100 mm every 7, 14, and 28 days. The bar chart illustrates the compressive strength of various concrete over different curing periods of 7, 14, and 28 days.

Figure 2 indicates the compressive strength attained at intervals 7, 14, and 28 days. Highest compressive strength (25.2 MPa) was attained by 12 M mix with 20% replacement by incorporating fine sand alongside bottom ash (12FB20). 12 M mix with 0% replacement of fine sand with bottom ash also showed similar compressive strength (24.6 MPa). Compressive strength was less for the mix 08FB60 and the rate of gain in strength between 14 and 28 days was considerably high in 10FB40. It can be observed that the resistance to crushing was higher in 12 M mix with 0 and 20% replacement of fine sand with bottom ash [12FB20].

4.1 Durability Characteristics of GPM with and Without Replacement

4.1.1 Sorptivity Test

Concrete that is going to be exposed to hostile conditions, such as sulphates and chloride ions, may have trouble absorbing water [60]. This test aims to determine the susceptibility of unsaturated concrete to water seeping through. It measures the rate at which liquids, such as water, are absorbed into unsaturated concrete by the use of capillary suction. Concrete mixture is used to create moulded cylinder specimens, which are then moist cured for 28 days. From the cylinder, a 2-inch (50-mm) slice is removed, and it is then sealed in a container for a minimum of 15 days and allowed to evaporatively cool for 3 days at 122 °F (50 °C) and 80% relative humidity. After that, the slice is covered with epoxy on both the exterior and the top before submerging it in water. Over time, the mass change of the slice is monitored. Figure 3 illustrates the sorptivity test results for solutions of 8, 10, and 12 M, highlighting the relationship between sorptivity and concentration.

4.1.2 Water Absorption Test

Figure 4 shows the relationship between water absorption (%) and the percentage of sand replaced with pond ash for different molarities. The x-axis represents the percentage replacement of sand with pond ash and the y-axis represents the water absorption percentage. For all three molarities, water absorption rises as pond ash replaces sand by a percentage. The water absorption for 0% replenishment is roughly 6% for 8 M and roughly 7% for both 10 and 12 M. Water absorption rises with replacement percentage; for all three molarities, it reaches around 14% at 60% replacement. Since there is not much difference in water absorption between the various molarities, it appears that the replacement percentage affects water absorption more than the molarity.

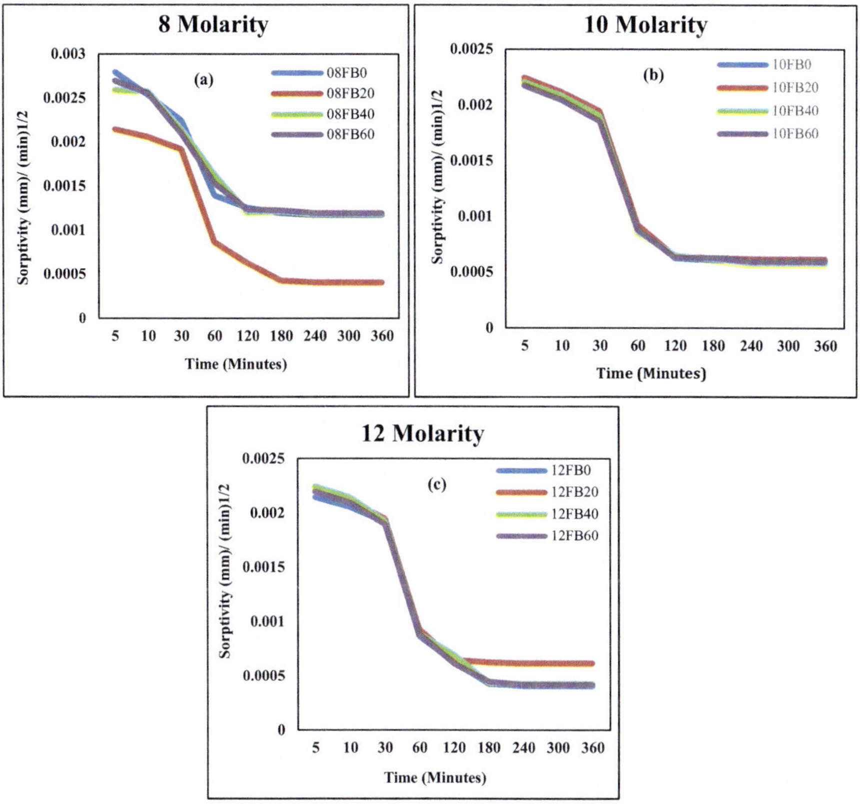

Fig. 3 Sorptivity for (**a**) 8 molar GPM mixes; (**b**) 10 molar; (**c**) 12 molar GPM mixes

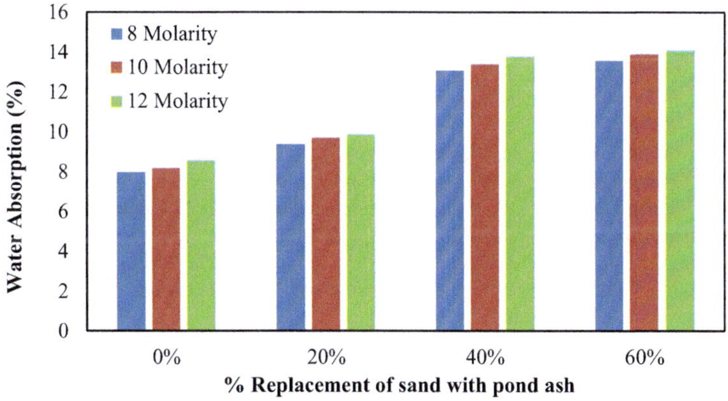

Fig. 4 Water absorption of GPM mixes for 8, 10, and 12 M

5 Conclusion

Fly ash is capable of being used as the primary binder to achieve GPM with a compressive strength of up to 15, 20, and 24 MPa at an ambient molarity of 8, 10, and 12 M, respectively. W/B ratio slightly influences the strength realized while all other factors remain static. In comparison to the 8 and 10 M mixes, the 12 M mixes' workability was significantly less prosperous. As the molarity of the mixtures grew, the compressive strength remained constant while the other parameters increased. Increased molarity of GPM mixtures results in increased strength and lower workability. The effect of substituting pond ash for sand in fine aggregate shows that substituting up to 20% of sand with pond ash results in a significant increase in strength.

The test results on durability of GPM, when pond ash is used in place of sand using the water absorption and sorptivity tests, indicate that the GPM mixtures produced are highly durable. Because the GPMs taken into consideration in this work had an alkaline character, they were resistant to corrosive action. The sorptivity test indicates that while the rate of capillary absorption increases initially, it subsequently declines with time. An increase in NaOH concentration from 8 to 12 M was found to result in an increase in water absorption. For each concentration from 8 to 12 M, an increase in the percentage of river sand replaced with pond ash was found to result in an increase in water absorption.

References

1. Varadharajan, S., Kirthanashri, S. V., Maurya, N., Bishetti, P., Shukla, B. K., & Bharti, G. (2023). Utilization of Fly ash in concrete: A state-of-the-art review. *Lecture Notes in Civil Engineering, 281*, 189–194. https://doi.org/10.1007/978-981-19-4731-5_17
2. Varadharajan, S., Kirthanashri, S. V., Dwivedi, A., Irfan, D., Shukla, B. K., Kumar, A., Bishetti, P., & Bharti, G. (2023). Use of fly ash and metakaolin to manufacture sustainable concrete. In *AIP conference proceedings* (Vol. 2800). AIP Publishing.
3. Davidovits, J. (1999). Chemistry of geopolymeric systems, terminology. *Geopolymer, 99*(292), 9–39.
4. Palomo, M. W., & Grutzeck, M. T. B. (1999). Alkali-activated Fly ashes, a cement for the future. *Cement and Concrete Research, 29*(8), 1323–1329.
5. Xu, H., & van Deventer, J. S. J. (2000). The geopolymerization of alumino silicate minerals. *International Journal of Mineral Processing, 59*(3), 247–266.
6. Van Jaarsveld, J. G. S., van Deventer, J. S. J., & Lorenzen, L. (1997). The potential use of Geopolymeric materials to immobilize toxic metals: Part I. Theory and applications. *Minerals Engineering, 10*(7), 659–669.
7. Barbosa, V. F. F., MacKenzie, K. J. D., & Thaumaturgo, C. (2000). Synthesis and characterisation of materials based on inorganic polymers of alumina and silica: Sodium Polysialate polymers. *International Journal of Inorganic Materials, 2*(4), 309–317.
8. Swanepoel, J. C., & Strydom, C. A. (2002). Utilization of fly ash in a geopolymeric material. *Applied Geochemistry, 17*(8), 1143–1148.
9. van Jaarsveld, J. G. S., van Deventer, J. S. J., & Lukey, G. C. (2002). The effect of composition and temperature on the properties of Fly ash and kaolinite based Geopolymers. *Chemical Engineering Journal, 89*(1–3), 63–73.

10. Cheng, T. W., & Chiu, J. P. (2003). Fire-resistant geopolymer produced by granulated blast furnace slag. *Minerals Engineering, 16*(3), 205–210.
11. Gourley, J. T. (2003). Geopolymers: Opportunities for environmentally friendly construction materials. In *Materials 2003 conference: Adaptive materials for a modern society*. Institute of Materials Engineering Australia.
12. Fernández-Jimenez, & Palomo, A. (2003). Characterisation of fly ash potential reactivity as alkaline cements. *Fuel, 82*(18), 2259–2265.
13. Mohamad, N., Muthusamy, K., Embong, R., Kusbiantoro, A., & Hashim, M. H. (2022). Environmental impact of cement production and solutions: A review. *Materials Today: Proceedings, 48*, 741–746.
14. Karadumpa, C. S., & Pancharathi, R. K. (2024). Study on energy use and carbon emission from manufacturing of OPC and blended cements in India. *Environmental Science and Pollution Research, 31*(4), 5364–5383.
15. Sarfraz, S., Sherif, Z., Jolly, M., & Salonitis, K. (2024). Benchmarking of energy consumption and CO2 emissions in cement production: A case study. In *TMS annual meeting & exhibition* (pp. 53–65). Springer Nature Switzerland.
16. Xiang, Q., Pan, H., Ma, X., Yang, M., Lyu, Y., Zhang, X., et al. (2024). Impacts of energy-saving and emission-reduction on sustainability of cement production. *Renewable and Sustainable Energy Reviews, 191*, 114089.
17. Ige, O. E., Von Kallon, D. V., & Desai, D. (2024). Carbon emissions mitigation methods for cement industry using a systems dynamics model. *Clean Technologies and Environmental Policy*, 1–19.
18. Hasanbeigi, A., et al. (2012). Emerging energy-efficiency and CO2 emission-reduction Technologies for Cement and Concrete Production. *Journal of Industrial Ecology, 16*(4), 526–542.
19. Hardjito, D., & Rangan, B. V. (2005). *Development and properties of low-calcium fly ash-based geopolymer concrete*. Research Report GC1, Faculty of Engineering, Curtin University of Technology, Perth, Australia.
20. Palomo, A., Grutzeck, M. W., & Blanco, M. T. (1999). Alkali-activated fly ashes: A cement for the future. *Cement and Concrete Research, 29*(8), 1323–1329.
21. Duxson, P., Provis, J. L., Lukey, G. C., & van Deventer, J. S. J. (2007). The role of inorganic polymer technology in the development of 'green concrete'. *Cement and Concrete Research, 37*(12), 1590–1597.
22. Fernández-Jiménez, A., & Palomo, A. (2005). Composition and microstructure of alkali activated fly ash binder: Effect of the activator. *Cement and Concrete Research, 35*(10), 1984–1992.
23. Kondolf, G. M. (1997). Hungry water: Effects of dams and gravel mining on river channels. *Environmental Management, 21*(4), 533–551.
24. Padmalal, D., & Maya, K. (2014). *Sand mining: Environmental impacts and selected case studies*. Springer.
25. Kondolf, G. M. (1994). Geomorphic and environmental effects of instream gravel mining. *Landscape and Urban Planning, 28*(2–3), 225–243.
26. Ashraf, M. A., Maah, M. J., & Yusoff, I. (2011). Sand mining effects, causes, and concerns: A case study from Bestari Jaya, Selangor, Peninsular Malaysia. *Scientific Research and Essays, 6*(6), 1216–1231.
27. Anbarashan, M., Balachandran, N., Mathevet, R., Barathan, N., & Uma Maheswari, P. (2024). An evaluation of coastal sand dune flora of Cuddalore District, Tamil Nadu, India: Perspectives for conservation and management. *Geology, Ecology, and Landscapes, 8*(2), 208–221.
28. Kommanamanchi, V., Duddu, S. R., & Chennarapu, H. (2024). Evaluation of geotechnical characterization of recycled sand as a sustainable replacement for natural sand. *Journal of Material Cycles and Waste Management*, 1–19.
29. Li-ping, G., Xiang-peng, F., Jian-dong, W., Ling-yi, Z., & Bang-cheng, L. (2024). High ductility cementitious composites incorporating seawater and coral sand (SCS-HDCC) for offshore engineering: Microstructure, mechanical performance and sustainability. *Cement and Concrete Composites, 147*, 105414.

30. Torres, A., Brandt, J., Lear, K., & Liu, J. (2017). A looming tragedy of the sand commons. *Science, 357*(6355), 970–971.
31. Zhen, H., Xiong, Z., Song, Y., Li, L., Qiu, Y., Zou, X., Chen, B., Chen, D., Liu, F., & Ji, Y. (2024). Early mechanical performance of glass fibre-reinforced manufactured sand concrete. *Journal of Building Engineering, 83*, 108440.
32. Shyam Prakash, K. S., & Rao, C. B. (2014). Study on the properties of concrete with manufactured sand as replacement to natural sand. *International Journal of Civil Engineering Research, 5*(4), 19–31.
33. Divya, S., & Praveenkumar, S. (2024). An integrated evaluation of graphene-based concrete mixture with copper slag and quarry dust using response surface methodology. *Journal of Building Engineering, 86*, 108876.
34. Rajamma, R., Ball, R. J., Tarelho, L. A. C., Allen, G. C., Labrincha, J. A., & Ferreira, V. M. (2009). Characterisation and use of biomass fly ash in cement-based materials. *Journal of Hazardous Materials, 172*(2–3), 1049–1060.
35. Ahmad, S., Ghazi, M. S. A., Syed, M., & Al-Osta, M. A. (2024). Utilization of fly ash with and without secondary additives for stabilizing expansive soils: A review. *Results in Engineering, 22*, 102079.
36. Singh, M., & Siddique, R. (2013). Effect of coal bottom ash as partial replacement of sand on properties of concrete. *Resources, Conservation and Recycling, 72*, 20–32.
37. Joshi, R. C., & Lohita, R. P. (1997). *Fly ash in concrete: Production, properties and uses* (Vol. 2). CRC Press.
38. Shukla, B. K., Gupta, A., Gowda, S., & Srivastav, Y. (2023). Constructing a greener future: A comprehensive review on the sustainable use of fly ash in the construction industry and beyond. *Materials Today: Proceedings, 93*(3), 257–264.
39. Provis, J. L., & Bernal, S. A. (2014). Geopolymers and related alkali-activated materials. *Annual Review of Materials Research, 44*, 299–327.
40. Davidovits, J. (1991). Geopolymers: Inorganic polymeric new materials. *Journal of Thermal Analysis and Calorimetry, 37*(8), 1633–1656.
41. Nath, P., & Sarker, P. K. (2014). Effect of GGBFS on setting, workability and early strength properties of fly ash geopolymer concrete cured in ambient condition. *Construction and Building Materials, 66*, 163–171.
42. Kumar, M., Srivastva, V., & Agarwal, V. C. (2015). Suitability of pond ash as partial replacement of fine aggregate in PPC concrete. *Journal of Multidisciplinary Engineering Science and Technology, 2*(8), 2156–2158.
43. Lakhiar, M. T., Bai, Y., Wong, L. S., Paul, S. C., Anggraini, V., & Kong, S. Y. (2022). Mechanical and durability properties of epoxy mortar incorporating coal bottom ash as filler. *Construction and Building Materials, 315*, 125677.
44. Lal, D., Chatterjee, A., & Dwivedi, A. (2019). Investigation of properties of cement mortar incorporating pond ash–an environmental sustainable material. *Construction and Building Materials, 209*, 20–31.
45. Minjigmaa, A., Zolzaya, T., Bayanjargal, E., Davaabal, B., & Temuujin, J. (2013). Characterization of ash pond ashes from 3rd thermal power plant by SEM/EDX and XRD methods. *Mongolian Journal of Chemistry, 14*, 61–65.
46. Attarde, S., Marathe, S., & Sil, A. (2014). Utilization of fly ash in construction industries for environment management. *International Journal of Environmental, 3*(2), 117–121.
47. Shukla, B. K., Bharti, G., Sharma, P. K., Sharma, M., Rawat, S., Maurya, N., Srivastava, R., & Srivastav, Y. (2024). Sustainable construction practices with recycled and waste materials for a circular economy. *Asian Journal of Civil Engineering., 25*, 5255–5276. https://doi.org/10.1007/s42107-024-01111-y
48. Varadharajan, S., Kirthanashri, S. V., Singh, K., Irfan, D., Shukla, B. K., & Kumar, A. (2023). Replacement of fine aggregate with imperial smelting furnace slag for manufacture of eco-friendly concrete. In *AIP conference proceedings* (Vol. 2800). AIP Publishing.
49. Sarkar, R., Singh, N., & Das, S. K. (2007). Effect of addition of pond ash and fly ash on properties of ash—Clay burnt bricks. *Waste Management & Research, 25*(6), 566–571.

50. Maurya, N., Srivastav, Y., Rawat, S., Sharma, M., Srivastava, R., Sihag, P., & Shukla, B. K. (2024). Augmenting concrete performance and sustainability with recycled glass: A critical examination of material characteristics and construction applications. In *AIP conference proceedings* (Vol. 3050). AIP Publishing.
51. Bakharev, T. (2005). Resistance of geopolymer materials to acid attack. *Cement and Concrete Research, 35*(4), 658–670.
52. Davidovits, J. (1994). Properties of geopolymer cements. In *First international conference on alkaline cements and concretes* (Vol. 1, pp. 131–149).
53. McLellan, B. C., Williams, R. P., Lay, J., Van Riessen, A., & Corder, G. D. (2011). Costs and carbon emissions for geopolymer pastes in comparison to ordinary Portland cement. *Journal of Cleaner Production, 19*(9–10), 1080–1090.
54. Hurukadli, P., Kumar, A., Shukla, B. K., & Bharti, G. (2023). Experimental study on the influence of available absorbent as internal curing agent on cement mortar. In *AIP conference proceedings* (Vol. 2800). AIP Publishing.
55. Neville, A. M. (1995). *Properties of concrete* (Vol. 4, p. 1995). Longman.
56. Shi, C., Jiménez, A. F., & Palomo, A. (2011). New cements for the 21st century: The pursuit of an alternative to Portland cement. *Cement and Concrete Research, 41*(7), 750–763.
57. Turner, L. K., & Collins, F. G. (2013). Carbon dioxide equivalent (CO2-e) emissions: A comparison between geopolymer and OPC cement concrete. *Construction and Building Materials, 43*, 125–130.
58. Bharti, G., Hurukadli, P., Shukla, B. K., Sihag, P., Jagudi, S., & Tripathi, A. (2023). Environmental impact analysis and utilization of copper slag for stabilising black cotton soil. In *Materials Today: Proceedings*. https://doi.org/10.1016/j.matpr.2023.04.305.
59. Varadharajan, S., Kirthanashri, S. V., Kumar, H., Irfan, D., Shukla, B. K., Kumar, A., Bishetti, P., & Bharti, G. (2023). Use of silica fume concrete to enhance mechanical properties and environmental benefits. In *AIP conference proceedings* (Vol. 2800). AIP Publishing.
60. Shukla, B. K., Sharma, P. K., Bharti, G., Gupta, A., Singh, A., Patel, T., Jaiswal, B. P., & Singh, C. K. (2024). A methological evaluation of mixed designs for self-healing concrete. *Lecture Notes in Civil Engineering, 509*, 51–73. https://doi.org/10.1007/978-981-97-3153-4_4

Slope Stability Analysis of Rainfall-Induced Landslide, Malin, Maharashtra

Bawane Krishna Yeshwantrao and Naveen James

1 Introduction

The village of Malin, located in Maharashtra's Pune district on the Deccan Plateau, experienced a devastating landslide in July 2014, highlighting the area's susceptibility to natural disasters. Malin has steep slopes and is mainly made up of basalt rock formations, which increase the risk of landslides during the annual monsoon rains. The aftermath of the 2014 landslide vividly shows how such events can severely affect local communities and infrastructure.

Rainfall-triggered landslides are a significant global hazard, often caused by changes in groundwater conditions, particularly in sloped terrain where the groundwater is usually deep. This creates negative pressures in the soil, known as matric suction, which help stabilize unsaturated soil slopes by maintaining effective stresses [1, 2]. However, intense rainfall can drastically change these conditions. When water seeps into the soil, matric suction decreases, leading to positive pore pressures that can raise the groundwater table and increase soil density. These changes weaken the soil's shear strength, making it more prone to landslides [3].

Our research focuses on understanding how rainfall influences slope stability in Malin using advanced geotechnical and hydrological modeling techniques. By combining analytical tools like GeoStudio's SEEP/W and SLOPE/W modules, we aim to simulate and predict how slopes respond to different levels of rainfall. These predictive capabilities are crucial for developing effective strategies to mitigate landslide risks and enhance resilience against future disasters.

B. K. Yeshwantrao (✉) · N. James
Indian Institute of Technology Ropar, Rupnagar, Punjab, India
e-mail: 2022cem1024@iitrpr.ac.in; naveen.james@iitrpr.ac.in

S. Pathak et al. (eds.), *Intelligent Infrastructure and Smart Materials*,
https://doi.org/10.1007/978-3-031-92421-7_16

2 Site Characteristics

Malin village is located in the Pune district of Maharashtra at coordinates 19°9′40″N and 73°41′18″E on the Deccan Plateau, known for its basalt rock formations. These rocks, especially the porous types, are prone to weathering and erosion, particularly during heavy monsoon rains. The area experiences a tropical monsoon climate, which contributes to frequent landslides, especially on the steep slopes of 20–40°. The local vegetation includes forests, farmlands, and grasslands, all of which affect how water moves and impacts soil stability. Landslides are often triggered by heavy rains and human activities like deforestation and construction, making the area particularly vulnerable. Figures 1 and 2 show the aerial image of Malin village before and after the landslide, which was taken from Google Earth Pro.

Fig. 1 Malin village before landslide

Fig. 2 Malin village after landslide

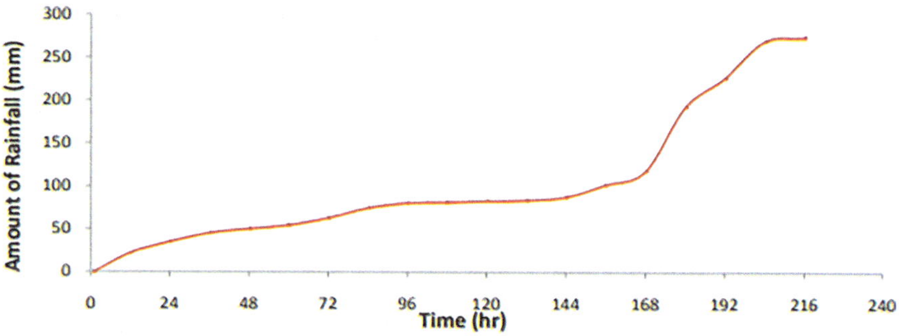

Fig. 3 Rainfall data of Malin site. (From "Forensic analysis of Malin landslide in India" by P. Ering, R. Kulkarni, Y. Kolekar, S.M. Dasaka and G.L.S. Babu, 2015, International Symposium)

3 Rainfall Characteristics

The primary cause of the Malin landslide was continuous rainfall, which occurred from July 21 to July 30, 2014 (Fig. 3). Malin village does not have its own rain gauge stations, so the rainfall data presented in the following graph is sourced from nearby areas and literature. The prolonged rain during this period led to soil saturation and instability, which triggered the landslide.

4 Material Properties

The material properties of soils from the Malin landslide area, sourced from Meshram (2016) [4], highlight significant factors influencing slope stability. The soils, primarily silty clay, show well-graded grain size distributions. The density ranges from 1.34 to 1.5 g/cm^3, with maximum dry density of 1.55 g/cm^3 in the slide area and 1.34 g/cm^3 above the slide. The cohesion values vary between 37.26 and 57.86 kPa, with higher internal friction angles above the slide (41°) compared to the slide area (25.6°), indicating better shear resistance in the soils above. Key properties such as liquid limit (53% in the slide area, 46% above), plasticity index (25.8 in the slide area, 10.4 above), and moisture content (28.8% in the slide area, 25% above) suggest that the soils in the slide area are more plastic and capable of higher water retention, critical for understanding slope behavior during rainfall-induced events. These data are instrumental in evaluating slope stability and the potential for future landslides under varying moisture conditions.

5 Methodology

5.1 *Geometric and Hydrological Modeling*

The study focused on the Malin slope's stability under varying rainfall conditions using GeoStudio SEEP/W and SLOPE/W modules. The geometry of the slope was meticulously reconstructed based on the topographic and soil properties [2]. The model encompassed the critical slide area and adjacent zones, extending from the toe to the hilltop to capture the full hydrological and mechanical behavior of the slope.

5.2 *Steady-State Seepage Analysis*

To establish initial groundwater conditions, a steady-state seepage analysis was conducted. The pressure head boundary conditions were defined with the water table set at 50 m at the toe and 100 m at the hilltop from the base of the slope. This configuration aimed to replicate the natural groundwater flow pattern. The water table was marked using these boundary conditions to simulate the hydrostatic pressure and seepage flow accurately. This steady-state analysis provided the baseline hydraulic head distribution throughout the slope, essential for initializing the transient analysis (Fig. 4).

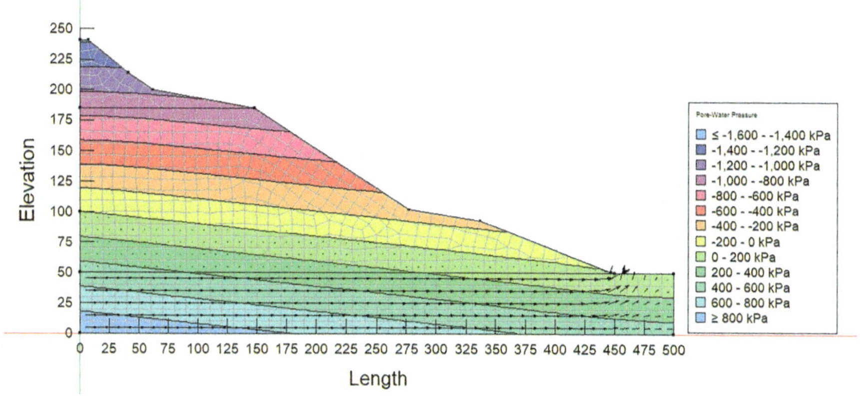

Fig. 4 Steady-state analysis results

5.3 *Transient Seepage Analysis*

The transient seepage analysis was conducted to simulate the daily impact of rainfall on the Malin slope from July 21 to July 30, 2014. The initial conditions for this analysis were established from the steady-state seepage results obtained on July 21, providing a baseline representation of the slope's hydraulic conditions. For each day, rainfall data were sourced from local meteorological records and applied as a unit flux boundary condition to accurately model rainfall infiltration. The analysis commenced with a 58 mm rainfall on July 22, directly influencing the slope's moisture conditions. Subsequently, each day's analysis used the results from the previous day as the starting point, thereby reflecting the cumulative effects of successive rainfall events. This sequential approach continued through July 23 to July 29, with daily rainfall amounts added as a unit flux. The final analysis on July 30 simulated the critical rainfall event that triggered the slope failure, considering the compounded effects of the continuous rainfall [3]. Throughout this period, rainfall was applied uniformly across the slope surface, assuming homogenous infiltration characteristics.

6 Results and Discussion

The results of the seepage and the slope stability analyses of the Malin slope are presented in terms of variation of pore water pressures, rainfall infiltration, formation of slip surfaces, and the corresponding factors of safety against slope failure for the actual rainfall events, and for different rainfall intensities and their duration.

The transient analyses monitored changes in volumetric water content and pore water pressure at various depths, capturing the progressive increase in soil moisture and rise in the groundwater table, which were critical in assessing the slope's response to prolonged rainfall (Fig. 5).

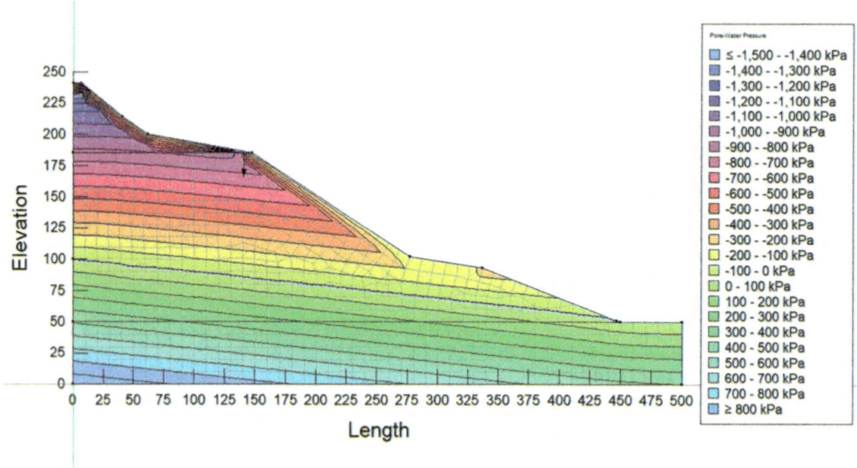

Fig. 5 Transient state analysis results

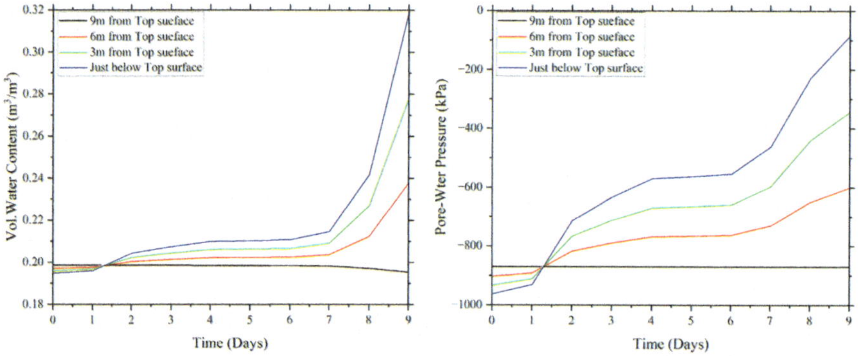

Fig. 6 Variation of volumetric water content and pore water pressure with time at four different depths

To thoroughly evaluate the impact of rainfall on the Malin slope, volumetric water content and pore water pressure were monitored at various depths over the study period. Daily measurements showed notable changes, capturing how water infiltrated and was stored within the soil layers of the slope.

(a) **Variation of pore water pressures and volumetric water content, rainfall infiltration**: From Fig. 6, volumetric water content over time exhibited notable increases across various soil depths during the rainfall period from July 21 to July 30, 2014. These increases were indicative of moisture accumulation within the slope due to continuous rainfall infiltration, which caused instability in the soil slope. Pore water pressure measurements illustrated rising pore pressures at different depths with rainfall (Fig. 6).

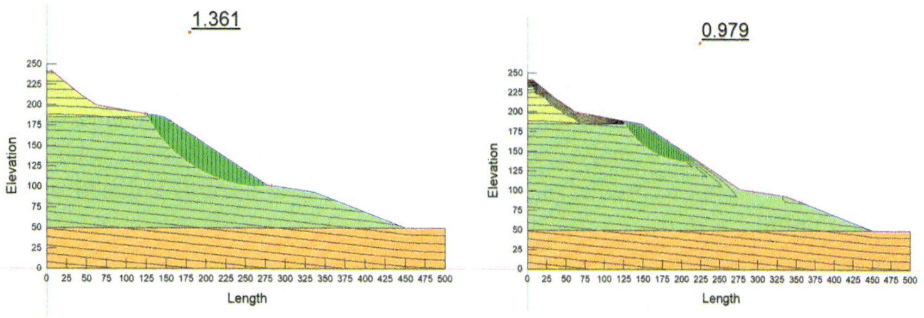

Fig. 7 FOS on July 21, i.e. before rainfall and FOS on July 30, i.e. after rainfall

The graphs displayed a progressive buildup of pore pressure, reflecting a decrease in stability of the slope. These findings were crucial in understanding the internal forces acting on the slope and their role in triggering potential instability.

(b) **Slip surfaces and the corresponding factors of safety**: The integration of slope stability analysis with transient seepage results provided a comprehensive assessment of the Malin slope's stability under rainfall conditions by using Morgenstern-Price method [5]. Following the initial stability analysis conducted after the steady-state seepage analysis, which established a baseline Factor of Safety (FoS) of 1.36, daily analyses were performed from July 22 to July 30, 2014 (Fig. 7). Each day's stability assessment incorporated updated pore water pressures and water content distributions derived from the transient seepage analyses.

On July 21, the stability analysis began with the steady-state conditions adjusted for rainfall-induced changes in pore pressure. Subsequent stability analyses from July 22 to July 30 considered the cumulative effects of rainfall from preceding days, progressively weakening the slope. By July 30, the final stability analysis coincided with the day of slope failure, where the FoS decreased significantly to 0.976. This sharp decline in FoS indicated imminent failure due to the substantial reduction in soil strength resulting from prolonged rainfall infiltration and saturation. The detailed integration of seepage dynamics with slope stability assessments underscored the critical role of rainfall in slope destabilization, highlighting the importance of continuous monitoring and modeling in predicting such geotechnical hazards.

7 Conclusion

This study highlights the critical impact of continuous rainfall on the stability of the Malin slope, leading to its eventual failure. Through daily monitoring of volumetric water content and pore water pressure from July 21 to July 30, 2014, significant changes were observed in the soil layers. These measurements demonstrated how sustained rainfall infiltrated the slope, resulting in increased moisture accumulation and rising pore pressures. The progressive buildup of pore pressure and water content significantly weakened the slope, as evidenced by the decreasing Factor of Safety (FoS), which dropped from 1.36 to 0.976 by the time of the landslide on July 30. This comprehensive analysis underscores the importance of continuous monitoring and modeling to predict and mitigate geotechnical hazards induced by prolonged rainfall.

References

1. Fredlund, D. G., & Rahardjo, H. (1993). *Soil mechanics for unsaturated soils* (p. 517). Wiley.
2. Dey, N., & Sengupta, A. (2018). Effect of rainfall on the triggering of the devastating slope failure at Malin, India. *Natural Hazards, 94*(3), 1391–1413. https://doi.org/10.1007/s11069-018-3483-9
3. Griffths, D. V., & Lu, N. (2005). Unsaturated slope stability analysis with steady infiltration or evaporation using elastic-plastic finite element. *International Journal for Numerical and Analytical Methods in Geomechanics, 29*, 249–267.
4. Meshram, S. G. (2016). *Geotechnical investigation and analysis of Malin landslide*. Master's thesis, Savitribai Phule Pune University, India.
5. Morgenstern, N. R., & Widger, R. A. (1978). The shear strength of unsaturated soils. *Canadian Geotechnical Journal, 15*(3), 313–321.

Experimental Assessment of Ultra-High-Performance Fiber-Reinforced Concrete for Bridge Applications

Gaurav Bharti, Bishnu Kant Shukla, and Pushpendra Kumar Sharma

1 Introduction

Ultra-high-performance fiber-reinforced concrete (UHPFRC) is a sophisticated type of concrete that consists mainly of a large quantity of Ordinary Portland Cement (OPC) and silica fume as binders [1–3]. It also contains steel fibers, a very low Water/Binder (W/B) ratio, and additives such as water reducers. UHPFRC differs from conventional concrete by using steel fiber reinforcement and high-quality plasticizers, resulting in a substantial improvement in its ability to resist cracking and the spread of cracks. This innovative material exhibits a compressive strength four to six times greater than normal concrete and is at least 100 times more durable [4]. These characteristics result in a longer service life and enhanced sustainability for concrete structures. The use of steel fiber reinforcement in UHPFRC contributes to its superior properties, providing advanced resistance to cracking and crack propagation. Experimental studies have shown that UHPFRC can achieve very high compressive strengths irrespective of the curing treatment applied. For instance, the average 28-day compressive strength of submerged and air-treated UHPFRC has been observed to be 180 MPa and 126.17 MPa, respectively. Compared to high-performance concrete (HPC), UHPFRC demonstrates superior strength, durability,

G. Bharti
Department of Civil Engineering, JSS Academy of Technical Education,
Noida, Uttar Pradesh, India

B. K. Shukla (✉)
Department of Civil Engineering, JSS Academy of Technical Education,
Noida, Uttar Pradesh, India

School of Civil Engineering, Lovely Professional University, Phagwara, Punjab, India

P. K. Sharma
School of Civil Engineering, Lovely Professional University, Phagwara, Punjab, India

S. Pathak et al. (eds.), *Intelligent Infrastructure and Smart Materials*,
https://doi.org/10.1007/978-3-031-92421-7_17

and long-term stability, making it a valuable material in the construction industry [5]. The global demand for more sustainable infrastructure has driven the development and use of UHPFRC. Traditional concrete, despite its widespread use, poses significant environmental challenges, including high CO_2 emissions during production. In contrast, UHPFRC consumes fewer materials and energy, thus contributing less CO_2 to the atmosphere [6, 7]. This aligns with the goals of sustainable construction by reducing the environmental footprint. Moreover, conventional steel-reinforced concrete faces issues such as reinforcement corrosion, leading to reduced longevity and increased maintenance costs. UHPFRC, with its enhanced properties, addresses these issues effectively, making it a promising material for future construction projects [8–10].

1.1 Background of the Study

Concrete is the most widely used man-made material globally, with about 7 km^3 produced annually as of 2006. This vast usage highlights its significant impact on our lives since its invention. The journey toward more sustainable infrastructures is essential, and UHPFRC technology supports this vision remarkably. Ordinary concrete production accounts for about 5% of global CO_2 emissions, equivalent to the emissions of an entire country like Russia. UHPFRC, by consuming fewer materials and energy, contributes less to CO_2 emissions and supports sustainability. Traditional steel-reinforced concrete, although designed to enhance strength, is plagued by reinforcement corrosion, leading to cracking and reduced longevity. UHPFRC, with its advanced resistance to cracking and long-term durability, presents a significant improvement [11–14]. These benefits, along with its superior mechanical properties, have spurred extensive research into HPC. The primary objectives of this project are to analyze the composition of UHPFRC and perform various tests to evaluate its mechanical and physical properties. To achieve this, several key tests will be conducted. The Flow Table test will assess the workability and consistency of the concrete mix. The Compressive Strength test will measure the ability of UHPFRC to withstand axial loads, providing a crucial indicator of its structural performance. The Four Point Load test will evaluate the flexural strength and toughness of the concrete, essential for understanding its behavior under bending stresses [15]. Lastly, the Tensile Strength test will determine the material's resistance to tension, highlighting its capacity to endure stretching forces without failure. Through these tests, the project aims to comprehensively assess the mechanical and physical characteristics of UHPFRC, paving the way for its optimized application in construction [16, 17].

Additionally, the project aims to prepare review and research papers based on the analysis of existing studies and experimental work conducted onsite. UHPFRC's tensile strength can be significantly enhanced, potentially doubling when transitioning from normal to high-strength concrete. This allows for the reduction in the thickness of concrete slabs, although this reduction is limited by the increase in modulus. While the construction cost of UHPFRC is higher than that of normal concrete pavement, the enhanced mechanical properties and potential for reduced thickness may offset the additional costs. There is a need to develop design methodologies for UHPFRC pavements under Indian conditions, along with guidelines and standards for design, construction, testing, and long-term performance monitoring. Several significant projects currently employ UHPFRC, showcasing its advanced capabilities in various infrastructure applications [18–20]. Notable among these are the bridges on the Samruddhi Mahamarg, which leverage UHPFRC's superior strength and durability to enhance their structural integrity and lifespan. Viaducts on NH17 at the Kashedi Ghat Section also utilize UHPFRC, benefiting from its high compressive strength and resistance to environmental stressors. Additionally, UHPFRC is being employed in projects along the Mumbai Goa (NH66) National Highway, where its robust properties contribute to the overall resilience and safety of the roadway infrastructure. Furthermore, the material has been crucial in the repair and rehabilitation of rigid pavement panels on the Mumbai-Pune Expressway, demonstrating its effectiveness in extending the service life and performance of existing concrete structures. These projects exemplify the practical and transformative applications of UHPFRC in modern construction.

Challenges Despite its numerous benefits, UHPFRC faces several implementation challenges. These include fiber dispersion, alignment, and orientation within the concrete matrix, corrosion of steel fibers, fiber balling and segregation during mixing, specialized construction procedures, cost considerations, and the need for appropriate mix design and proportioning. Addressing these challenges is crucial for optimizing the performance and reliability of UHPFRC in various construction applications [21].

In summary, UHPFRC represents a significant advancement in concrete technology, offering superior mechanical properties and sustainability benefits. However, successful implementation requires overcoming several technical challenges and developing appropriate design methodologies and standards.

2 Literature Review

UHPFRC has garnered significant attention due to its superior mechanical properties and potential applications in modern construction. Various studies have explored different aspects of UHPFRC, including material properties, design procedures, and practical applications. A comprehensive review of UHPFRC emphasizes its material properties and design procedures, which serve as foundational knowledge for subsequent research in this field [1]. An evaluation has been conducted to determine the appropriateness of UHPFRC for the restoration and strengthening of concrete structures. The assessment has emphasized the efficacy of UHPFRC in improving the structural integrity [2]. The study examined the effect of silica fume and steel fiber on the mechanical features of high-strength concrete, revealing substantial enhancements in performance [3]. This is also reinforced by conversations on the attributes of ultra-high-performance 'ductile' concrete and its influence on sustainable building [3]. A study has investigated the impact of silica fume on the permeability and microstructure of high-strength concrete, demonstrating improved durability and decreased porosity [4]. Furthermore, researchers have investigated the permeability and porosity properties of concrete reinforced with steel fibers, confirming the advantages of including fibers [5]. Research investigating the impact of superplasticizer and curing circumstances on the characteristics of concrete, both with and without fiber, has demonstrated that the inclusion of fibers greatly enhances mechanical qualities across different curing conditions [6]. Insights into the progress in knowledge and design codes for high-performance fiber-reinforced concrete emphasize the advancements in this field [7]. The tensile strength behavior of ultra-high-performance concrete (UHPC) and UHPFRC has been focused on, highlighting the superior performance of UHPFRC in tension compared to conventional concrete [9]. Discussions on the application of UHPFRC in bridge construction underscore its potential in infrastructure projects [10]. Reviews of the mechanical performance, durability, sustainability, and implementation challenges of UHPC provide a holistic view of the material's capabilities and limitations [11]. Case studies on fiber-reinforced concrete emphasize its practical applications and benefits [12]. The impact of vibration on fresh concrete of conventional and UHPFRC has been examined, revealing how vibration affects the properties and performance of the concrete [13]. Experimental studies on the mechanical properties of UHPFRC confirm its superior performance in various mechanical tests [14]. Reviews of the properties of UHPFRC focus on its application in construction and the challenges associated with its use [15]. Investigations into the impact of nano-silica and micro-steel fiber on the engineering properties of UHPC demonstrate enhanced mechanical and microstructural properties [16]. Experimental research on UHPC contributes valuable data on its mechanical behavior under different conditions [17]. Standards for fresh concrete are essential for ensuring consistency and quality in concrete production [18]. Reviews of the cost impact of UHPC in durability aspects highlight the economic considerations in using UHPFRC [19]. Reviews of the use of UHPFRC in bridge restoration and new structure in Switzerland provide practical insights into its application in real-world projects [20]. Discussions on the mix design and flexural response of UHPFRC offer guidance on optimizing the concrete mix for specific applications [21]. Investigations into the mechanical and microstructural properties of steel fiber–reinforced concrete (SFRC) provide further evidence of the benefits of fiber incorporation [22]. Several other studies have also contributed to the understanding of UHPFRC, including research on its application in strengthening black cotton soil, internal curing agents, and energy-efficient residential building design [23–26]. Further investigations into the behavior of UHPFRC under different loading conditions, such as the effect of fiber reinforcement on the deflection of prestressed concrete beams, provide deeper insights into its structural capabilities [27, 28]. The impact of substituting industrial wastes with fine aggregate in concrete has also been reviewed, highlighting sustainable construction practices [29]. These studies collectively underscore the versatile applications and superior performance of UHPFRC in various construction scenarios. Further investigations into the behavior of UHPFRC under different loading conditions, such as the impact of fiber reinforcement on the deflection of prestressed concrete beams, provide deeper insights into its structural capabilities [30, 31]. The impact of substituting industrial wastes with fine aggregate in concrete has also been reviewed, highlighting sustainable construction practices [32]. Studies on the performance of fiber-reinforced polymer laminates in strengthening prestressed concrete beams further illustrate the advancements in the field [33, 34]. The extensive body of research on UHPFRC highlights its remarkable mechanical properties, durability, and suitability for a wide range of construction applications. Continued research and development are essential to fully realize the potential of UHPFRC in enhancing modern infrastructure.

3 Material and Methodology

3.1 Materials

UHPFRC is composed of OPC, silica fume, fine aggregate, water, steel fibers, and a high-range water reduction agent. The typical concrete mix design includes 2% steel fibers by volume.

3.1.1 Cement

The primary binder in UHPFRC is OPC. UHPFRC requires more than twice as much cement (700 kg/m^3) compared to conventional strength concrete. Interestingly, due to the low water content in the mixture, not all the cement in the UHPFRC matrix hydrates. Anhydrate cement grains can serve as high elastic modulus reinforcements in the matrix, while hydrated cement serves as a bonding agent.

3.1.2 Sand

Sand acts as an aggregate in UHPFRC, and the durability of UHPFRC is increased by removing coarse particles. Quartz sand, with its high stiffness and favorable paste-aggregate interfaces, is often suggested. Typically, the mean particle size is less than 1 mm, although UHPFRC has been developed with maximum particle sizes ranging from 8 to 16 mm. The sand limits the cement matrix to increase strength, and a type of quartz sand that is not chemically vigorous in the cement hydration reaction at room temperature is often utilized.

3.1.3 Silica Fume

Silica fume, a highly reactive pozzolanic material, is essential for both HPC and UHPFRC. With a diameter of 0.2 μm, silica fume is the smallest component of UHPFRC. It improves the rheological properties of UHPFRC, fills gaps between cement grains, and generates hydration products through pozzolanic action. The pozzolanic reaction, which occurs when silica hydrates react with $Ca(OH)_2$ (portlandite), results from the hydration of OPC. This reaction produces C-S-H hydrates using portlandite as a raw material, leading to the elongation of C-S-H chains. Furthermore, the presence of silica fume enhances the microstructure, compacity, and mechanical strength of UHPFRC.

3.1.4 Steel Fibers

UHPC is inherently brittle because of its exceptionally high strength and uniformity. However, its brittleness may be mitigated by including steel fibers, which enhance its ductility. The predominant dimensions of steel fibers often employed are 13 mm (0.5 in.) in length and 0.20 mm (0.008 in.) in diameter. The fibers in the UHPFRC pass through small fissures, carrying tensile loads and helping to make it more flexible. The alignment of the fibers relative to the plane of cracking has a considerable impact on this malleable behavior. It is crucial to use caution when mixing and placing UHPFRC to avoid the formation of fiber clusters and to guarantee the uniform distribution of fibers inside each component. The workability of a concrete mix including fibers is influenced by the size of the coarse particles and the fibers present in the mix. Within UHPFRC, the lack of large particles results in the size of the fibers exerting the most significant influence on the flowability of the concrete. Fig. 1 depicts the arrangement of steel fibers in UHPFRC.

Fig. 1 Cross-section of a UHPFRC sample showing steel fiber orientation

Table 1 Optimum content of steel fiber in different erosion environments

Environmental erosion conditions	Optimum volume ratio (%)
Water	2
Freeze-thaw cycles	2
NaCl	1
Na_2SO_4	1
H_2SO_4	2
NaOH	2

Fig. 2 Superplasticizer carried on gantry

With increasing fiber size, the workability of UHPFRC blends declines. The most prevalent content for UHPFRC is 2% fiber volume, which also happens to be the most affordable content found [15]. According to Qinyuan Zhang et al. [23], various steel fiber percentages can be opted for varying conditions, as shown in Table 1.

3.1.5 Superplasticizer

Superplasticizers can effectively enhance the workability of UHPC by counteracting its low water-to-binder ratio (SP) [11]. They enhance the fluidity of UHPFRC mixes by dispersing cement and silica fume. Superplasticizers aid in achieving reduced water/cement and water/binder ratios (the latter consisting of both cement and silica fume) while maintaining the workability of the mixture. The optimal quantity of superplasticizer is rather substantial, comprising around 1.6% of the cement content, due to the use of low water-to-cement and water-to-binder ratios in UHPFRC. Superplasticizers of the third generation, specifically polycarboxylates and polycarboxylate ethers, are commonly used due to their high productivity and absence of a specific limit for low water-to-cement ratios [15]. The superplasticizers utilized on the construction site were melamine based and were introduced using an appropriate gantry, as seen in Fig. 2.

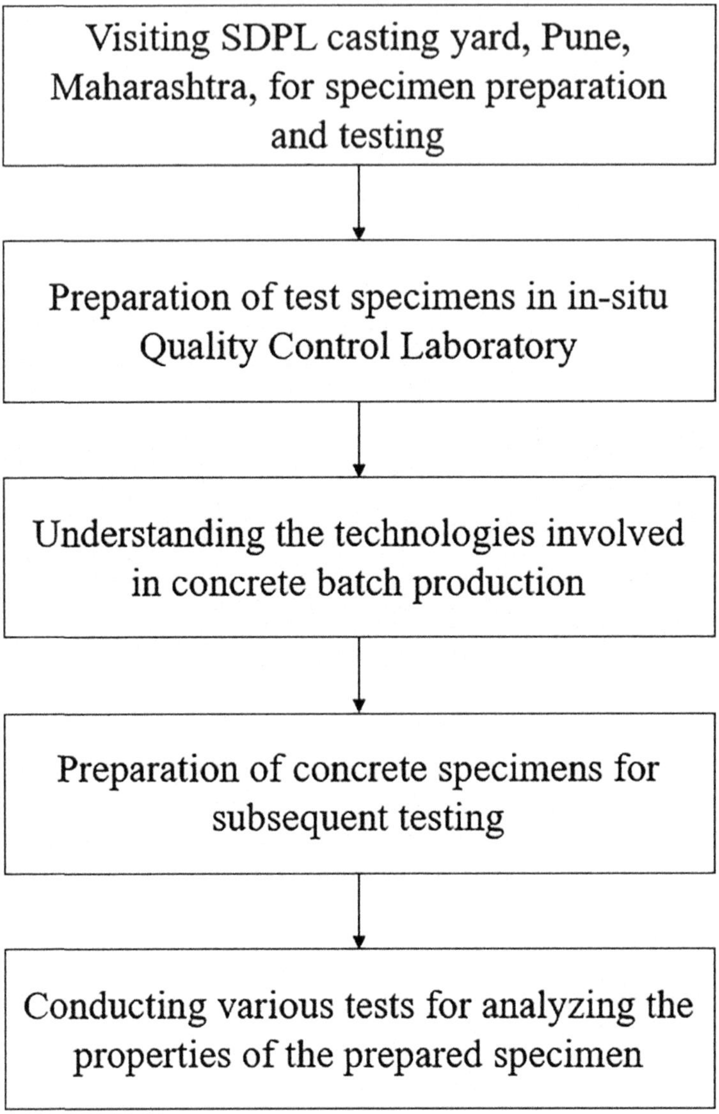

Fig. 3 Flowchart of methodology adopted in the present study

The methodology adopted for this research, as shown in Fig. 3, involved several systematic steps to ensure comprehensive analysis and accurate results. Initially, we visited the SDPL casting yard in Pune, Maharashtra, to prepare and test the specimens. Following this, the preparation of test specimens was conducted in an in situ quality control lab to maintain controlled conditions and high standards. Understanding the technologies involved in concrete batch production was crucial, as it provided insights into the materials and processes used. Subsequently, concrete specimens were prepared for testing, ensuring they met the required specifications. Various tests were then conducted to analyze the properties of the prepared specimens, covering aspects such as compressive strength, flexural strength, and workability. Finally, the tested samples were collected for further analysis and documentation, completing the methodology cycle and ensuring that all relevant data was captured accurately for comprehensive evaluation.

4 Technology and Chemical Processes in Concrete Production

Concrete production involves a variety of advanced technologies and materials to create the final product, ensuring it meets specific structural and durability requirements. Key technologies commonly used in concrete production include mixing, cement production, admixtures, reinforcement, formwork, and curing.

4.1 Mixing Technology

Concrete mixing technology is essential for blending cement, aggregates (such as sand and gravel), water, and often additional additives. Two primary mixing technologies are utilized: batch plants and continuous mixing plants. Batch plants mix concrete ingredients in specified proportions batch by batch, allowing for greater control over the mixture, making them ideal for smaller-scale concrete production. In contrast, continuous mixing plants, also known as ready-mix plants, produce concrete in a continuous flow, suitable for large-scale projects requiring a high volume of concrete.

4.2 Cement Production

Cement, a crucial component of concrete, is produced through a series of technological processes. Quarrying involves extracting limestone, clay, and other raw materials from quarries and mines. These materials are then crushed and finely ground before being blended in specific proportions to produce cement powder. The blended raw materials are heated in a kiln at high temperatures, around 1450 °C (2642 °F), to produce clinker, which is then ground into a fine powder to form cement.

4.3 Admixtures

Admixtures are substances added to the concrete mixture to enhance its properties or improve its workability. Typical varieties of admixtures encompass water reducers, which decrease the water required for a specific concrete mix, thus enhancing its strength and longevity. Retarders are substances that delay the process of concrete setting, resulting in a longer period of time during which the concrete may be worked with. This is particularly beneficial in situations where the weather is hot. Accelerators expedite the process of setting and solidifying concrete, proving advantageous in low temperatures or situations when swift building is necessary. Regarding the slump test, there is a slight increase in the slump of vibrated conventional concrete compared with regular concrete. Additionally, the slump of Mixed UHPFRC increases compared with Casting UHPFRC.

4.4 Reinforcement Technology

Reinforced concrete is extensively utilized in construction and entails the incorporation of steel bars or mesh throughout the concrete to augment its ability to withstand tension. Steel reinforcement is the strategic placement of steel bars or mesh into the concrete formwork before pouring, in order to enhance strength and resistance against tensile pressures. Prestressed concrete applies tension to the reinforcing steel prior to the concrete being poured. Upon release, the steel exerts pressure on the concrete, thus improving its structural integrity. Regarding fatigue strength, the inclusion of fibers rises fatigue strength by about 90% and 70% of the static strength after 2×106 cycles for nonreverse and complete reversal loading, respectively [12].

4.5 Formwork Technology and Curing Technology

Formwork refers to temporary structures used to shape and support the poured concrete until it hardens. Various formwork technologies exist, including customary timber formwork, steel formwork, and modular formwork systems. These technologies ensure that the concrete keeps the desired shape and integrity during the curing process. Curing is the process of maintaining favorable conditions (temperature, moisture) for the concrete to develop its desired strength and durability. Different curing methods, such as water curing, steam curing, and chemical curing, are employed depending on the project requirements. Proper curing ensures that the concrete achieves its maximum strength and extends its lifespan, making it a critical step in the concrete production process. Using steam curing treatment is required to receive all the improved benefits associated with UHPFRC. According to some experts, the conventional heat curing procedure follows the initial curing for 48 h at 90 °C. Heat curing increases durability, encourages high density and high strength, and lessens delayed shrinkage and creep effects. To eliminate the potential of delayed ettringite development, it must also be completed only after the concrete has fully set, necessitating thorough understanding of the setting period. UHPC's overall porosity appears to be influenced by the curing method used on the substance. Untreated UHPC has a total porosity of around 8.4%; however after heat treatment, it has a total porosity of around 1.5%. Hence, UHPC's greater endurance is due to its unconnected pore structure [15]. For the performed experiment, cubes were treated in an accelerated curing tank for 48 h and a temperature of 90 °C was maintained.

5 Results and Discussion

5.1 Flexural Tensile Test

The flexural tensile strength test was conducted on prismatic specimens of 150 mm by 150 mm by 500 mm. The test followed the beam technique outlined in the ASTM C1018 standard, with force applied at one-third intervals along the specimen. Two prismatic specimens were generated for each series. The testing was conducted using a closed-loop deflection-controlled loading frame with a capacity of 250 kN and a loading rate of 0.5 mm/min. Simultaneously, the test captured the deflections of the bottom surface and the supports of the beams at their mid-span. The net mid-span deflection of the beam was determined by subtracting the average of the support deflections from the mid-span deflection. The load-deflection curves for each specimen were also produced and shown graphically [3]. The flexural strength and ductility characteristics of UHPFRC are about 250 times greater than those of conventional concrete [15]. Figure 4 illustrates the load-deflection diagram for UHPFRC during bending, with the normal phases clearly highlighted. Additionally, Fig. 5 presents the failure produced from a point loading test. UHPFRC's behavior under flexure loading may be described using three stages.

1. The material exhibits linear elastic behavior until it reaches its original breaking strength.

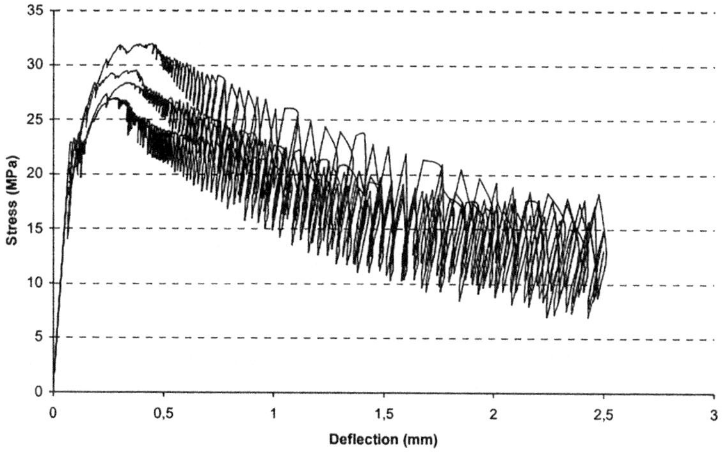

Fig. 4 Four-point flexural tests: stress versus deflection

Fig. 5 Failure obtained after four-point loading test

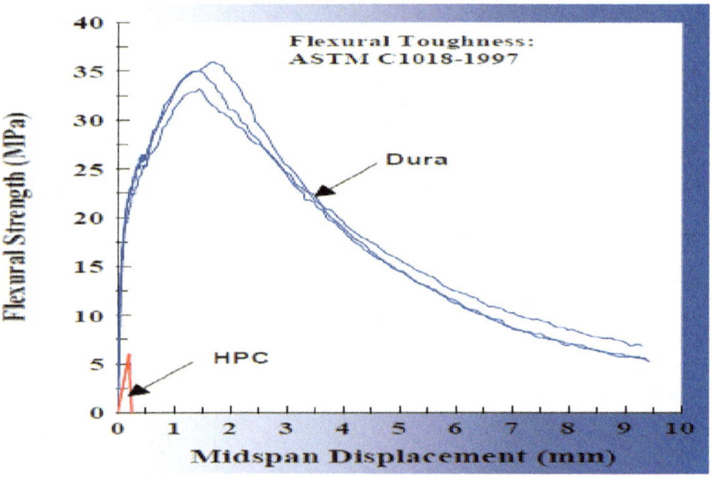

Fig. 6 The mid-span deflection with respect to the flexural strength in HPC and UHPFRC

2. A period of displacement hardening occurs prior to reaching the maximum load.
3. A phase of deflection softening occurs after reaching the maximum load.

The experiment performed on the specimen, produced by mixing the constituent materials for 1.5 h, was conducted in accordance with ASTM C1018 standard. It was performed using a Universal Testing Machine with a capacity of 200 kN. The flexural strength was 27 MPa which is much higher than that of ordinary concrete. It is universally accepted that addition of steel fibers leads to an overall increase of ultimate strength in concrete beams. Figure 6 shows the comparison between HPC and UHPFRC for the flexural values and the midspan deflection of the concrete

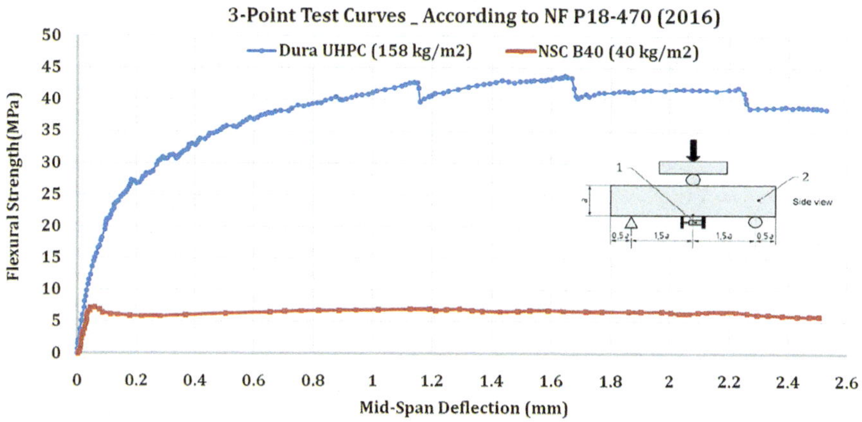

Fig. 7 The mid-span deflection with respect to flexural strength in NSC and UHPFRC

specimen. Figure 7 shows the comparison between normal strength concrete (NSC) and UHPFRC for the flexural values and the midspan deflection of the specimens visible in the figure. UHPFRC has a much greater loading at first crack than NSC and HPC and hence leads to a greater flexural toughness as visible in the figure.

5.2 *Compressive Strength*

The ratio of water to cement was 0.26. Steel fiber might slightly affect the water requirement for concrete mixing. The concrete compressive strength specimens were cubic in shape, measuring 150 mm × 150 mm × 150 mm. Each group consisted of three identical specimens, which were used to ensure consistent findings. The steel fiber concentration in this investigation was varied at 0, 1, and 2% [22]. For the purpose of evaluating compressive strength and elastic modulus, three cylindrical specimens measuring 150 mm in diameter and 300 mm in height were utilized. The compressive strength was determined by calculating the average test results of three specimens. The cube that exhibited failure is seen in Fig. 8.

The compressive strength test performed in a Compression Testing Machine with a capacity of 2000 kN, the failure occurred at 141 MPa which is at least two to three times stronger than NSC, as shown in Fig. 9.

5.3 *Flow Table Test*

Flowability (ASTM-C230M 2008). Through mixing and before casting, a straightforward flow table test in accordance with ASTM-C230M (2008) is used to gauge the UHPFRC's consistency, as shown in Fig. 10. In general, compressive strength declines as flowability rises. At flow values between 170 and 190 mm, a significant

Fig. 8 Failure obtained in compressive strength test on standard specimen

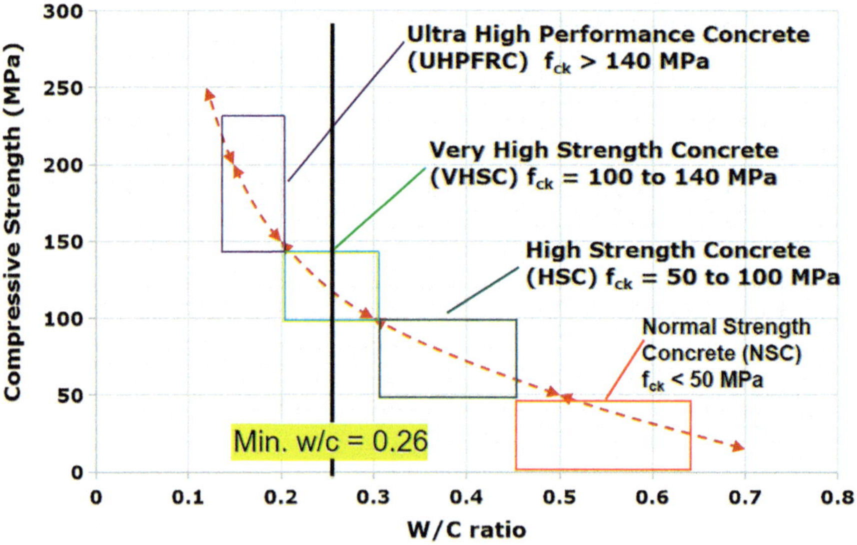

Fig. 9 Graph showing the relationship between w/c ratio and compressive strengths of various types of concrete

Fig. 10 Flow table test apparatus for workability

variance has been noted. This is mostly caused by the variety of cement sources offered by local vendors [4].

The ability of concrete to handle, transport, and place between the formwork with minor segregation is called the workability of concrete. Flow table test was employed using metal cone of 50 mm, mold diameter of 130 mm, and 20 blows in accordance with ASTM C230. A flow diameter of 230 mm was obtained signifying a relatively higher consistency of UHPFRC than ordinary concrete.

5.4 *Tensile Strength*

The average tensile strength values for UHPC (which contains no fiber) in experimental direct tensile strength testing range from 7 to 10 MPa. Depending on the fiber qualities, the slope of the falling branch may vary in the UHPFRC tensile stress-crack opening diagram (i.e., amount and orientation) [10]. The UHPFRC is one of the concrete types included in the HPFRCC group. Its extra benefit of having a very dense low-permeable matrix is in addition to its typical high-performing attributes. Figure 7 compares the uniaxial tensile behavior of normal concrete, SFRC, and UHPFRC. The findings unmistakably demonstrate that, in comparison to other cementitious materials, UHPFRC has superior tensile strength and strain-hardening behavior [15]. As per study, adding 3% fiber to mortar raises its tensile strength by nearly 2.5 times linked to mortar without reinforcement [13].

Tensile Strength is the capacity of concrete to resist tension/stretched tight. It is generally known that tensile strength. UHPFRC shows usable structural ultimate tensile strength (>8 MPa). The split tensile strength performed on cylindrical specimens in accordance with ASTM C496 yielded a value of 12 MPa.

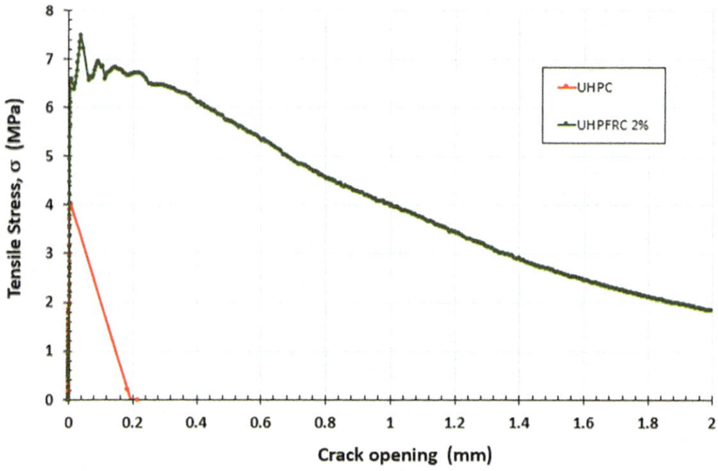

Fig. 11 UHPC versus UHPFRC at (2% fiber)

An extremely dense, low-permeable matrix adds another benefit to its regular high-performance qualities. Figure 11 compares the uniaxial tensile behavior of normal concrete, SFRC, and UHPFRC. In comparison to other cementitious materials, the results unmistakably demonstrate that UHPFRC has better tensile strength and strain-hardening tendency. The investigation used UHPC and UHPFRC, which exhibited average maximum tensile strengths of 4.0263 MPa (for UHPC) and 6.5851 MPa (for UHPFRC 2%). Additionally, the correlating fracture opening lengths were measured to be 0.0078 mm (for UHPC) and 0.0068 mm (for UHPFRC 2%).

6 Conclusion

The comprehensive evaluation of UHPFRC through standardized tests, including flexural tensile strength, compressive strength, flow table test, and tensile strength, reveals significant improvements over normal and HPC. UHPFRC exhibits exceptional strength and ductility in flexural tensile strength tests, with performance approximately 250 times greater than that of normal concrete. The load-deflection behavior is characterized by linear elastic behavior, displacement-hardening, and deflection-softening phases, demonstrating superior resistance to cracking and deformation under flexural loads. Compressive strength tests showed UHPFRC's failure at 141 MPa, which is two to three times stronger than conventional concrete, enhanced by the inclusion of steel fibers. The flow table test highlighted UHPFRC's high workability with a flow diameter of 230 mm, indicating better flowability compared with ordinary concrete. Tensile strength tests confirmed remarkable improvements, with average values ranging from 7 to 10 MPa for UHPC and up to 12 MPa for UHPFRC, significantly enhanced by steel fibers. Adding 3% fiber increased tensile strength by nearly 2.5 times compared to mortar without reinforcement. These findings confirm UHPFRC's superior mechanical properties, making it an ideal material for high-performance and durable concrete structures. UHPFRC's enhanced flexural and compressive strength, combined with excellent workability and tensile properties, position it as a valuable solution for modern construction challenges, particularly in bridge construction. Future research and development should focus on optimizing mix designs and exploring further applications to maximize UHPFRC's benefits in various structural contexts.

References

1. Buttignol, T. E. T., Sousa, J. L. A. O., & Bittencourt, T. N. (2017). Ultra high-performance fiber-reinforced concrete (UHPFRC): A review of material properties and design procedures. *Revista IBRACON de Estruturas e Materiais, 10*, 957–971.
2. Bandara, S., Wijesundara, K., & Rajeev, P. (2023). Ultra-high-performance fibre-reinforced concrete for rehabilitation and strengthening of concrete structures: A suitability assessment. *Buildings, 13*(3), 614.
3. Köksal, F., Altun, F., Yiğit, İ., & Şahin, Y. (2008). Combined effect of silica fume and steel fiber on the mechanical properties of high strength concretes. *Construction and Building Materials, 22*(8), 1874–1880.
4. Voo, Y. L., & Foster, S. J. (2010). Characteristics of ultra-high performance 'ductile' concrete and its impact on sustainable construction. *The IES Journal Part A: Civil & Structural Engineering, 3*(3), 168–187.
5. Chaudhary, S. K., & Sinha, A. K. (2020). Effect of silica fume on permeability and microstructure of high strength concrete. *Civil Engineering Journal, 6*(9), 1697–1703.
6. Zeng, W., Zhao, X., Zou, B., & Chen, C. (2023). Topographical characterization and permeability correlation of steel fiber reinforced concrete surface under freeze-thaw cycles and NaCl solution immersion. *Journal of Building Engineering, 80*, 108042.
7. Aruntaş, H. Y., Cemalgil, S., Şimşek, O., Durmuş, G., & Erdal, M. (2008). Effects of super plasticizer and curing conditions on properties of concrete with and without fiber. *Materials Letters, 62*(19), 3441–3443.
8. Walraven, J. C. (2009). High performance fiber reinforced concrete: Progress in knowledge and design codes. *Materials and Structures, 42*(9), 1247–1260.
9. Faisal, F. W., & Ashour, S. A. (1992). Mechanical properties of high-strength fiber reinforced concrete. *ACI Material Journal, 89*(5), 449–455.
10. Kusumawardaningsih, Y., Fehling, E., Ismail, M., & Aboubakr, A. A. M. (2015). Tensile strength behavior of UHPC and UHPFRC. *Procedia Engineering, 125*, 1081–1086.
11. Foster, S. J., & Voo, Y. L. (2015). UHPFRC as a material for bridge construction: Are we making the most of our opportunities? *Concrete in Australia, 41*(2), 1–22.
12. Abbas, S. M. L. N., Nehdi, M. L., & Saleem, M. A. (2016). Ultra-high performance concrete: Mechanical performance, durability, sustainability and implementation challenges. *International Journal of Concrete Structures and Materials, 10*, 271–295.
13. Ragavendra, S., Reddy, I. P., & Dongre, A. R. C. H. A. N. A. A. (2017). Fibre reinforced concrete-a case study. In *Proceedings of the Architectural Engineering Aspect for Sustainable Building Envelopes, Hyderabad, India* (pp. 10–11).
14. Sudarshan, N. M., & Chandrashekar Rao, T. (2017). Vibration impact on fresh concrete of conventional and UHPFRC. *International Journal of Applied Engineering Research, 12*(8), 1683–1690.
15. Gangwar, S., Mishra, S., & Sharma, H. K. (2019). An experimental study on mechanical properties of ultra-high-performance fiber-reinforced concrete (UHPFRC). In *Sustainable con-*

struction and building materials: Select proceedings of ICSCBM 2018 (pp. 873–886). Springer Singapore.
16. Tayeh, B. A., Aadi, A. S., Hilal, N. N., Bakar, B. A., Al-Tayeb, M. M., & Mansour, W. N. (2019). Properties of ultra-high-performance fiber-reinforced concrete (UHPFRC)—A review paper. In *AIP conference proceedings* (Vol. 2157, p. 020040). AIP Publishing LLC.
17. Hakeem, I. Y., Amin, M., Abdelsalam, B. A., Tayeh, B. A., Althoey, F., & Agwa, I. S. (2022). Effects of nano-silica and micro-steel fiber on the engineering properties of ultra-high performance concrete. *Structural Engineering and Mechanics, 82*(3), 295–312.
18. Zhang, X., & Zhang, H. (2019). Experimental research on ultra-high performance concrete (UHPC). In *IOP conference series: Materials science and engineering* (Vol. 562, p. 012045). IOP Publishing.
19. Harrison, T. (2004). *Standards for fresh concrete*. BSI British Standards Institution.
20. Kravanja, G., Mumtaz, A. R., & Kravanja, S. (2024). A comprehensive review of the advances, manufacturing, properties, innovations, environmental impact and applications of ultra-high-performance concrete (UHPC). *Buildings, 14*(2), 382.
21. Bertola, N., Schiltz, P., Denarié, E., & Brühwiler, E. (2021). A review of the use of UHPFRC in bridge rehabilitation and new construction in Switzerland. Frontiers. *Built Environment, 155*.
22. Sudarshan, N. M., & Rao, T. C. (2018). Mix design and flexural response of ultra high performance fiber reinforced concrete (UHPFRC). *Materials Today: Proceedings, 5*(11), 25420–25429.
23. Zhang, Q., Xu, W., Sun, Y., & Ji, Y. (2022). Investigation on mechanical and microstructure properties of steel fiber reinforced concrete. *Advances in Materials Science and Engineering*.
24. Bharti, G., Shukla, B. K., Kumar, A., Nagaraju, M. B., Bansal, V., & Varadharajan, S. (2023). Impact of copper and steel slag blend in strengthening mechanical properties of black cotton soil. In *AIP conference proceedings* (Vol. 2800). AIP Publishing.
25. Shukla, B. K., Bharti, G., Sharma, P. K., Sharma, M., Rawat, S., Maurya, N., & Srivastav, Y. (2024). Sustainable construction practices with recycled and waste materials for a circular economy. *Asian Journal of Civil Engineering., 25*, 5255. https://doi.org/10.1007/s42107-024-01111-y
26. Hurukadli, P., Kumar, A., Shukla, B. K., & Bharti, G. (2023). Experimental study on the influence of available absorbent as internal curing agent on cement mortar. In *AIP conference proceedings* (Vol. 2800). AIP Publishing.
27. Shukla, B. K., Parashar, B., Bharti, G., Sharma, P. K., & Kumar, A. (2023). A sustainable design of energy efficient and environment friendly residential building for tropical climate. In *AIP conference proceedings* (Vol. 2800). AIP Publishing.
28. Bharti, G., Shukla, B. K., Kumar, A., & Hurukadli, P. N. (2023). Stabilization of black cotton soil using steel slag. In *AIP conference proceedings* (Vol. 2800). AIP Publishing.
29. Bharti, G., Shukla, B. K., Kumar, A., Nagaraju, M. B., Bansal, V., & Varadharajan, S. (2023). Effect of implementation of waste shredded plastic on construction of flexible pavement. In *AIP conference proceedings* (Vol. 2800). AIP Publishing.
30. Hurukadli, P., Kumar, A., Shukla, B. K., & Bharti, G. (2023). A critical review on behavior of fiber reinforced pre stressed concrete. In *AIP conference proceedings* (Vol. 2800). AIP Publishing.
31. Hurukadli, P., Bharti, G., Shukla, B. K., Garg, P., & Tripathi, A. (2023). Deflection of reinforced concrete beams: An analytical study on the impact of steel fibers. *Materials Today: Proceedings*. https://doi.org/10.1016/j.matpr.2023.03.321.
32. Parikshit, N. H., Garg, P., Bharti, G., Tripathi, A., & Shukla, B. K. (2023). Analytical study on the effect of steel fiber in the deflection of pre-stressed concrete beams. In *IOP conference series: Earth and environmental science* (Vol. 1110, p. 012027). IOP Publishing.
33. Varadharajan, S., Kirthanashri, S. V., Samim, M., Shukla, B. K., & Bharti, G. (2023). A review on influence of replacing industrial wastes with fine aggregate in concrete. *Lecture Notes in Civil Engineering, 281*, 183–188.
34. Hurukadli, P., Bharti, G., & Shukla, B. K. (2023). Behavior of fiber reinforced polymer laminates strengthening prestressed concrete beams. *Materials Today: Proceedings, 78*, 731–737.

Modelling of Temporal Clear Water Scour Depth Around Bridge Piers Using XGBoost and SVM-PSO

Prince Gaurav, G. Lohith Reddy, A. Baranwal, and B. S. Das

1 Introduction

Scour is the process by which sediment particles are removed from the water stream bed through the erosive action of flowing water, carrying sediment away from hydraulic structures. The scour around the foundation of a bridge pier is a critical consideration in bridge design and construction, as many bridge failures have been attributed to scour or undermining of hydraulic structures such as piers, abutments, and spur dikes. Accurate scour depth estimation around these structures is crucial for the safe and cost-effective design of hydraulic structures, determining the depth of their foundations beneath the water stream bed. Scouring around bridge piers results from bed sediment movement caused by the obstruction of the bridge pier in the flow direction [1]. This obstruction accelerates the flow, forming vortices around the bridge pier that erodes sediment from its vicinity. Ettema [3] identified three scour phases: initial, intermediate, and equilibrium [16]. However, determining the criteria for reaching equilibrium has been a subject of controversy and is somewhat arbitrary. Underestimating scour depth may lead to designing shallow foundations, risking exposure of the foundation to the flow and compromising bridge safety. Furthermore, overestimating local scour depth results in an uneconomical bridge pier design. Scouring around a bridge pier occurs due to sediment erosion from the bed foundation in flow conditions.

Tison [4] laid the groundwork for subsequent investigations into bridge pier scour, contributing to an enhanced comprehension of the factors influencing equilibrium scour depth around different bridge pier shapes. Various studies have been performed to advance our knowledge of scouring around the bridge pier and develop effective mitigation strategies. Noteworthy contributions include insights

P. Gaurav · G. L. Reddy · A. Baranwal (✉) · B. S. Das
National Institute of Technology, Patna, Bihar, India

S. Pathak et al. (eds.), *Intelligent Infrastructure and Smart Materials*,
https://doi.org/10.1007/978-3-031-92421-7_18

into the significance of bridge pier shape, sediment particle size, and flow velocity, as demonstrated in studies by Kothyari et al. [1], Melville and Chiew [5], Sheppard et al. [6], Rathod and Manekar [7], Baranwal and Das [8]. Various non-circular shapes of bridge piers exhibit distinct effects on equilibrium scour depth. Consequently, meticulous consideration of pier shape is imperative during bridge design.

Moreover, many researchers have provided remedial measures, as evidenced by the work of Garg et al. [9], Vijayasree and Eldho [10]; Zhang et al. [11], and Singh et al. [12]. These measures aim to diminish the impact of scour on bridge piers by implementing protective structures, managing sediment transport, and designing pier shapes that are less vulnerable to scouring. The river current exhibits unpredictability during events and rapid changes in discharge can occur. Due to the brief duration of the maximum flood discharge, predictions of scour depth using steady flow relations for the maximum flood effluent result in overly deep estimates. Consequently, designing hydraulic foundations based on these maximum scour depth estimations for the design discharge may lead to inaccuracies. The primary objective of this study is to formulate a model for calculating the temporal variation of scour depth around the circular shape of bridge piers placed at a streambed level. This study models scour depth around bridge piers using two soft-computing techniques—XGBoost and SVM-PSO—under unsteady flow conditions. Existing datasets and empirical scour depth equations for unsteady flows were compiled for analysis. The findings may assist researchers and bridge engineers in predicting local scour depth under varying temporal conditions.

2 Literature Review

Many researchers have done extensive work to develop a relationship to predict the scour depth around a bridge pier, such as Shen et al. [13], Kothyari et al. [1], Chang et al. [14], Lee and Sturm [15], Pandey et al. [16], and Kumar et al. [42]. Mia and Nago [17] employed a sediment transport equation on a cylinder-shaped pier in controlled bed conditions to predict local scour depth over time, reaching equilibrium as the bed-shear stress approached critical levels. Lee and Sturm [15] explored the influence of relative sediment size on pier scour depth, employing three sediment sizes and diverse bridge pier designs at different geometric scales. Das et al. [18] utilized ultrasonic Doppler velocimetry to focus on the equilibrium scour hole around circular piers, emphasizing the significance of the horseshoe vortex phenomenon. The impact of flow unsteadiness on bridge pier scour has been explored by Kim et al. [19]. However, these investigations have predominantly focused on scouring around piers with uniform cross-sections and consistent pier heights. In recent years, various machine learning (ML) approaches have been explored for modelling bridge pier scouring. Bateni et al. [20] conducted a study employing ANFIS to model scour depth around bridge piers. The findings indicated that pier diameter (b) exerts a more substantial influence on equilibrium scour depth than other independent parameters. The generalized regression neural network (GRNN) model exhibited superior performance in modelling the scour depth and emphasized the effectiveness of the GRNN model in capturing the complexities of scour processes. Still, it does not provide specific insights into its applicability under distinct flow regimes [21]. Support vector regression tuned radial basis function (SVR-RBF) has superiority in predicting the scour depth and identifying that flow depth (y) and pier width (b) are crucial influencing factors in modelling the scour depth around the bridge pier. Artificial neural network (ANN) models exhibited the highest level of accuracy in predicting scour depth based on laboratory data, particularly under clearwater scouring (CWS) and live bed scouring (LBS) conditions. However, the same level of performance was not replicated when applied to field data, and it was observed that ANN models may not be as effective in real-world scenarios. According to Talpur et al. [22], Genetic programming (GP) yields viable results in estimating suspended sediment load (SSL). However, most studies indicate that ANN is a more potent tool than GP. Utilizing a support vector machine (SVM) demonstrates precise outcomes, particularly when employing carefully chosen kernels. Notably, hybrid models yield superior results to models solely based on artificial intelligence. Etemad-Shahidi et al. [23] emphasize the importance of careful data selection and empirical models based on CWS and LBS conditions to enhance the accuracy of scour depth prediction models. Their study reveals that a genetic function-based model outperformed ANN and regression-based models, particularly in predicting clear-water scour depth. Particle swarm optimization (PSO) demonstrated superior performance to other scour depth models, especially for clear-water scour datasets. The study identified the parameters ratio of pier width to flow depth (b/y) and the ratio of median grain size to flow depth (d_{50}/y) as the most influential factors for accurately predicting scour depth in laboratory and field data [24]. Aksoy et al. [25]; Pandey et al. [26] attributed scour hole creation to the development of an armour layer in non-uniform gravel bed conditions. Chou and Pham [27] utilized the classification and regression trees (CART) model to predict the monthly SSL of the Haraz River. As a result, the CART model demonstrated commendable performance across various evaluation criteria, surpassing the capabilities of the ployed data-driven techniques to predict SSL. The findings indicated that the ANFIS models exhibited superior predictive performance to the ANN, FL, and LSSVM models for SSL. Hamidifar et al. (2021) investigated equations for estimating critical flow velocity ratios, proposing enhanced models for CWS conditions. Garg et al. [29] conducted experiments in tandem arrangements, developing a scour depth model through multivariable regression analysis have modelled a local scour depth using Sequential minimal optimization regression (SMOREG), GBM, k-nearest neighbors (k-NN), and RF models. All the selected empirical equations examined in this study overestimate the scour depth. The SMOREG techniques were substantially closer to observation values compared to empirical equations. Rathod and Manekar [7] performed a study to predict the local scour depth around the bridge pier by using the GEP approach. These findings establish GEP as a robust tool for practical engineering applications, though the authors emphasize the importance of parameter sensitivity analysis during model development. The research provides valuable guidance for selecting modelling techniques and key input variables in future scour prediction studies. Nil et al. [30] applied SVM techniques to predict scour depth under CWS and LBS conditions around bridge piers, achieving satisfactory results. In the previous literature, few studies have been presented on modelling scour depth using soft computing techniques (SCT) such as XGBoost and SVM-PSO under temporal conditions. In this research work, an attempt is made to develop a temporal scour depth predictive model by using XGBoost and SVM-PSO ML approach to predict local scour depth to prevent the collapse of bridge piers, contributing to boosting the economy of the country and saving lives of people.

2.1 Governing Previous Empirical Equations

Numerical and experimental measurements of temporal scour depth around bridge piers have been conducted, focusing predominantly on laboratory-based investigations. Numerous researchers have formulated equations for calculating scour depth around the circular shape of bridge piers under temporal conditions, considering diverse flow behaviour and bed materials [31–34]. Researchers extensively utilize these equations to compute scour depth around bridge piers.

Oliveto and Hager [33]:

$$d_s / y = \left\{0.068\sigma_g^{-0.5} F_d^{1.5} (\log T) Z_r\right\} / y, \text{where } F_d = \frac{V}{(g' d_{50})^{0.5}}, z_R = (yb2)1/3, t_R = \frac{z_R}{\left(\sigma^{1/3} g' d_{50}\right)^{0.5}}, T = \frac{t}{t_R} \tag{1}$$

Lança et al. [33]:

$$d_s / y = \left\{\left[2.3\frac{y}{b^{1/3}}\right]\left[1 - e^{\left\{-a_1\left(\frac{v_t}{b}\right)^{a2}\right\}}\right]\right\} / y$$

$$\text{If } 0.50 \le \frac{y}{b} \le 1.45, 60 < \frac{b}{d_{50}} \le 100 \tag{2a}$$

$$d_s / y = \left\{\left[2.3\frac{y}{b^{1/3}}\right] 0.55 \left[1 - e^{\left\{-a_1\left(\frac{v_t}{b}\right)^{a2}\right\}}\right]\right\} / y$$

$$\text{If } 0.50 \le \frac{y}{b} \le 1.45, \frac{b}{d_{50}} > 500 \tag{2b}$$

$$d_s / y = \left\{ \left[2.3 \frac{y}{b^{1/3}} \right] 5.8 \left(\frac{b}{d_{50}} \right)^{-0.38} \left[1 - e^{\left\{ -a_1 \left(\frac{v_t}{b} \right)^{a_2} \right\}} \right] \right\} / y$$

$$\text{If } 0.50 \le \frac{y}{b} \le 1.45, 100 < \frac{b}{d_{50}} \le 500 \tag{2c}$$

$$d_s / y = \left\{ [2.6][0.55] \left[1 - e^{\left\{ -a_1 \left(\frac{v_t}{b} \right)^{a_2} \right\}} \right] \right\} / y$$

$$\text{If } \frac{y}{b} > 1.45, 500 < \frac{b}{d_{50}} \tag{2d}$$

$$d_s / y = \left\{ [2.6] \left[5.8 \left(\frac{b}{d_{50}} \right)^{-0.38} \left[1 - e^{\left\{ -a_1 \left(\frac{v_t}{b} \right)^{a_2} \right\}} \right] \right] \right\} / y$$

$$\text{If } \frac{y}{b} > 1.45, 100 < \frac{b}{d_{50}} \le 500 \tag{2e}$$

$$d_s / y = \left\{ [2.6] \left[1 - e^{\left\{ -a_1 \left(\frac{v_t}{b} \right)^{a_2} \right\}} \right] \right\} / y$$

$$\text{If } \frac{y}{b} > 1.45, 60 < \frac{b}{d_{50}} \le 100 \tag{2f}$$

where $a_1 \left(b / d_{50} \right) = 1.22 \left(\frac{b}{d_{50}} \right)^{-0.764}$ and $a_2 \left(b / d_{50} \right) = 0.09 \left(\frac{b}{d_{50}} \right)^{0.244}$.

Franzetti et al. [32]:

$$d_s / y = 2.57\left\{\begin{array}{l}\left[1-0.675e^{-1.451\left(\frac{y}{b}\right)}\right]\\ \left[0.849\left(\left(\frac{b}{d_{50}}\right)^{1.815}e^{-2.990}\left(\frac{b}{d_{50}}\right)^{0.235}\right)\right]\\ \left[0.740\left(\mathrm{e}^{-0.066\sigma_g^{3.923}}+0.416\right)\right]\left[1-2.017\left(1-\frac{V}{V_c}\right)^3\right]\\ \left[1-\mathrm{e}^{-0.0837T^{0.231}}\right]\end{array}\right\}, \text{where } T=\frac{t}{t_r} \quad (3)$$

Tang et al. [34]:

$$\frac{d_s}{y} = \sigma^{-1}\left(\frac{y^{0.15}\left(\frac{b}{d_{50}}\right)^{0.25}}{d^{0.4}}\right)\left(\frac{V}{V_{sc}}-1\right)^{0.5}\ln\left(1+0.006\Delta\frac{Vt}{d}\right) \quad (4)$$

where $V_{sc} = \left(0.462\frac{\bar{D}}{d_*}\right)^{0.06} V_c$, $\Delta = \frac{\rho_s - \rho}{\rho}$, $\bar{D}$ = mean value pier width, d_* = effective value of pier width.

3 Methodology of Present Study

Various equations have been formulated, each tailored for different data set ranges and specific parameters. Melville and Chiew [5] incorporated flow depth (y) and flow intensity correction factors in their equations, while Oliveto and Hager [33] gave a scour depth predictive empirical equation that included time (t). Furthermore, Pandey et al. [16] use the densiometric Froude number (F_{rd}) and standard deviation of the bed materials (σ_g) to develop their empirical scour depth model. All these input parameters are essential to develop the empirical temporal models [35].

3.1 Influential Input Parameters for Temporal Scour Depth Modelling Under CWS Condition

The local scour depth (d_s) developed around the bridge pier, characterized by bridge pier width (b) and size of bed sediment (d_{50}), can be represented using the following expression. This scour depth (d_s) increases with time (t) under CWS with approach velocity (V) and water depth (y) [5]. The following functional relationship describes scour depth:

$$d_s = f\left(b,\rho,\mu,g,d_{50},\sigma_g,y,V,V_c,F_r,t\right) \tag{5}$$

where b = pier diameter, μ = dynamic viscosity of fluid, ρ = density of fluid, $g = 9.81\ \text{m/s}^2$, d_{50} = size of bed sediment, σ_g = geometric standard deviation of bed materials, y = flow depth, V = approach mean velocity, V_c = particle incipient velocity, F_r = Froude number, and t = time.

Oliveto and Hager [33], Franzetti et al. [33], and Tang et al. [34]combined non-dimensional factors to create an improved predictive model for scour depth. The relationship for the SDR (d_s/y) is expressed through the following relation in terms of non-dimensional input parameters:

$$\frac{d_s}{y} = f\left(\frac{b}{y},\frac{V}{V_c},F_r,\frac{d_{50}}{b},\sigma,\frac{Vt}{b}\right) \tag{6}$$

3.2 Datasets Collection

Validation of ML approaches i.e., XGBoost and SVM-PSO for predicting local scour depth requires field or experimental datasets. The present study did not conduct its experimental study; instead, existing previous field or experimental datasets from various sources, as outlined in Table 1, were compiled. Specifically, 501 experimental datasets from different literatures [25, 26, 31, 36–39] under CWS conditions (i.e., $V/V_c \leq 1$) were collected. The components of CWS datasets include five non-dimensional input parameters such as b/y, V/V_c, F_r, d_{50}/b, σ_g, and Vt/b and one observed output parameter, denoted as SDR (d_s/y). To model the temporal scour depth, 75% of the datasets were used for training purposes, while 25% of data sets were used for testing purposes.

3.3 XGBoost Approaches

In this research, artificial intelligence techniques, specifically XGBoost and SVM-PSO, were employed for scour depth modelling by utilizing previously published datasets on local scour depth around bridge piers. The introduction of the boosting concept aimed to enhance the capabilities of a "weak learner," transforming it into a more effective learner through certain modifications. From a statistical perspective, this process resembles improving a "good hypothesis" from a relatively "poor hypothesis." Jason Brownlee defines a "weak hypothesis" as a model with slightly better performance than random chance. Boosting iteratively filters observations, retaining those easily handled by the weak learner while emphasizing challenging

Table 1 Different parameters range for CWS collected from the literature

Authors and years	b/y	V/V_c	F_r	d_{50}/b	σ_g	Vt/b	d_s/y
Yanmaz and Altinbilek [36]	0.280–1.480	0.430–0.770	0.250–0.290	0.010–0.020	1.130–1.280	3.00–6.00	0.136–0.231
Melville and Chiew [5]	0.050–5.080	0.400–1.000	0.120–0.880	0.004–0.080	1.010–3.150	3.33–250.50	0.051–0.576
Mia and Nago [17]	0.140–1.500	0.700–0.900	0.120–0.400	0.002–0.060	1.200–1.290	2.33–313.63	0.080–0.317
Chang et al. [14]	5.000–6.666	0.805–0.845	0.592–0.632	0.007–0.010	1.200–3.000	19.00–56.00	0.202–0.950
Molinas [37]	0.089–1.275	0.205–0.999	0.110–0.724	0.003–0.095	1.150–3.700	8.00–30.00	0.005–0.214
Sheppard et al. [6]	0.086–5.352	0.609–0.970	0.071–0.387	0.001–0.007	1.210–1.510	18.83–616.00	0.066–0.642
Grimaldi	0.360–0.500	1.000–1.000	0.210–0.330	0.005–0.017	1.440–1.460	96.00–148.35	0.131–0.187
Lança et al. [31]	0.200–2.000	0.278–0.347	0.259–0.317	0.002–0.017	1.360–1.360	168.00–330.00	0.121–0.549
Lopez et al.	0.250–0.810	0.500–0.860	0.165–0.336	0.017–0.018	1.320	2.00–145.20	0.080–0.189
Fael et al. [38]	0.250–1.000	0.960	0.220	0.004–0.017	1.360	150.00–316.32	0.143–0.348
Aksoy et al. [25]	0.213–0.227	0.450–0.560	0.259–0.326	0.017–0.086	1.390	6.68	0.075–0.149
Pandey et al. [26]	0.550–0.700	0.640–0.900	0.138–0.203	0.003–0.004	1.220	13.53–17.50	0.106–0.230
Yang et al. [39]	0.135–0.477	0.530	0.099–0.099	0.004–0.013	1.300	0.07–12.41	0.057–0.105

cases. XGBoost, built on gradient boosting, extends this framework with advanced features: a sparse-aware algorithm for automatic missing data handling, a block structure enabling parallelized tree construction, and support for continuous model updates with new data. A key innovation is its use of Taylor-expanded loss functions to enhance generalization. The gradient boosting-based algorithm introduces a regularization term to the objective function, effectively reducing model complexity to prevent overfitting. The objective function is expressed as follows:

$$\text{obj}(\phi) = \sum_{i=1}^{n} l(y_i, \hat{y}_l) + \sum_{k} \Omega(f_k) \tag{7}$$

$$\Omega(f) = \gamma T + 1/2\lambda\omega^2 \tag{8}$$

The XGBoost algorithm incorporates key components in its objective function formulation in the presented framework. The XGBoost algorithm executes as depicted:

$$\text{obj}(\phi) = \sum_{i=1}^{n} \left[l\left(y_i, \hat{y}_l^{(t-1)}\right) + g_i f_t(x_i) + 1/2 h_i f_t^2(x_i) \right] + \Omega(f_t) \tag{9}$$

where Eqs. (10) and (11) are the first-order and second-order spin-offs of the loss function individually.

$$g_i = \alpha_{\hat{y}_l^{(t-1)}} l\left(y_i, \hat{y}_l^{(t-1)}\right) \tag{10}$$

$$h_i = \alpha^2_{\hat{y}_l^{(t-1)}} l\left(y_i, \hat{y}_l^{(t-1)}\right) \tag{11}$$

3.4 SVM-PSO Approaches

Kennedy and Eberhart [40] pioneered particle swarm optimization (PSO), a metaheuristic where each swarm member (particle) combines individual memory and swarm-wide knowledge to converge toward optimal solutions. Each particle tracks its personal best solution, assessed via a fitness function, and adjusts its velocity based on both its own experience and neighboring particles' positions. The algorithm iteratively refines solutions through this collaborative exploration of the search space. In each epoch, every particle is updated with two crucial "best" values, p_{best} and g_{best}. SVM-PSO demonstrates superior error globalization capabilities compared to SVM, neural networks, and fuzzy inference systems.

The following formulations give the general equation for PSO:

$$v_{i+1} = \omega v_i + c_1 r_1 \left(p_{\text{best}} - x_i\right) + c_2 r_2 \left(g_{\text{best}} - x_i\right) \tag{12}$$

$$x_{i+1} = x_i + v_{i+1} \tag{13}$$

In the context of the PSO algorithm, the position of each particle is denoted by x_i, and v_i represents its velocity. The particle's best value is captured by p_{best}, while the global best value is indicated as g_{best}. Random numbers between 0 and 1, designated as r_1 and r_2, come into play, along with acceleration coefficients c_1 and c_2. The PSO searches to optimize the input datasets by increasing the input specification until the maximum limit of input parameters is reached.

3.5 *Gamma Test (GT)*

To identify the most influential input parameters for temporal scour depth modeling using XGBoost and SVM-PSO machine learning approaches, the Gamma Test (GT) was employed. It is crucial to select the optimal range of input values, and the value of the GT ranges from 0 to 1 [41, 42]. A total of 26 experiments were conducted, out of which ten experiment results are presented in Table 2. The V_{ratio} is defined as follows:

$$V_{\text{ratio}} = \frac{r}{\sigma^2(y)} \tag{14}$$

The V_{ratio}, denoted as $\sigma^2(y)$, represents the yield variance (y). Its utility lies in its effectiveness in comparing yields from different sources. When the V_{ratio} approaches zero, the real yield exhibits stability. Conversely, when the V-ratio is close to one,

Table 2 Different combinations of input parameters for CWS condition using Gamma Test

S. no.	Input parameters combination	Mask	Gamma value	Gradient	Standard error	V_{ratio}	Nearest neighbour
1	(b/y, V/V_c, d_{50}/b, σ_g, Vt/b)	110,111	−0.0115	0.4053	0.0083	−0.0461	10
2	(b/y, F_r, d_{50}/b, σ_g, Vt/b)	101,111	−0.0115	0.4053	0.0083	−0.0461	10
3	(b/y, V/V_c, F_r, σ_g)	111,010	−0.0001	0.2663	0.0037	−0.0004	8
4	(b/y, V/V_c, d_{50}/b, Vt/b)	110,101	0.0030	0.2440	0.0056	0.0120	10
5	(b/y, V/V_c, F_r, σ_g, Vt/b)	111,011	0.0033	0.2026	0.0059	0.0132	8
6	(b/y, V/V_c, d_{50}/b, σ_g, Vt/b)	110,111	0.0038	0.1915	0.0034	0.0154	12
7	(b/y, V/V_c, F_r, σ_g, Vt/b)	111,011	0.0050	0.1858	0.0035	0.0199	10
8	(b/y, V/V_c, F_r, σ_g)	111,010	0.0052	0.2305	0.0056	0.0207	12
9	(b/y, V/V_c, F_r)	111,000	0.0066	0.3233	0.0053	0.0264	8
10	(b/y, V/V_c, d_{50}/b, Vt/b)	110,101	0.0073	0.1949	0.0034	0.0290	12

the yield, in the context of the smooth model, resembles irregular disturbance. The Win-Gamma software was utilized to apply the GT [43]. The authors assert the high efficacy of this method, suggesting its applicability to other non-linear hydraulic modelling problems.

4 Results and Discussions

4.1 Statistical Indices

The mean absolute percentage error (MAPE) represents the average magnitude of errors a model generates, measuring the average deviation of predictions.

$$\text{MAPE} = \frac{\sum_{i=1}^{n}\left|\frac{O_i - P_i}{O_i}\right|}{n} \tag{15}$$

where P_i is the predicted value, O_i is the experimental value, and n is the number of trials.

Root mean square error (RMSE) holds significance in modelling as it quantifies the extent of error present during the modelling process.

$$\text{RMSE} = \sqrt{\frac{1}{n}\sum_{i=1}^{n}\left(P_i - O_i\right)^2} \tag{16}$$

The coefficient of determination (R^2) statistical index is the degree to which alterations of one variable can be done by changes in another variable when an event outcome is predicted. The equation for R^2 is as follows:

$$R^2 = \frac{\left(n\left(\Sigma O_i P_i\right) - \left(\Sigma P_i\right)\left(\Sigma P_i\right)\right)^2}{\left(n\Sigma O^2_i - \left(\Sigma O_i\right)^2\left(n\Sigma P^2_i - \left(\Sigma P_i\right)^2\right)\right.} \tag{17}$$

4.2 Development of Temporal Scour Depth Model around Bridge Pier Using XGBoost Machine Learning Technique

For the XGBoost model development, the 501 available datasets were randomly split into 75% for training and 25% for testing to predict scour depth. Six input parameter combinations were trained and tested for each of the seven input parameters (b/y, V/V_c, F_r, d_{50}/b, σ_g, Vt/b) used to model scour depth. Three statistical

Table 3 The finding of training and testing datasets for non-dimensional input parameters (b/y, V/V_c, F_r, d_{50}/b, σ_g, Vt/b) by utilizing the XGBoost machine learning approaches

	XGBoost ML approaches optimal parameter		Training data			Testing data		
S. no.	Learning rate	Max depth	MAPE	RMSE	R^2	MAPE	RMSE	R^2
1	0.1	6.0	1.57	0.0051	0.95245	12.69	0.0294	0.96777
2	0.4	7.0	1.34	0.0049	0.96192	12.84	0.0305	0.96533
3	0.3	5.0	1.46	0.0050	0.94188	13.43	0.0305	0.96527
4	0.4	9.0	1.10	0.0047	0.94138	13.72	0.0306	0.96520
5	0.2	8.0	1.32	0.0049	0.95618	13.06	0.0308	0.96478
6	0.1	8.0	1.45	0.0050	0.96176	13.41	0.0308	0.96464
7	0.4	10.0	1.12	0.0047	0.95129	13.72	0.0310	0.96428
8	0.4	6.0	1.25	0.0048	0.95168	13.85	0.0314	0.96336
9	0.4	8.0	1.22	0.0047	0.93123	14.48	0.0324	0.96102
10	0.4	5.0	1.37	0.0048	0.95616	16.09	0.0382	0.94577

indices, namely MAPE, RMSE, and R^2, were employed to calculate the models and compare their performance with existing models. The Root Mean Square Error (RMSE) metric, which measures the square root of averaged squared differences between predicted and observed values, was used to evaluate model accuracy. An error analysis was conducted to determine the optimal XGBoost model, and the results, presented in Table 3, indicate that a learning rate of 0.1 and a maximum depth of six instances yield superior accuracy and lower error compared to increasing the number of instances.

4.3 Development of Temporal Scour Depth Model around Bridge Pier Using SVM-PSO Machine Learning Technique

In SVM-PSO modelling, 75% and 25% of datasets are randomly selected to model temporal scour depth for training and testing stages, respectively. For various swarm sizes, C_1 and C_2 are the coefficients; the results are presented in Table 4.

To evaluate the effectiveness of the developed soft computing XGBoost and SVM-PSO models and compare their performance with existing models, three statistical indices, MAPE, RMSE, and R^2, were employed. TA value of zero for this index signifies optimal model performance. In Table 4, the best combination for the optimal parameter of SVM-PSO is observed at swarm size (C_1) = 0.3, C_2 = 3.7, P = 25.0, C = 132.0, γ = 3.7, ε = 0.6, and d = 3.0. At this parameter, the values of statistical indices such as MAPE, RMSE, and R^2 are 17.070 (lower), 0.03414 (lower), and 0.94844 (higher), respectively, which provides good results among the other selected optimal parameter combinations.

Table 4 The finding training and testing datasets for non-dimensional input parameters (b/y, V/V_c, F_r, d_{50}/b, σ_g, Vt/b) by utilizing the SVM-PSO machine learning approaches

	SVM-PSO ML approaches optimal parameter							Training data			Testing data		
S. no.	C_1	C_2	P	C	γ	ε	d	MAPE	RMSE	R^2	MAPE	RMSE	R^2
1	0.3	3.7	25.0	132.0	3.7	0.6	3.0	7.231	0.02085	0.949848	17.070	0.03414	0.94844
2	2.4	1.6	25.0	380.0	35.0	2.3	3.0	7.207	0.02081	0.949849	17.075	0.03416	0.94838
3	0.8	3.6	25.0	48.0	39.0	2.6	3.0	7.110	0.02068	0.949853	17.149	0.03417	0.94833
4	0.8	3.3	25.0	749.0	76.0	9.9	3.0	7.241	0.02087	0.949847	17.107	0.03418	0.94830
5	3.4	0.6	25.0	225.0	47.0	3.4	3.0	7.072	0.02063	0.949854	17.153	0.03419	0.94827
6	1.3	2.8	25.0	625.0	97.0	1.8	3.0	6.806	0.02024	0.949865	17.537	0.03479	0.94644
7	1.8	2.3	25.0	375.0	24.0	9.7	3.0	9.300	0.02581	0.949642	17.176	0.03540	0.94456
8	2.4	1.7	25.0	214.0	83.0	0.9	3.0	4.522	0.01748	0.949925	22.398	0.03990	0.92955
9	1.0	3.0	25.0	250.0	3.0	0.3	3.0	4.525	0.01747	0.949925	22.550	0.03997	0.92931
10	0.9	3.1	25.0	253.0	4.8	1.7	3.0	4.496	0.01746	0.949925	22.660	0.03998	0.92928

C_1 cognitive factor; C_2 social factor; d degree

5 Comparison of the Present Developed Models with the Existing Scour Depth Empirical Models

The scatter plots of predicted SDR and observed SDR of XGBoost and SVM-PSO models are shown in Fig. 1a, b. The proposed XGBoost model performed well in predicting SDR values, as all the predicted SDR values lie near the best-fit line. For SDR values 0.5–2.0, the predicted XGBoost ML model overpredicted SDR for a few values, and for SDR values of 3.0–4.5, the XGBoost model predicted precisely and approximately same SDR values, as shown in Fig. 1a. Figure 1b indicates that the SVM-PSO model overpredicted a few SDR values for a range of 0.5–1.0 and underpredicted SDR values between 2.7 and 3.5 this model also shows that SVM-PSO model predicted the SDR values well and nearby at the 3.0–4.5 observed SDR range. Figure 2 compares a scatter plot of the predicted SDR values of presently developed XGBoost and SVM-PSO models with the SDR prediction values of four existing empirical equations from different researchers.

The predicted SDR value of the XGBoost and SVM-PSO models lies near the best-fit line. Among existing researchers, Oliveto and Hager [33] predicted a better SDR value. Lança et al. [31] overpredicted an SDR value of up to 1.5, and Franzetti et al. [32] and Tang et al.'s [34] model underpredicted an SDR value of 0.5–2.5. Oliveto and Hager [33], Franzetti et al. [32], and Tang et al. [34] empirical models very poorly predicted SDR value for the present study, which may be because their models may be developed for a different range of datasets compared to the present selected datasets.

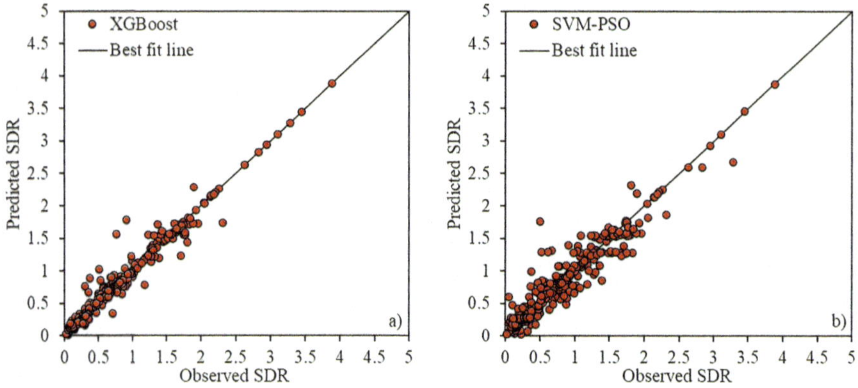

Fig. 1 Scatter plot of observed versus predicted SDR: (**a**) XGBoost model and (**b**) SVM-PSO model

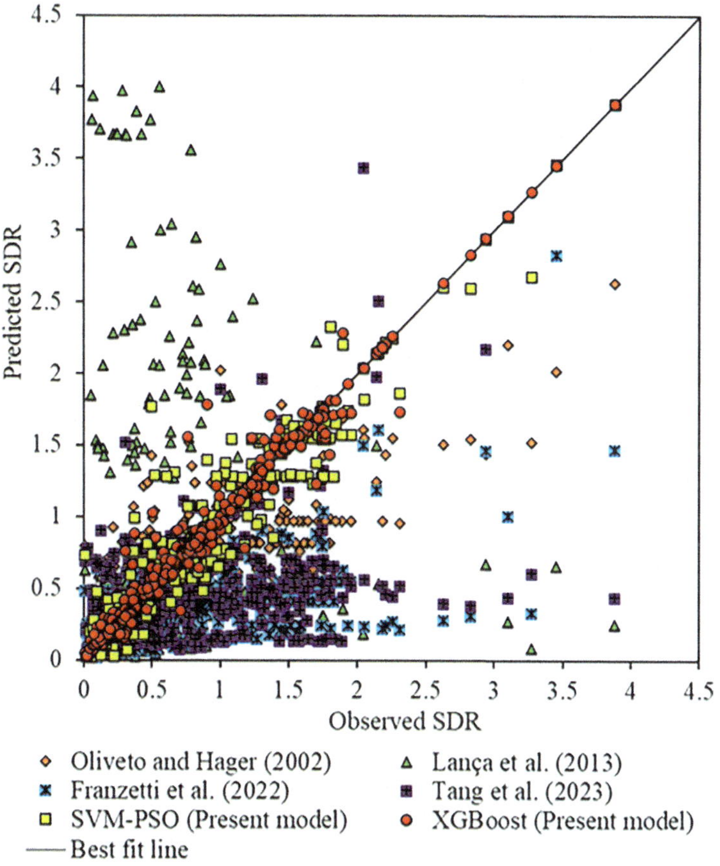

Fig. 2 Scatter plot of predicted versus observed SDR estimated using the existing empirical equation of different researchers and present developed model

6 Conclusion

In the present study, 501 datasets for CWS under temporal conditions were collected, of which 75%, i.e., 325, and 25%, i.e., 126 datasets were used for training and testing purposes, respectively. The following conclusion has been drawn from the present study:

- The GT analysis reveals that scour depth around bridge piers is predominantly influenced by seven key parameters: pier width (b), flow depth (y), Froude number (F_r), approach flow velocity (V), particle incipient velocity (V_c), the geometric standard deviation of the bed material (σ_g), and the mean particle size (d_{50}).
- For CWS, it was found that at learning rate = 0.1 and max depth = 6.0, the XGBoost model predicted better results than other combinations with lower MAPE and RMSE values and a high R^2 value.

- This study achieved strong predictive performance, with the XGBoost model demonstrating a coefficient of determination (R^2) of 0.97 for SDR estimation, slightly outperforming the SVM-PSO model's R^2 of 0.95. Both developed models showed robust accuracy, with mean absolute percentage error (MAPE) values below 17%, indicating reliable predictive capability for scour depth ratio assessment.
- Oliveto and Hager's [33] model predicted better results for SDR values up to 2.0 among all the selected existing empirical models developed by various previous researchers.
- Recent studies have revealed that existing empirical models based on temporal data exhibit significant limitations in accurately predicting scour depth ratio (SDR) values. These shortcomings highlight the need for further research to develop an improved predictive model with enhanced accuracy and reliability for engineering applications.

A comparison between the XGBoost and SVM-PSO models for predicting the scour depth ratio (SDR) reveals that the XGBoost model demonstrates superior predictive accuracy, as confirmed by error analysis metrics. This suggests that XGBoost may be a more reliable approach for SDR estimation compared to the SVM-PSO hybrid model.

References

1. Kothyari, U. C., Garde, R. C. J., & Ranga Raju, K. G. (1992). Temporal variation of scour around circular bridge piers. *Journal of Hydraulic Engineering, 118*(8), 1091–1106.
2. Baranwal, A., Das, B. S., & Setia, B. (2023). A comparative study of scour around various shaped bridge pier. *Engineering Research Express, 5*(1), 015052.
3. Ettema R (1980) *Scour at bridge piers*. Master Thesis, School of Eng., University of Auckland, Auckland, New Zealand
4. Tison, L. J. (1940). Erosion autour de piles de pont en riviere. Annales des Travaux publics de Belgique, 41(6), 813–871. (in French).
5. Melville, B. W., & Chiew, Y. M. (1999). Time scale for local scour at bridge piers. Journal of Hydraulic Engineering, 125(1), 59–65.
6. Sheppard, D. M., Odeh, M., & Glasser, T. (2004). Large scale clear-water local pier scour experiments. Journal of Hydraulic Engineering, 130(10), 957–963.
7. Rathod, P., & Manekar, V. L. (2022). Gene expression programming to predict local scour using laboratory and field data. ISH Journal of Hydraulic Engineering, 28(2), 143–151. https://doi.org/10.1080/09715010.2020.1846144
8. Baranwal, A., & Das, B. S. (2024). Live-bed scour depth modelling around the bridge pier using ANN-PSO, ANFIS, MARS, and M5Tree. Water Resources Management, 38(12), 4555–4587.
9. Garg, V., Setia, B., & Verma, D. V. S. (2005). Reduction of Scour around a bridge pier by multiple collar plates. ISH Journal of Hydraulic Engineering, 11(3), 66–80. https://doi.org/10.1080/09715010.2005.10514802
10. Vijayasree, B. A., & Eldho, T. I. (2016). Experimental study of scour around bridge piers of different arrangements with same aspect ratio. In Scour and Erosion - Proceedings of the 8th International Conference on Scour and Erosion, ICSE 2016 (pp. 889–895). CRC Press/Balkema.

11. Zhang, W., Wu, C., Li, Y., Wang, L., & Samui, P. (2021). Assessment of pile drivability using random forest regression and multivariate adaptive regression splines. Georisk, 15(1), 27–40. https://doi.org/10.1080/17499518.2019.1674340
12. Singh, N. B., Devi, T. T., & Kumar, B. (2022). The local scour around bridge piers—a review of remedial techniques. ISH Journal of Hydraulic Engineering, 28(S1), 527–540. https://doi.org/10.1080/09715010.2020.1752830
13. Shen, H. W., Schneider, V. R., & Karaki, S. (1969). Local scour around bridge piers. *Journal of the Hydraulics Division, 95*(6), 1919–1940.
14. Chang, W. Y., Lai, J. S., & Yen, C. L. (2004). Evolution of scour depth at circular bridge piers. Journal of Hydraulic Engineering, 130(9), 905–913.
15. Lee, S. O., & Sturm, T. W. (2009). Effect of sediment size scaling on physical modeling of bridge pier scour. Journal of Hydraulic Engineering, 135(10), 793–802.
16. Pandey, M., Sharma, P. K., Ahmad, Z., & Karna, N. (2018). Maximum scour depth around bridge pier in gravel bed streams. Natural Hazards, 91, 819–836.
17. Mia, M. F., & Nago, H. (2003). Design method of time-dependent local scour at circular bridge pier. Journal of Hydraulic Engineering, 129(6), 420–427.
18. Das, S., Das, R., & Mazumdar, A. (2013). Circulation characteristics of horseshoe vortex in scour region around circular piers. Water Science and Engineering, 6(1), 59–77. https://doi.org/10.3882/j.issn.1674-2370.2013.01.005
19. Kim, I., Fard, M. Y., & Chattopadhyay, A. (2015). Investigation of a bridge pier scour Prediction Model for Safe Design and Inspection. Journal of Bridge Engineering, 20(6), 04014088. https://doi.org/10.1061/(asce)be.1943-5592.0000677
20. Bateni, S. M., Borghei, S. M., & Jeng, D. S. (2007). Neural network and neuro-fuzzy assessments for scour depth around bridge piers. Engineering Applications of Artificial Intelligence, 20(3), 401–414.
21. Firat, M., & Gungor, M. (2009). Generalized regression neural networks and feed forward neural networks for prediction of scour depth around bridge piers. Advances in Engineering Software, 40(8), 731–737.
22. Talpur, N., Salleh, M. N. M., & Hussain, K. (2017). An investigation of membership functions on performance of ANFIS for solving classification problems. *IOP Conference Series: Materials Science and Engineering, 226*, 012103. https://doi.org/10.1088/1757-899X/226/1/012103
23. Etemad-Shahidi, A., Bonakdar, L., & Jeng, D. S. (2015). Estimation of scour depth around circular piers: Applications of model tree. *Journal of Hydroinformatics, 17*(2), 226–238. https://doi.org/10.2166/hydro.2014.151
24. Baranwal, A., Das, B. S., Khatua, K. K., Khuntia, J. R., & Devi, K. (2025). Modelling of live bed scour depth around bridge pier using machine learning techniques. In River Flow 2024 (pp. 880–888). CRC Press.
25. Aksoy, A. O., Bombar, G., Arkis, T., & Guney, M. S. (2017). Study of the time-dependent clear water scour around circular bridge piers. Journal Of Hydrology And Hydromechanics, 65(1), 26.
26. Pandey, M., Azamathulla, H. M., Chaudhuri, S., Pu, J. H., & Pourshahbaz, H. (2020). Reduction of time-dependent scour around piers using collars. Ocean Engineering, 213, 107692.
27. Chou, J. S., & Pham, A. D. (2014). Hybrid computational model for predicting bridge scour depth near piers and abutments. Automation in Construction, 48, 88–96. https://doi.org/10.1016/j.autcon.2014.08.006
28. Hamidifar, H., Zanganeh-Inaloo, F., & Carnacina, I. (2021). Hybrid scour depth prediction equations for reliable design of bridge piers. Water (Switzerland), 13(15), 1–14. https://doi.org/10.3390/w13152019
29. Garg, R. K., Chandra, S., & Kumar, A. (2022). Analysis of bridge failures in India from 1977 to 2017. Structure and Infrastructure Engineering, 18(3), 295–312. https://doi.org/10.1080/15732479.2020.1832539
30. Nil, Baranwal, A., & Das, B. S. (2023). Clear-water and live-bed scour depth modelling around bridge pier using support vector machine. Canadian Journal of Civil Engineering, 50(6), 445–463.

31. Lança, R., Fael, C., Maia, R., Pêgo, J. P., & Cardoso, A. H. (2013). Clearwater scour at pile groups. *Journal of Hydraulic Engineering, 139*(10), 1089–1098. https://doi.org/10.1061/(ASCE)HY.1943-7900.0000770
32. Franzetti, S., Radice, A., Rebai, D., & Ballio, F. (2022). Clear water scour at circular piers: A new formula fitting laboratory data with less than 25% deviation. Journal of Hydraulic Engineering, 148(10), 04022021.
33. Oliveto, G., & Hager, W. H. (2002). Temporal evolution of clear-water pier and abutment scour. Journal of Hydraulic Engineering, 128(9), 811–820.
34. Tang, H., Liu, Q., Zhou, J., Guan, D., Yuan, S., Tang, L., & Zhang, H. (2023). Process-based design method for pier local scour depth under clear-water condition. Journal of Hydraulic Engineering, 149(12), 06023009.
35. Baranwal, A., Kumar, S., Saurabh, S., & Das, B. S. (2025). Estimation of bridge pier scour depth under clear water and live bed scouring conditions using MARS and ETR machine learning technique. Iranian Journal of Science and Technology, Transactions of Civil Engineering, 1–27.
36. Yanmaz, A. M., & Altinbilek, H. D. G. A. (1991). Study of time-dependent local scour around bridge piers. Journal of Hydraulic Engineering, 117(10), 1247–1268.
37. Molinas, A. (2004). Bridge scour in non-uniform sediment mixtures and in cohesive materials: Synthesis report (No. FHWA-RD-03-083). United States. Federal Highway Administration. Office of Infrastructure Research and Development.
38. Fael, C., Lança, R., & Cardoso, A. (2016). Effect of pier shape and pier alignment on the equilibrium scour depth at single piers. International Journal of Sediment Research, 31(3), 244–250. https://doi.org/10.1016/j.ijsrc.2016.04.001
39. Yang, Y., Melville, B. W., Macky, G. H., & Shamseldin, A. Y. (2020). Temporal evolution of clear-water local scour at aligned and skewed complex bridge piers. Journal of Hydraulic Engineering, 146(4), 04020026.
40. Kennedy, J., & Eberhart, R. (1995). Particle swarm optimization (PSO). Proceedings of ICNN'95-International Conference on Neural Networks, 4(1), 1942–1948.
41. Agalbjorn, S., Koncar, N., & Jones, A. J. (1997). A note on the gamma test. Neural Computing & Applications, 5, 131–133.
42. Kumar, A., Baranwal, A., & Das, B. S. (2023). Modelling of clear water scour depth around bridge piers using M5 tree and ANN-PSO. AQUA—Water Infrastructure, Ecosystems and Society, 72(8), 1386–1403. https://doi.org/10.2166/aqua.2023.225
43. Durrant, P. J. (2001). winGamma: a non-linear data analysis and modelling tool with applications to flood prediction. Cardiff University.

Circular Economy and Construction Industry: A Case Study Approach for Adaptive Reuse in Construction and Demolition Waste

Kavana, B. Gowri Shenoy, and Anoop Kumar Shukla

1 Introduction

Construction and demolition waste (CDW) is one of the largest waste streams in the world due to human activities, mainly concrete, metals, wood, and plastics from construction. Sustainable management of CDW reduces environmental impacts like depletion of resources and associated pollution. Circular Economy (CE) provides viable possibilities to replace the traditional linear model with the economical use of resources and addresses the challenges. It focuses on waste reduction, reuse, and recycling of materials, and maintaining or restoring natural systems [1].

Implementing the circular economy principle in construction will have the environmental impact of reusing and reducing construction waste. Moreover, the Circular economy will reduce the demand for new materials and construction costs as well as reduce the demand for virgin materials. The mismanagement of Construction and Demolition Waste (CDW) is a global issue that has led to increased landfills as well as soil and groundwater contamination [2]. Recycling and handling facilities for construction waste are not well maintained despite rising construction and demolition waste per year due to a lack of infrastructure and regulation. Disposal in landfills and by incineration increases the overall environmental footprint, emitting greenhouse gases due to the lack of recycling opportunities. Hence, addressing all the issues will encourage the circular economy principle, ensuring efficient resources and maximizing waste reduction.

The objective of this study is to evaluate the effects of adaptive reuse of construction and demolition (C&D) waste in the construction industry to achieve a circular economy. This study addresses the three questions below:

Kavana · B. Gowri Shenoy · A. K. Shukla (✉)
Manipal School of Architecture and Planning, Manipal Academy of Higher Education, Manipal, Karnataka, India
e-mail: anoop.shukla@manipal.edu; gowri.shenoy@manipal.edu

S. Pathak et al. (eds.), *Intelligent Infrastructure and Smart Materials*,
https://doi.org/10.1007/978-3-031-92421-7_19

RQ 1: How does adaptive reuse reduce the need for new materials in construction?
RQ 2: How do real-life cases demonstrate the benefits of adaptive reuse?
RQ 3: How does adaptive reuse contribute to achieving the principles of a circular economy in the construction industry?

However, addressing the above questions will allow us to explore various types of construction and demolition waste materials for adaptive reuse and achieve sustainability. The scope highlights identifying new reuse pathways such as concrete, wood, and metal, emphasizing the benefits and resource conservation [3].

1.1 The Challenges of Construction and Demolition Waste

Construction and demolition waste in the construction industry generates enormous waste, such as wood, bricks, metal, concrete, and plastics [4]. A lot of these materials are dumped instead of being recycled or reused. As cities grow, finding areas to dispose of or to dump becomes difficult. However, the waste dumped in landfills will release harmful toxins that will impact environmental pollution. As a result, it is a common practice to dump the C&D waste illegally, which is not a sustainable solution. Thirty-six percent of the world's solid waste is generated due to rapid urban growth, infrastructure development, and non-maintained buildings, such as innovating or tearing down structures like buildings, roads, and bridges. C&D waste from buildings, roads, and bridges will include brick, wood, metal, concrete, glass, plastics, and hazardous materials like asbestos and lead-based paint [5].

However, the quantity of waste generated depends on factors like the size of the project, the materials used, and construction methods. Managing these waste materials involves separating them, recycling them, and reusing them, which will reduce environmental issues and promote sustainability and innovation in the construction industry [6].

Hence, to address these issues, the circular economy can be a good principle to adopt in achieving sustainability and efficient reuse of materials in the construction industry [7].

1.2 Principles of Circular Economy for the Construction Sector

The principle of Circular economy in the construction sector works toward minimizing waste and maximizing the recycling and reuse of materials. This helps reduce environmental destruction and extract the resources from the traditional linear economic model, i.e., take, make, and dispose [8].

The construction industry is very resource-intensive, which leads to large amounts of waste and greenhouse gas emissions. Thus, a transition to a circular economy

Table 1 Strategies of circular economy

Sl. No	Strategies of circular economy	Contribution to the construction industry	References
1.	Design for disassembly and reuse	Designing buildings that allow for gradual deconstruction and reuse at the end-of-life and material maintenance ensures the materials used can be re-extracted and reused, reducing the impact	[9]
2.	Minimize waste to achieve resource efficiency	Material efficiency can be achieved by retaining the maximum material value by keeping them in the closed loop	[10]
3.	Reduce and recycle for material circularity	Reuse and recycling are key elements of the circular economy as they involve recovering resources from secondary building materials and avoiding the consumption of new raw materials	[10]
4.	Lifecycle assessment	It identifies opportunities for applying circular economy principles to buildings. This involves considering how recycling and reuse can be integrated at every phase of the lifecycle of a building	[10]
5.	Economic opportunities and environmental consequences	It not only minimizes the environmental impact of production and consumption but also provides economic opportunities through a market for secondary materials and savings on waste disposal and raw materials	[11]

(CE) is needed to make this sector sustainable. The CE model encourages reuse, recycling, repair, recovery, and refurbishment to reduce resource consumption and improve conservation [9] and complements SDG-9, 11, 12, and 13. It offers a way to reduce waste, cut costs, and create sustainable cities adapting to future challenges.

The strategies of the circular economy are mentioned in Table 1.

The construction sector's introduction of the circular economy is essential for sustainable progress. Shifting to a circular economy faces challenges but offers benefits like decreased landfills, conserving raw materials, and improved environmental sustainability.

1.3 Framework in Achieving Circular Economy

The construction sector's transition toward a circular economy (CE) is growing, with greater attention to addressing environmental impacts, resource efficiency, and longevity [12]. Construction has been an incredibly unsustainable process, consuming large amounts of scarce resources and producing vast amounts of waste, making it necessary to shift to circular practices. The Construction Project-Level Circular Economy Assessment Framework explains the transition from the linear to the circular economy model. This framework will propose a systematic definition of the implementation of CE principles for different aspects of construction projects. Also, it comprehensively approaches sustainability, covering all aspects of resource

Table 2 Indicators and their categories of CE for adaptive reuse

Sl. No	Organization-level indicators categories	Relevant indicators	References
1.	Input related	1. Material circularity	[13]
		2. Energy circularity	
		3. Water circularity	
		4. Land circularity	
		5. Economic circularity	
2.	Process related	1. Circular economy-driven building design	
		2. Circular economy-driven building construction	
		3. Managing logistics	
		4. Managing supply chain	
		5. Circular economy services	
		6. Adoption of technologies	
		7. Circular economy-driven innovations	
		8. Adoption of policies and regulations	
3.	Output related	1. Waste circularity	
		2. Emission circularity	
		3. Social circularity	
		4. Economic circularity	

management, from input and process efficiencies to output results. This highlights the importance of a framework of construction project-level indicators for circular economy practices in the construction sector. The three subcategories of indicators are Input Related, Process Related, Output Related, and People Related [13]. These include input indicators that measure the circularity of materials, energy, water, and land use. Indicators of process focus on circular designing, construction techniques, and supply chain management; at the same time as output indicators, for example, waste and emissions reductions, along with economic and social benefits created. Furthermore, indicators concerning people emphasize the contribution of human resources and stakeholders in putting circularity into practice. They are interconnected to maintain the circular flow of materials, energy, and resources within the economy, thus reducing waste and environmental impact while promoting economic and social benefits, as mentioned in Table 2.

1.4 The Importance of Adaptive Reuse Strategies of Construction and Demolition Waste

Adaptive reuse is widely recognized as one of the most powerful tools to implement circular economy principles in the built environment [14]. Using waste building materials and structures involves breaking down existing ones for reuse and rebuilding. Adaptive reuse allows for the use of construction waste and vastly decreases the excavation of precious natural resources [15]. Reuse is conceptualized through maximizing value from conserving and reusing the existing materials by which energy is saved, i.e., transforming the existing structures [16].

For example, structural elements like beams or floor slabs can be reused in some other buildings, reducing CO_2 emissions and energy (manpower, transport, etc. will be reduced). Converting old warehouses to new offices or out-of-use factories to habitable spaces will save resources and suitable materials from landfills. Hence, old warehouse adaptive reuse will align with sustainability, i.e., reuse can reconstruct a better urban environment that is ecologically friendly and economically viable [17]. In line with this, circular economy indicators will help evaluate adaptive reuse to achieve a life cycle assessment of materials or structures instead of demolition.

2 Methodology

Adaptive reuse of CDW to achieve CE has been done through a case study approach in this study (Table 3). This approach allows for a specific analysis of existing structures by identifying materials, reuse patterns, integration into new construction, and elaborating the process to reduce waste generation and enhance sustainability in the building lifecycle.

2.1 *The Mill*

The Mill uses the concept of adaptive reuse of an old industrial mill into a restaurant (Fig. 1 and Table 4). Wood, bricks, concrete, and metals were reused in furniture, flooring, and structural elements. They adopted local materials to reduce costs and achieve a circular economy. The project preserved land, reduced demolition waste, and created material circularity.

Table 3 Case studies and their approach toward CE

Sl. No	Case study	Typology	Approach to circular economy
1.	The Mill, Manipal	Restaurant	Local reuse, adaptive strategies
2.	Kachra Mane, Bengaluru	Residence	Waste repurposing, environmental focus
3.	Upcycle House, Denmark	Residence	Significant carbon reduction and material reuse
4.	Hines Headquarters, Seattle	Office	Sustainable retrofitting, diverse material adaptation
5.	The Collage House, Mumbai	Residence	Creative reuse, aesthetic-functional integration
6.	Manav Sadhna, Ahmedabad	Cultural center	Low-cost construction, community-driven reuse

Fig. 1 Open sitting area of the restaurant

Table 4 Adaptive reuse in The Mill

Case study	Organization-level indicator categories	Indicators achieved
The Mill, Manipal	Input related	Material circularity: wood, concrete, and bricks bought about 50% of an overall reduction in cost
		Land circularity: Reuse the existing mill structure
		Economic circularity: Reduction in transportation cost and support of local economy
	Process related	Circular economy-driven building construction: Adapting the existing structure for modern use aligns with this indicator as it preserves the building while enhancing its functionality
		Managing logistics and supply chain: The use of locally sourced materials minimizes logistical complexity and supply chain distances
	Output related	Waste circularity: Reusing materials and repurpose process

2.2 *Kachra Mane*

Kachra Mane, also known as "Garbage House," is a unique residential project that utilizes construction and demolition waste (Fig. 2 and Table 5). Recycled glass, reclaimed wood, and metal scrap were used from the surroundings of the demolished building's structural and aesthetic components. It also incorporates sustainable systems, such as water management and passive energy strategies, to help reduce its environmental footprint. It reduces waste and encourages maximum efficiency with local resources.

Fig. 2 Interior of the Kachra Mane

Table 5 Adaptive reuse in Kachra Mane

Case study	Organization-level indicator categories	Indicators achieved
Kachra Mane	Input related	Energy circularity: Passive design strategies
		Water circularity: Grey water recycling, rainwater harvesting system
	Process related	Circular economy-driven building design: Commitment to sustainability, integrating renewable energy solutions
		Circular economy services: The project considers the entire lifecycle of materials and systems, focusing on durability and ease of maintenance to boost the lifecycle of the elements
	Output related	Waste circularity: House design emphasizes minimizing waste using recycled materials and effective waste management systems
		Economic circularity: By using locally sourced and reclaimed materials

2.3 *Upcycle House*

A house in Denmark, the Upcycle House, responds to the need for recycling and repurposing to reduce carbon emissions (Fig. 3 and Table 6). The project involves recycling bricks, metals, and papers for constructing walls and structural components with cork tiles reused from champagne corks. Integrating energy-efficient systems in design reduces 60% of the carbon footprint through material conservation. The principle of adaptive reuse is achieved in Upcycle House through material and economic circularity.

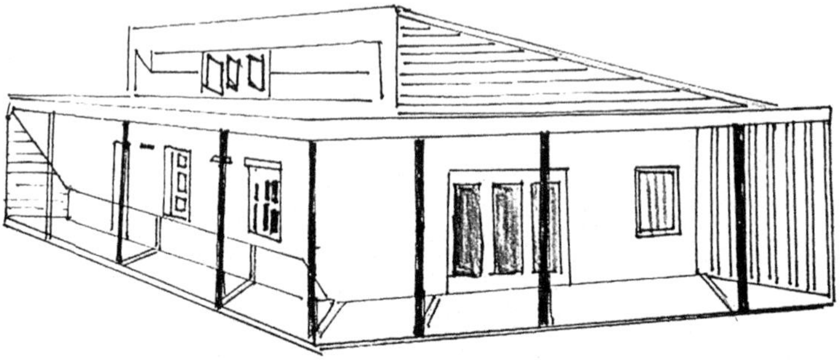

Fig. 3 Exterior of the Upcycle House

Table 6 Adaptive reuse in Upcycle House

Case study	Organization-level indicator categories	Indicators achieved
Upcycle House	Input related	Material circularity: Recycled materials like bricks and refurbished metal components
	Process related	Circular economy-driven building design: Commitment to sustainability by incorporating energy-efficient systems and sustainable construction practices
	Output related	Waste circularity: Reusing construction waste materials

2.4 *Hines Headquarters*

At the Hines Headquarters in Seattle, modular building components were repurposed through reconfiguration, making the project cost-effective by enabling the reuse of materials that typically end up in landfills (Fig. 4 and Table 7). The sustainable elements included recycled textiles, reused glass, remilled lumber for carpets, entrance glazing, and accent walls. By focusing on local sourcing networks, transportation-linked emissions were reduced, and the circular economy was promoted. Contributing to minimizing environmental impact through resource efficiency and modularity, the design features an adaptive element that decreases the overall carbon footprint.

Fig. 4 Interiors of Hines Headquarters

Table 7 Adaptive reuse in Hines Headquarters

Case study	Organization-level indicator categories	Indicators achieved
Hines Headquarters	Input related	Material circularity: Recycled textiles, reused glass, re-milled lumber
	Process related	Circular economy-driven building design: Design emphasizes environmental sustainability, with features such as adaptive reuse of materials, efficient energy systems
	Output related	Waste circularity includes strategies for maximizing recycling

2.5 *Collage House*

In Mumbai, the Collage House incorporates waste items through features like ornamental wall patterns made from rejected tiles and bricks (Fig. 5 and Table 8). Reused timber sashes, second-hand metal panels, and rubble are converted into load-bearing and non-load-bearing wall systems, forming curtain walls and cladding. The project respects waste by using these materials and promotes local sourcing and the aesthetic potential of adaptive reuse. Collage House is an example of circularity that uses background and foreground elements by repurposing existing resources to meet the needs.

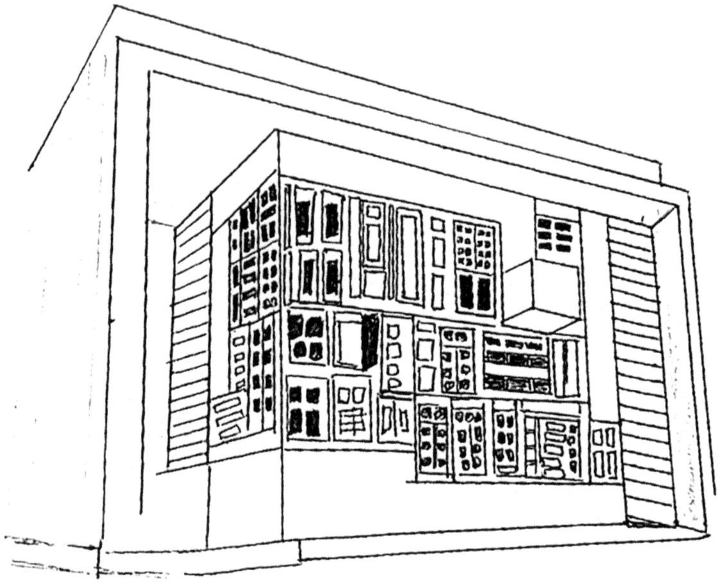

Fig. 5 Exterior of the Collage House

Table 8 Adaptive reuse in Collage House

Case study	Organization-level indicator categories	Indicators achieved
Collage House	Input related	Material circularity: Rejected tiles and bricks, reused timber sashes, second-hand metal panels, and rubble
	Process related	Circular economy-driven building design: Rainwater harvesting and natural cooling strategies
	Output related	Waste circularity: By utilizing local materials and focusing on efficient construction practices

2.6 *Manav Sadhna Activity Center*

A community-centric project at Manav Sadhna Activity Center in Ahmedabad used locally sourced recycled materials such as flash bricks, glass bottles, and timber for several structural purposes and decorative applications (Fig. 6 and Table 9). The center acts as a resource for children and social gatherings in the community, promoting environmental and social sustainability. By thoughtfully maximizing efficiency while minimizing waste, its practical yet purposeful approach to sustainable practices can yield community benefits far beyond its own.

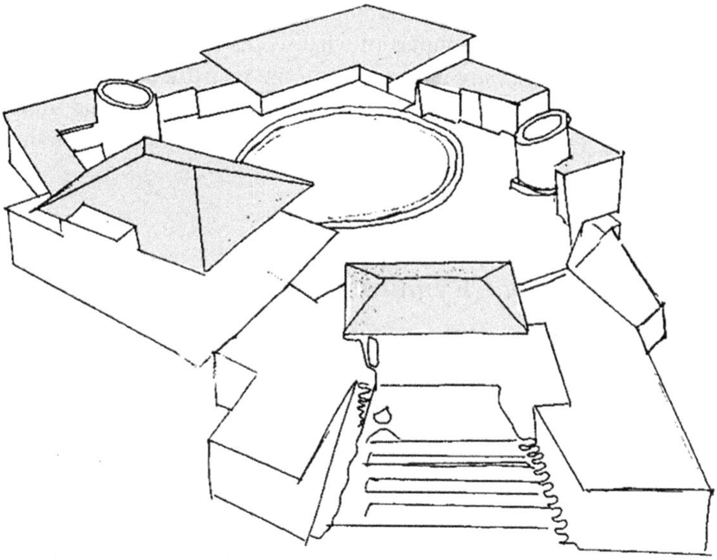

Fig. 6 Manav Sadhna Center

Table 9 Adaptive reuse in Manav Sadhna

Case study	Organization-level indicator categories	Indicators achieved
Manav Sadhna	Input related	Material circularity: Flash bricks, glass bottles and timber
	Process related	Circular economy-driven building design: Designing for longevity and adaptability reduces the need for frequent repairs or replacements
	Output related	Social circularity: Improving living conditions and providing community benefits

3 Results and Discussion

Adaptive reuse for construction and demolition waste can be studied through a case study approach using various methods across residential and community habitats. Reusing materials like timber, bricks, steel, and glass shows the potential of waste cycles through a circular economy approach. The analysis shows a shift toward implementing reuse methods, including deconstruction, selective dismantling, and upcycling. Case studies discussed in this study explain these key sustainable practices that prevent environmental impacts by reducing the use of new materials, transportation emissions, and increasing energy efficiency. Promoting tangible retention, these case studies provide proven examples of conserving resources and preventing overcrowding of landfills. Additionally, these adaptive reuse techniques are in support of the circular economy. It promotes the recovery of materials, which is the idea that buildings are material banks for future use. The case studies offer inspiring examples of what works and what doesn't, indicating the need for circular economy approaches in the construction sector. This study helps develop an understanding of the transformative potential and limitations of adaptive reuse as a more resource-efficient approach to construction and demolition waste generation.

3.1 Achievements in CE Indicators

The following indicators can be used to achieve the principle of circular economy as discussed in case studies:

3.1.1 Input Related

Material Circularity, as shown in case studies like the Upcycle House, The Mill, and Collage House, reused bricks and metals and innovatively used newspaper wool as insulation, resulting in 60+% carbon emission reduction. It shows the potential of adaptive reuse in decreasing the pressure on new materials. Energy circularity showcases the role of passive cooling strategies and sustainable energy solutions to demonstrate the role of adaptive reuse in energy efficiency, as shown in Kachra Mane.

3.1.2 Process Related

In most cases, applied design approaches allow sustained use and adaptability of reused materials, budget-friendly maintenance, and extended life of components. As seen in The Mill and Hines Headquarters, effective logistics and supply chain management derived nearly all materials from in-state, significantly reducing transportation emissions.

3.1.3 Output Related

Reusing old tiles, bricks, and timber, case studies such as Collage House, Kachra Mane, and Manav Sadhna have reduced the amount of waste in landfills by 85%. All the examples used in this study demonstrate cost savings from material reuse and local material sourcing while generating new circular economy opportunities for the secondary materials market.

Table 10 Material matrix

Material	Structural					Interior		Landscaping		Temporary Structures	
	Foundation	Walls	Roof	Columns	Slab	Flooring	Decorative	Paving	Garden structures	Formwork	Scaffolding
Concrete	Yellow	Yellow	Red	Yellow	Yellow	Green	Yellow	Yellow	Yellow	Red	Red
Brick	Green	Green	Red	Red	Yellow	Yellow	Yellow	Green	Green	Red	Red
Wood	Yellow	Yellow	Yellow	Yellow	Yellow	Green	Green	Green	Green	Green	Green
Metal	Yellow	Yellow	Green	Green	Green	Yellow	Green	Yellow	Green	Green	Green
Ceramic	Red	Yellow	Red	Red	Yellow	Green	Green	Yellow	Yellow	Red	Red
Glass	Red	Yellow	Red	Red	Red	Green	Green	Red	Yellow	Red	Red
Plastic	Red	Red	Red	Red	Red	Yellow	Green	Yellow	Yellow	Yellow	Yellow
Stone	Green	Yellow	Yellow	Green	Green	Yellow	Yellow	Green	Green	Red	Red
Green- Highly Reusable Yellow- Reusable with modifications Red- Limited reuse											

3.2 *Material Matrix*

Materials can be re-utilized in various construction applications. It helps to find the best materials for adaptive reuse by categorizing concrete, brick, wood, metal, and other materials according to their use. These materials can be used in designing structures and spaces both indoors and outdoors, including landscapes that offer new possibilities and eco-friendly solutions. However, we can propose a color-coding system that explains the identification of areas where materials can be reused and align with the circular economic principles. Green color suggests materials like wood or brick that are utilized again with minimal modification, such as in garden fixtures and foundation work. The red color suggests that material reusability is limited, and some methods must be proposed for reuse. The yellow color suggests material with minor alteration for reuse (Table 10). This matrix benefits decision-makers, architects, and other construction specialists by supporting the increase of materials' lifespan through waste reduction and promoting a more circular economy in the building industry.

3.3 *Challenges in Adaptive Reuse*

Construction waste will not be ready for immediate reuse without significant processing. Newer technologies for recycling and material suitability assessment are still in process, restricting efficiency [18]. Adaptive reuse adoption on a larger scale would be nearly impossible without proper supportive policies [19]. However, the current study highlights local sourcing, integrated design, and community engagement to achieve cost-effective techniques and a sustainable environment.

3.4 Future Scope

Despite an expected lifetime, the performance of recovered materials can be further explored to assess the accuracy and safety of reusing these products. Moreover, various technologies, notably artificial intelligence and robotics, can also transform how construction waste is sorted and recycled, facilitating more efficient systems [20]. Policy support is also significant in contributing to adaptive reuse practices by creating regulatory environments and encouraging overcoming logistical issues toward a more sustainable construction industry, which needs further exploration [21].

4 Conclusion

The study explores the adaptive reuse approach to achieve the circular economy principle through case studies focusing on construction and demolition waste (CDW). The study also mentions waste reduction through material reuse, which helps decrease carbon emissions, resource depletion, and landfills. Additionally, this study focuses on reducing costs and creating new markets for secondary materials and environmental and financial benefits. Adaptive reuse is a key approach to supporting sustainability in the construction industry, focusing on achieving Circular economy goals. Exploration of material reuse in different scenarios is essential. Also, developing advanced methods for assessing the suitability of CDW would allow for safer and more appropriate reuse in different construction applications. Moreover, a more detailed study into the utilization of emerging technologies for recycling and adaptive reuse can be explored further to achieve a circular economy at the global level extensively [22]. Further, the construction industry can adopt the circular economy principle in future projects. Urban planners and architects can implement these principles by designing nature-adapted buildings and infrastructure. Also, this study will help policymakers minimize waste through the circular economy principle and achieve a sustainable built environment [23].

References

1. Mazur, Ł. (2021). Circular economy in housing architecture: Methods of implementation. *ACTA SCIENTIARUM POLONORUM - Architectura Budownictwo, 20*(2), 65–74. https://doi.org/10.22630/aspa.2021.20.2.15
2. Cervantes Puma, G. C., Salles, A., & Bragança, L. (2024). Exploring the potential of circular economy strategies in urban planning: A comparative analysis of successful case studies. In *Lecture notes in civil engineering* (pp. 491–500). Springer Science and Business Media Deutschland GmbH. https://doi.org/10.1007/978-3-031-57800-7_45

3. Joseph, H. S., Pachiappan, T., Avudaiappan, S., & Flores, E. I. S. (2022). A study on mechanical and microstructural characteristics of concrete using recycled aggregate. *Materials, 15*(21). https://doi.org/10.3390/ma15217535
4. Lee, S., Chang, H., & Lee, J. (2024). *Construction and demolition waste management and its impacts on the environment and human health: Moving forward sustainability enhancement*. Elsevier. https://doi.org/10.1016/j.scs.2024.105855
5. Banias, G. F., et al. (2022). Environmental assessment of alternative strategies for the management of construction and demolition waste: A life cycle approach. *Sustainability (Switzerland), 14*(15). https://doi.org/10.3390/su14159674
6. Chen, L., et al. (2024). *Conversion of waste into sustainable construction materials: A review of recent developments and prospects*. Elsevier. https://doi.org/10.1016/j.mtsust.2024.100930
7. Bhavsar, V., Sridharan, S. R., & Sudarsan, J. S. (2023). Barriers to circular economy practices during construction and demolition waste management in an emerging economy. *Resources, Conservation and Recycling Advances, 20*. https://doi.org/10.1016/j.rcradv.2023.200198
8. Benachio, G. L. F., Freitas, M. d. C. D., & Tavares, S. F. (2020). *Circular economy in the construction industry: A systematic literature review*. Elsevier. https://doi.org/10.1016/j.jclepro.2020.121046
9. Eberhardt, L. C. M., Birkved, M., & Birgisdottir, H. (2022). Building design and construction strategies for a circular economy. *Architectural Engineering and Design Management, 18*(2), 93–113. https://doi.org/10.1080/17452007.2020.1781588
10. López Ruiz, L. A., Roca Ramón, X., & Gassó Domingo, S. (2020). *The circular economy in the construction and demolition waste sector—A review and an integrative model approach*. Elsevier. https://doi.org/10.1016/j.jclepro.2019.119238
11. Ogunmakinde, O. E., Sher, W., & Egbelakin, T. (2021). *Circular economy pillars: A semi-systematic review*. Springer Science and Business Media Deutschland GmbH. https://doi.org/10.1007/s10098-020-02012-9
12. Guerra, B. C., & Leite, F. (2021). Circular economy in the construction industry: An overview of United States stakeholders' awareness, major challenges, and enablers. *Resources, Conservation and Recycling, 170*, 105617. https://doi.org/10.1016/j.resconrec.2021.105617
13. Jayakodi, S., Senaratne, S., Perera, S., & Bamdad, K. (2024). *Circular economy assessment using project-level and organisation-level indicators for construction organisations: A systematic review*. Elsevier B.V. https://doi.org/10.1016/j.spc.2024.05.025
14. Hamida, M. B., Remøy, H., Gruis, V., & Jylhä, T. (2023). Circular building adaptability in adaptive reuse: Multiple case studies in the Netherlands. *Journal of Engineering, Design and Technology*. https://doi.org/10.1108/JEDT-08-2022-0428
15. Lederer, J., Gassner, A., Kleemann, F., & Fellner, J. (2020). Potentials for a circular economy of mineral construction materials and demolition waste in urban areas: A case study from Vienna. *Resources, Conservation and Recycling, 161*, 104942. https://doi.org/10.1016/j.resconrec.2020.104942
16. Olumo, A., & Haas, C. (2024). Building material reuse: An optimization framework for sourcing new and reclaimed building materials. *Journal of Cleaner Production, 479*, 143892. https://doi.org/10.1016/j.jclepro.2024.143892
17. Guerra, B. C., et al. (2021). Circular economy applications in the construction industry: A global scan of trends and opportunities. *Journal of Cleaner Production, 324*, 129125. https://doi.org/10.1016/j.jclepro.2021.129125
18. Mlote, D. S., Budig, M., & Cheah, L. (2024). *Adaptability of buildings: A systematic review of current research*. Frontiers Media SA. https://doi.org/10.3389/fbuil.2024.1376759
19. Pintossi, N., Ikiz Kaya, D., & Pereira Roders, A. (2021). Assessing cultural heritage adaptive reuse practices: Multi-scale challenges and solutions in Rijeka. *Sustainability (Switzerland), 13*(7). https://doi.org/10.3390/su13073603
20. Dodampegama, S., Hou, L., Asadi, E., Zhang, G., & Setunge, S. (2024). *Revolutionizing construction and demolition waste sorting: Insights from artificial intelligence and robotic applications*. Elsevier B.V. https://doi.org/10.1016/j.resconrec.2023.107375

21. Varde, A. S., & Liang, J. (2023). *Machine learning approaches in agile manufacturing with recycled materials for sustainability*. Retrieved form http://arxiv.org/abs/2303.08291.
22. Dubale, M., Vasić, M. V., Goel, G., Kalamdhad, A., & Singh, L. B. (2023). Utilization of construction and demolition mix waste in the fired brick production: The impact on mechanical properties. *Materials, 16*(1). https://doi.org/10.3390/ma16010262
23. Ho, O., et al. (2024). A conceptual model for integrating circular economy in the built environment: An analysis of literature and local-based case studies. *Journal of Cleaner Production, 449*, 141516. https://doi.org/10.1016/j.jclepro.2024.141516

Experimental Analysis of Fly Ash and Bagasse Ash Integration as Sustainable Stabilisers for Improving the Engineering Properties of Black Cotton Soil

Gaurav Bharti, Amit Tripathi, Parikshit Hurukadli, B. Bharath, Pushpendra Kumar Sharma, and Bishnu Kant Shukla

1 Introduction

Black cotton soil, predominantly found in tropical and subtropical regions, is characterised by its high swelling and shrinkage properties, posing significant challenges for construction and agricultural activities. These expansive soils exhibit considerable volume changes with moisture variations, leading to ground heave and settlement issues that compromise the structural integrity of foundations, pavements, and other civil engineering structures [1]. Traditional methods of soil stabilisation often rely on chemical additives like lime and cement, which, although effective, come with environmental and economic drawbacks. Consequently, there is an increasing interest in exploring sustainable and cost-effective alternatives for soil stabilisation. One promising approach is the utilisation of industrial by-products such as fly ash and bagasse ash [2]. Fly ash, a by-product of coal combustion in thermal power plants [3, 4], is rich in pozzolanic materials that enhance the mechanical properties and stability of soils. Bagasse ash, derived from the combustion of sugarcane bagasse in sugar mills, also contains silica and other beneficial compounds that can improve soil properties. Both materials offer the dual benefits of waste utilisation and soil stabilisation, aligning with sustainable construction

G. Bharti · A. Tripathi · P. Hurukadli · B. Bharath
Department of Civil Engineering, JSS Academy of Technical Education,
Noida, Uttar Pradesh, India

P. K. Sharma
School of Civil Engineering, Lovely Professional University, Phagwara, Punjab, India

B. K. Shukla (✉)
Department of Civil Engineering, JSS Academy of Technical Education,
Noida, Uttar Pradesh, India

School of Civil Engineering, Lovely Professional University, Phagwara, Punjab, India

S. Pathak et al. (eds.), *Intelligent Infrastructure and Smart Materials*,
https://doi.org/10.1007/978-3-031-92421-7_20

practices and environmental conservation [5]. Despite the recognised potential of fly ash and bagasse ash in soil stabilisation, there is a notable gap in the literature concerning their combined effects on black cotton soil. Previous studies have primarily focused on the individual impacts of these ashes on various soil types, leaving a critical gap in understanding their synergistic effects when used together. Furthermore, most research has concentrated on short-term performance metrics, with limited attention to the long-term durability and environmental implications of using these by-products in soil stabilisation.

The future scope of this research is substantial. Investigating the combined application of fly ash and bagasse ash on black cotton soil can provide valuable insights into their complementary mechanisms and the resulting improvements in soil behavior. Additionally, exploring the long-term performance and environmental impact of these stabilised soils will be crucial for developing sustainable and resilient construction practices. This research can pave the way for innovative soil stabilisation techniques that not only enhance the engineering properties of problematic soils but also contribute to waste management and environmental sustainability. In summary, this study aims to address the existing literature gap by evaluating the combined effects of fly ash and bagasse ash on the stabilisation of black cotton soil. By doing so, it seeks to provide a comprehensive understanding of the potential benefits and challenges associated with this approach, thereby informing future research and practical applications in the field of geotechnical engineering.

2 Literature Review

In order to solve the major issues that expansive soils, and black cotton soil in particular, offer in the construction industry, a great deal of research has been done on the stabilisation of expansive soils utilising diverse industrial by-products and innovative materials. The results of significant investigations in this field are discussed in this review of the literature, with an emphasis on the merits of various stabilisation strategies and materials. Fly ash was investigated by Bhuvaneshwari et al. (2005) as a potential stabilising agent for expansive soils [6]. As a by-product of burning coal, fly ash is full of pozzolanic elements that improve the qualities of soil. Their research showed that fly ash is a good substitute for conventional chemical additions since it greatly enhances the mechanical qualities and stability of expanding soils. In order to stabilise black cotton soil, Yadu et al. (2011) evaluated the efficaciousness of fly ash and rice husk ash. Both components enhanced the soil's qualities, according to their research, although fly ash had a stronger impact on increasing the soil's compressive strength and lowering its flexibility index [7]. The potential of agricultural waste products, such as rice husk ash, in stabilising soil and advancing environmentally friendly building techniques is not fully highlighted by this study. Tiwari (2014) investigated the use of fly ash, coconut coir fibre, and broken glass to stabilise the soil for growing black cotton [8]. The combination of these minerals greatly increased the strength and durability of the soil, according to the study. In addition to improving the mechanical qualities of the soil, the addition of broken glass and coconut coir fibre offered a sustainable means of recycling waste products. Ashraf and Joseph (2014) used raw plastic bottles to test soil stabilisation. According to their research, adding plastic bottles to soil can significantly increase its stability and capacity to support loads [9]. This creative method turns plastic garbage into a useful resource for building, providing a long-term solution to the issue of plastic waste. Singh and Mali (2013) concentrated on using glass fibres to stabilise soil. Strength and durability were increased when glass fibres were added to the soil, suggesting that this method of improving the performance of poor soils is a viable one [10]. This study demonstrated how industrial waste materials, such glass fibres, may be used to stabilise soil. The behaviour of soil reinforced with plastic waste strips in terms of the California Bearing Ratio (CBR) was examined by Choudhary (2012). The results showed that adding plastic waste strips considerably raised the CBR value and improved the soil's ability to support loads [11]. This study backs the utilisation of recycled plastic materials in geotechnical applications, which have positive effects on the environment and the economy. Joseph and Muhammad Haneef (2014) investigated how plastic granules affected the characteristics of the soil. According to their findings, adding plastic granules strengthened the soil and decreased its flexibility [12]. The possibility of employing plastic trash for soil stabilisation is further highlighted by this study, which supports sustainable building practices. Michel and Aderinle (2014) looked at the stabilisation of clay soil with glass powder. The outcomes showed that glass powder effectively improved the mechanical characteristics of the soil, indicating that it is a good substitute for conventional stabilising agents [13]. This research encourages the recycling of industrial by-products by highlighting the potential use of glass waste in geotechnical engineering. Sivakumar Babu (2014) conducted research on the impact of combining plastic garbage with black cotton soil. According to the research, adding plastic garbage to the soil increased its strength and stability, making it a good alternative for stabilising the soil [14]. This study encourages the use of recycled materials in buildings as a long-term solution to problems with soil stabilisation. The stabilisation of black cotton soil using fibres and stone dust was investigated by Suresh (2009). The study discovered that the soil's compressive strength and durability were greatly increased by the combination of these components [15]. This study emphasises the possibility of stabilising soil with locally accessible resources like natural fibres and stone dust. Apart from these investigations, a number of Indian Standards (IS) include protocols for analysing and assessing soil characteristics. The processes for ascertaining the liquid and plastic limits of soils are described in IS 2720 (Part V) (1985) [16]. These experiments are critical to comprehending expanding soil behaviour and assessing stabilisation treatment efficacy. The procedures for figuring out a soil's specific gravity are outlined in IS 2720 (Part III) (1980) [17]. When evaluating the density and compaction properties of stabilised soils, this value is essential. IS 2720 (Part XL) (1977) outlines how to calculate a soil's free swell index [18]. This test aids in assessing expansive soils' capacity to swell, which is essential for stabilising the soil. Guidelines for estimating the moisture content and dry density of soils are given in IS 2720 (Part VIII) (1980) [19].

When evaluating the stability and compaction of stabilised soils, these characteristics are essential. The techniques for determining the grain size distribution of soils are described in IS 2720 (Part IV) (1975) [20]. Understanding the particle size distribution and how it affects soil behaviour requires the completion of this exam. The process for ascertaining the unconfined compressive strength (UCS) of soils is outlined in IS 2720 (Part X) (1973) [21]. Assessing the strength and stability of stabilised soils requires the use of this test. A lot of research has been done on the use of different industrial by-products and agricultural wastes for soil stabilisation in order to improve the geotechnical characteristics of problematic soils, such as black cotton soil [22–25].

The results of several investigations on the efficiency of substances including fly ash, sugarcane bagasse ash, fibres, and other additions in enhancing soil stability are summarised in this study. Sales and Lima (2010) looked into replacing sand in concrete with bagasse ash from Brazilian sugarcane. They discovered that adding bagasse ash to concrete may greatly increase its compressive strength, indicating that bagasse ash may have use in stabilising soil [26]. By investigating the recycling of sugarcane bagasse ash in the production of clay bricks, Faria et al. (2012) expanded on this research and showed increases in brick strength and sustainability [27]. Kharade (2014) concentrated on stabilising expansive soils with bagasse ash from the sugar sector. According to the study's findings, bagasse ash is a practical solution for stabilising soil since it successfully decreased the plasticity index and raised the UCS of black cotton soil [28]. An experimental investigation on stabilising black cotton soil using fly ash, coconut coir fibre, and broken glass was carried out by Tiwari and Mahiyar (2014). They discovered that by combining these materials, the soil's potential for swelling was decreased and its ability to support weight increased [29]. Chittaranjan et al. (2011) conducted a study on the use of agricultural wastes to stabilise soil. According to their research, substances like sawdust and rice husk ash can improve the resilience and strength of soils, providing a long-term fix for soil stabilisation [30]. The leaching behaviour and strength characteristics of black cotton soil stabilised by fly ash were investigated by Kommu et al. [31]. The findings demonstrated that the addition of fly ash strengthened the soil and decreased its permeability, making it more appropriate for building. Ratna et al. (2016) assessed how well Recron 3s fibre worked with lime to stabilise expansive soil. According to the study, the soil's strength and durability were greatly increased and its propensity for swelling was decreased by the addition of lime and Recron 3s fibre [32]. Shanthi and Sajida (2009) conducted research on the behaviour of soil reinforced with sisal fibre in black cotton. According to their findings, sisal fibre reinforcement increased the soil's overall stability and tensile strength [33]. The stress-strain behaviour of soil reinforced with polypropylene fibres was examined by Verma et al. [34]. The soil's ductility and load-bearing capability were found to be enhanced by polypropylene fibres, increasing the soil's resistance to deformation. Osinubi et al. (2009) investigated the use of bagasse ash as an additive in the lime stabilisation of black cotton soil [35]. The results showed that adding lime and bagasse ash together significantly decreased the soil's plasticity and increased its strength. Black cotton soil stabilisation using fibres and stone dust was the subject of an experimental investigation carried out by Suresh et al. [36]. Their findings demonstrated that the strength and compaction properties of the soil were enhanced by the addition of fibres and stone dust. Saini et al. (2019) concentrated on employing sugarcane bagasse ash to stabilise the soil for black cotton [37]. According to the study's findings, bagasse ash significantly decreased the soil's potential for swelling and increased the soil's ability to support loads. Black cotton soil's strength qualities were examined by Kiran and Kiran (2013), who used stabilisers such as bagasse ash. Their findings showed that the strength and durability of the soil were greatly enhanced by the use of chemical additives together with bagasse ash [38]. The stability and durability of sawdust ash-lime stabilised black cotton soil were investigated by Obeta et al. (2019). According to the study, adding lime and sawdust ash to the soil improved its resistance to weathering and compressive strength [39]. Salim et al. (2014) looked at utilising sugarcane bagasse ash to increase the bearing strength of compressed earth block bricks made from sandy loam soil [40]. Their research showed that bagasse ash increased the strength and longevity of the bricks, proving that it is a practical building material. Sowmiya et al. (2018) investigated the use of plastic straws, polyester, and LDPE to increase the bearing capacity of black cotton soil. The findings showed that these substances greatly decreased the soil's compressibility and increased its ability to support loads [41]. The effects of plastics in soil, particularly the effects of polyester textile fibres on soil invertebrates, were investigated by Selonen et al. [42]. The study's findings, which suggested that polyester fibres may modify soil characteristics and biodiversity, underlined the importance of giving plastic trash considerable thought when stabilising soil. Seshasai et al. (2016) concentrated on using fibres to stabilise expansive soil. According to their research, fibres increased the soil's stability and tensile strength, making them more appropriate for use in buildings [43]. Yilmaz and Karatas (2011) looked at how polypropylene fibre affected the tensile properties of soil mixes with a light cement content that contain clay [44]. According to the study, the addition of polypropylene fibres greatly increased the soil's tensile and compressive strength, increasing its resistance to deformation and suitability for use in building projects. The large amount of research that has been evaluated shows how different materials, such as agricultural wastes and industrial byproducts, may be used to stabilise problematic soils like black cotton soil. According to the studies, soils may be made significantly more geotechnically sound by adding materials like fly ash, sugarcane bagasse ash, fibres, and other additions. This increases the soils' resilience to environmental changes and increases their strength and longevity [45].

These results encourage ecologically friendly building methods and the effective use of waste materials by supporting the use of economical and sustainable materials for soil stabilisation. The re-examined studies offer a thorough grasp of the possible advantages and uses of different materials in expanding soil stabilisation. A number of industrial by-products, including plastic waste, glass fibres, fly ash, and rice husk ash, have been used to increase soil stability and mechanical qualities. These materials provide economical and environmentally friendly ways to stabilise soil, encourage waste product recycling, and lessen the negative effects of building on the environment. The best ratios and combinations of these materials should be investigated further in future studies to optimise their soil stabilisation efficacy. Long-term research is also required to assess the stability and environmental effects of stabilised soils. By utilising the results of this research, the building sector may solve structural and environmental issues by adopting more robust and sustainable soil stabilisation techniques.

3 Material and Methodology

3.1 Soil Made of Black Cotton

Black cotton soil, which is well recognised as an expansive soil in India, is regarded as one of the soils that present the greatest challenges in the field of civil engineering as shown in Fig. 1. It is something that cannot be avoided in many development projects because it covers around 15% of India's land area. The behaviour of this soil in terms of swelling and shrinking is controlled by a number of different variables, the most important of which is the high concentration of minerals such as montmorillonite and illite that it contains. Gujarat is the location from where the black cotton soil is obtained that is being utilised for this project.

Silicon dioxide (SiO_2), the most prevalent oxide at 56.80%, is essential for structural integrity and stability [44]. Fe_2O_3 (17.10%) and Al_2O_3 (10.90%) increase the material's strength, hardness, and magnetic characteristics. TiO_2 at 2.71% improves durability, whereas Na_2O at 2.14% and CaO at 3.85% affect melting point and reactivity as shown in Table 1 (constructed based on discussion presented in [39]). MnO, MgO, and K_2O fine-tune the material's characteristics, while trace levels of phosphorus pentoxide (P_2O_5) and sulphur trioxide (SO_3) alter chemical behaviour. Loss on ignition (LOI) at 3.19% shows volatile chemicals, providing a complete composition profile.

Fig. 1 Black cotton soil

Table 1 Chemical composition of BCS

Oxides	Composition (chemical) (%)
MnO	0.29
TiO_2	2.71
SiO_2	56.8
P_2O_5	<0.001
Fe_2O_3	17.1
Na_2O	2.14
Al_2O_3	10.9
L.O.I	3.19
MgO	1.8
CaO	3.85
K_2O	1.5
SO_3	<0.001

Fig. 2 Fly ash

3.2 *The Fly Ash*

Fly ash is a thin powder that is mostly made up of spherical glass particles. It is a by-product of the burning of coal in the power sector, as shown in Fig. 2. The reaction between fly ash and lime results in the formation of cementitious compounds, which in turn improves the strength and durability of construction materials. Fly ash is well known for its pozzolanic capabilities. The Municipal Corporation of Delhi (MCD) is the source of the fly ash that is being used in this project.

Table 2 Chemical composition of fly ash

Characteristics	Fly ash (%)
Sulphates	0.1
Phosphorus oxide	<1
Silica	55–65
Titanium oxide	<1
Aluminium oxide	22–25
Alkali oxide	<1
Iron oxide	5–7
Magnesium oxide	<1
Calcium oxide	5–7

Table 2 (constructed based on discussions presented in [46]) outlines the chemical composition of fly ash, presenting the percentages of various oxides and compounds it contains. Silica (SiO_2) is the most abundant component, ranging from 55% to 65%, indicating a high silica content which is typical for fly ash. Aluminium oxide (Al_2O_3) follows with a significant proportion of 22% to 25%, reflecting its alumina content. Iron oxide (Fe_2O_3) and calcium oxide (CaO) each account for 5–7%, highlighting the presence of iron and calcium. Minor components include sulphates at 0.1%, and trace amounts of phosphorus oxide, titanium oxide, magnesium oxide, and alkali oxides, all less than 1%. These minor components, despite their low percentages, can influence the reactivity and properties of the fly ash when used in various applications, such as in cement and concrete production. Overall, the table provides a comprehensive overview of the primary and secondary constituents of fly ash, essential for assessing its suitability and performance in different industrial uses.

3.3 Ash Made from Bagasse

A variety of percentages of bagasse ash will be mixed into the black cotton soil in addition to fly ash in order to determine the effect that bagasse ash has on the strength qualities of the soil, as shown in Fig. 3. Bagasse ash is produced by the sugarcane industry, which results in the production of waste by-products. The Indian Institute of Sugarcane Research (IISR) in Lucknow was the establishment from which the bagasse ash was procured for the purpose of this experiment. The purpose of this research is to investigate the ways in which fly ash and bagasse ash can work together to bring about improvements in the engineering features of black cotton soil.

Table 3 (prepared based on information outlined in [40]) presents the chemical composition of a mineral sample, highlighting the percentages of various oxides and compounds present. SiO_2 is the predominant component, constituting 72.9% of the sample, indicating a high silica content. Aluminium oxide (Al_2O_3) and iron oxide (Fe_2O_3) are present in smaller amounts, at 6.6% and 6.4%, respectively, suggesting the presence of alumina and iron. Other significant components include

Fig. 3 Bagasse ash

Table 3 Chemical composition of BA

Mineral	Percentage (%)
Fe_2O_3	6.4
CaO	2.9
Na_2O	1.2
SiO_2	72.9
P_2O_5	4.1
Loss of ignition	1
Al_2O_3	6.6
K_2O	2.2
MgO	3.1

calcium oxide (CaO) at 2.9%, magnesium oxide (MgO) at 3.1%, potassium oxide (K_2O) at 2.2%, and phosphorus pentoxide (P_2O_5) at 4.1%. Sodium oxide (Na_2O) is the least abundant among the major oxides, making up 1.2% of the sample. The LOI, which represents the amount of weight loss due to the release of volatile substances upon heating, is 1.0%, indicating the presence of some volatile compounds or water. Overall, the table provides a detailed breakdown of the mineral's composition, essential for understanding its properties and potential applications.

3.4 Methodology

The methodology for this study (Fig. 4) on the stabilisation of black cotton soil using fly ash and bagasse ash follows a systematic approach, as illustrated in the flow chart.

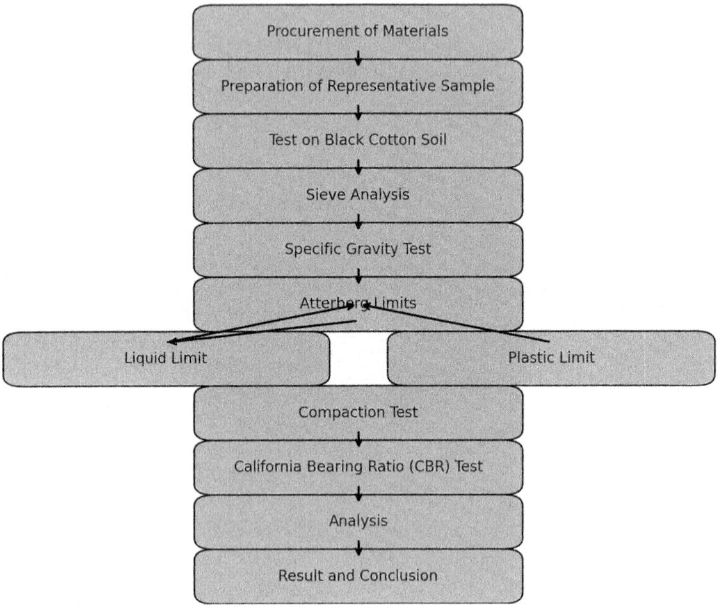

Fig. 4 Flow chart for methodology

The first step involves sourcing all necessary materials for the study, including black cotton soil, fly ash, and bagasse ash. Once the materials are procured, a representative sample of the black cotton soil is prepared. This ensures that the sample used for testing is uniform and representative of the soil type being studied. Various tests are conducted on the black cotton soil to determine its initial properties. This includes identifying its inherent characteristics and behaviour. Sieve analysis is performed to determine the particle size distribution of the soil. This helps in understanding the gradation and texture of the soil. The specific gravity of the soil is determined using appropriate methods. This property is essential for various calculations and understanding the soil's density relative to water. The liquid limit and plastic limit of the soil are determined to understand its plasticity and consistency limits. These tests help in classifying the soil and predicting its behaviour under different moisture conditions. Compaction tests are conducted to determine the optimal moisture content and maximum dry density (MDD) of the soil. This step is crucial for understanding how the soil can be compacted in the field to achieve desired properties. The CBR test is conducted to assess the soil's strength and its suitability for use in subgrade and subbase layers of pavements. This test provides a measure of the soil's resistance to penetration. The data collected from the various tests are analysed to understand the effects of adding fly ash and bagasse ash on the properties of black cotton soil. This involves comparing the results with the baseline properties of the untreated soil.

Finally, the results of the analysis are compiled, and conclusions are drawn regarding the effectiveness of fly ash and bagasse ash in stabilising black cotton soil. The study's findings are summarised, highlighting the improvements in soil properties and potential applications. Understanding the soil's properties related to its flexibility requires a comprehension of the Atterberg limits. Whereas the plastic limit test finds the water content at which the soil transitions from a semi-solid to a plastic state, the liquid limit test finds the water content at which the soil transitions from a plastic to a liquid state. These boundaries aid in the classification of the soil and the forecasting of its behaviour under different moisture regimes. The purpose of the compaction test is to ascertain how the dry density and soil moisture content relate to one another. This test offers information on the MDD and optimum moisture content (OMC), which may be used to determine how best to compress the soil in the field. To increase the soil's load-bearing capacity and decrease its compressibility, proper compaction is essential. The purpose of this test is to evaluate the soil's compressive strength in the absence of lateral confinement. It offers important details on the behaviour of the soil under axial stress and its capacity to sustain structures. An important consideration when developing foundations and other geotechnical constructions is the UCS. By using this technique, the research offers a thorough grasp of the efficacy of fly ash and bagasse ash as soil stabilisers by methodically evaluating their capacity to improve the characteristics of black cotton soil.

4 Result and Discussion

4.1 Test on Parent Soil

4.1.1 Specific Gravity

The specific gravity test used in this investigation is in accordance with IS: 2720 (Part 3) [17]. The specific gravity of soil is determined using a density bottle, and this information is useful for determining the unit weight and saturation level of wet soils. The parent soil's specific gravity was found to be 2.330.

4.1.2 Liquid Limit

The purpose of this test is to ascertain the soil's liquid limit in accordance with IS: 2720 (Part 5)—1985 [16]. The water content at which fine-grained soil behaves almost like a liquid yet has low shear strength is known as the liquid limit. In only 25 blows, its flow seals the groove in Casagrande's liquid limit mechanism. It was determined that the parent soil's liquid limit was 76%.

4.1.3 Plastic Limit

The water content at which soil loses its flexibility and behaves more like a brittle substance as it transitions to a semi-solid state is known as the "plastic limit". The water content at which soil would just start to disintegrate when spun into a thread about 3 mm in diameter is another definition for it. To ascertain the plastic limit in accordance with IS: 2720 (Part-5)—1985 (Reaffirmed—2006) [16], tests were carried out. It was determined that the parent soil's plastic limit was 34%.

4.1.4 Grain Size Analysis

The grain sieve analysis used in this work is in accordance with IS: 2720 (Part 4) (1985) [20], as shown in Fig. 5. This falls under the classification of MH.

4.1.5 Proctor Compaction Test

We performed standard Proctor test in order to find out the OMC of the parent soil. The OMC obtained for parent soil was 28%. The MDD obtained for parent soil was 1.455 g/cc.

4.1.6 CBR Test

CBR is the ratio, given in percentage terms, of the force per unit area needed to break through a standard 50 mm diameter circular plunger at a rate of 1.25 mm/min into a soil mass to the force needed to break through a standard material at the same

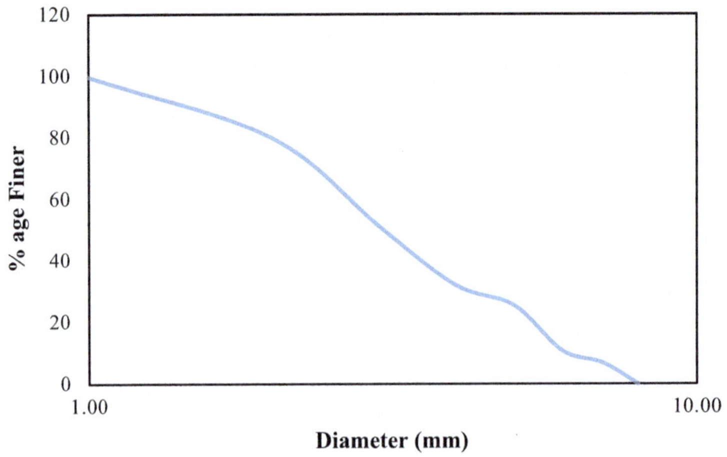

Fig. 5 Grain size analysis

rate. Typically, the ratio is calculated assuming a 1.5 mm penetration. The parent soil's CBR value was 1.5 mm.

4.2 Result on Treated Soil with Varying Percentage of Fly Ash and Bagasse Ash

Firstly, we have kept percentage of bagasse ash constant, that is, 8%, and then we go on increasing the percentage of fly ash from 2 to 4 to 6%, as shown in Fig. 6. Then in the next phase we have kept the percentage of fly ash constant and then we have increased the percentage of bagasse ash from 2 to 4 to 6 to 8%.

A specific gravity (SG) of 2.33 is found in the parent soil, which is known as black cotton soil (BCS). In general, the specific gravity of the soil mixes is reduced when admixtures such as bagasse ash (BA) and fly ash (FA) are added to the mixtures. The combination of 8% BA and 2% FA results in the lowest specific gravity of 2.21, which is the most noticeable decrease that can be observed. Different combinations also exhibit variable degrees of decrease, with specific gravity values ranging from 2.25 to 2.32, depending on the proportions of BA and FA. Other combinations also show varying degrees of reduction. Based on these findings, it is evident that the specific gravity of BCS may be efficiently decreased by integrating various amounts of BA and FA. This might be a key step in the process of customising soil parameters to satisfy specific engineering needs.

The figure displays the results of the liquid limit (LL) test conducted on BCS that contained different proportions of bagasse ash (BA) and fly ash (FA) admixtures, as shown in Fig. 7. The moisture level at which the soil transitions from a plastic to a liquid condition is referred to as the liquid limit. Based on the findings, it can be concluded that the incorporation of bagasse ash and fly ash into black cotton soil

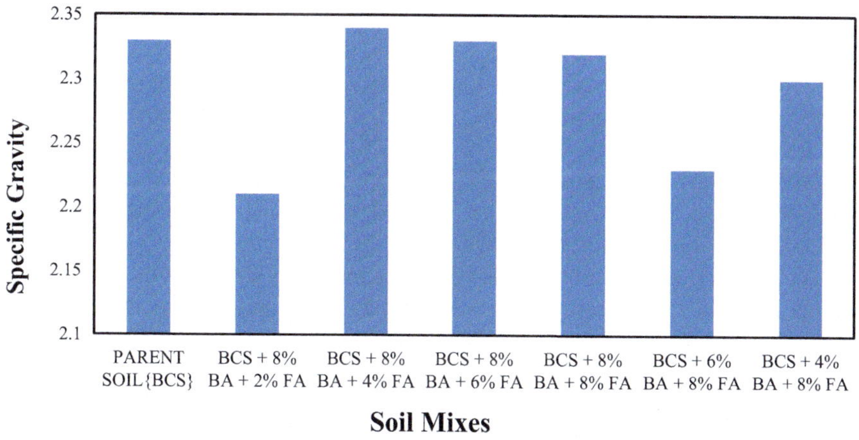

Fig. 6 Specific gravity for various combinations of BCS, FA, and BA

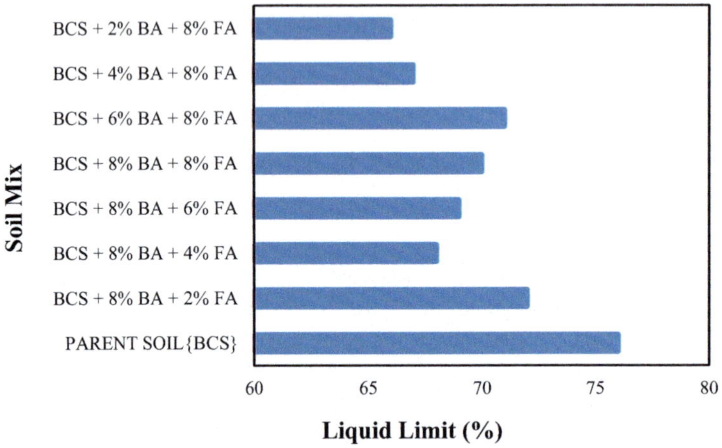

Fig. 7 Liquid limit for various combinations of BCS, FA, and BA

leads to a general decrease in the liquid limit of the soil, which is advantageous for the enhancement of soil stability. The liquid limit is reduced to 66% when 2% BA and 8% FA are combined, which is the combination that results in the greatest substantial drop in liquid limit. Based on this information, it appears that a greater proportion of fly ash in comparison to bagasse ash has the impact of reducing the flexibility of the soil. When the proportion of bagasse ash is increased, on the other hand, the liquid limit tends to increase as well, which indicates that the plasticity is increased. Consequently, the plasticity properties of the soil may be optimised by the use of a balanced mixture of bagasse ash and fly ash, which then results in improved performance in building applications.

Plastic limit for the parent soil was 34%. On adding 2% of fly ash and 8% of bagasse, it came out to be 39%, which was higher than that of the parent soil. After that, it decreased with the increasing percentage of the fly ash from 34% to 25% to 26% at percentages of 4 to 6 to 8% of the fly ash, as shown in Fig. 8. It was again high when we added 4% of the bagasse ash and 8% of fly ash.

OMC reading for the parent soil was 28%, but with the addition of admixture, the value of OMC was constantly at 27%, as shown in Fig. 9. It was least at 6% BA and 8% FA, and the value was 26%.

MDD for the parent soil was 1.455 gm/cc, but with the addition of admixture, the value decreased. It was highest at 8% BA and 6% FA with a value of 1.48 gm/cc, and the trend of the property is shown in Fig. 10.

The CBR values for soil at 2.5 and 5 mm penetrations show that the soil exhibits higher resistance at 2.5 mm penetration, starting at 2.5 and peaking at 7.71 before gradually declining to 3.83, as shown in Fig. 11. In contrast, the CBR values at 5 mm penetration begin lower at 2, reach a peak of 7.07, and then decline more sharply to 3.4. Both penetration depths follow a similar trend of an initial increase, a midpoint peak, and a subsequent decline, but the CBR values for 2.5 mm penetration consistently remain higher than those for 5 mm penetration. This indicates that the soil's load-bearing capacity is higher at shallower penetrations, with optimal strength observed around the midrange penetration points.

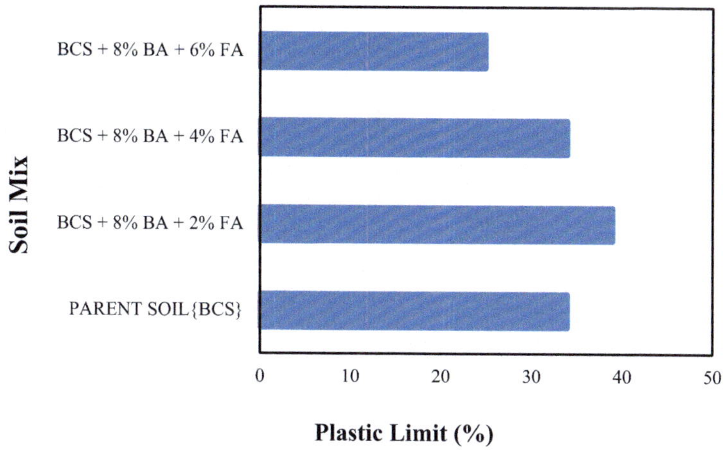

Fig. 8 Plastic limit for various combinations of BCS, FA, and BA

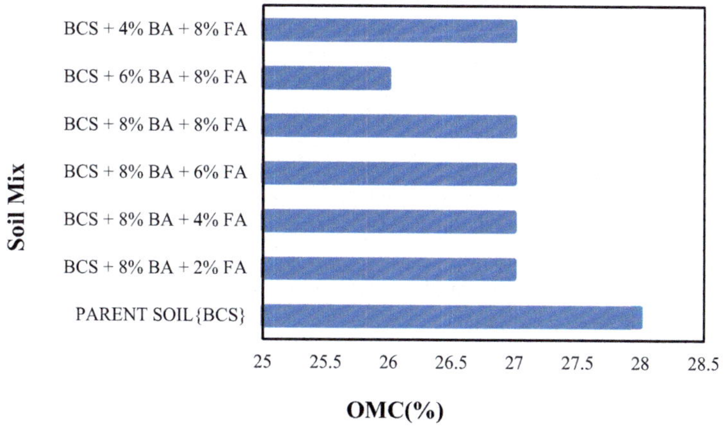

Fig. 9 OMC for various combinations of BCS, FA, and BA

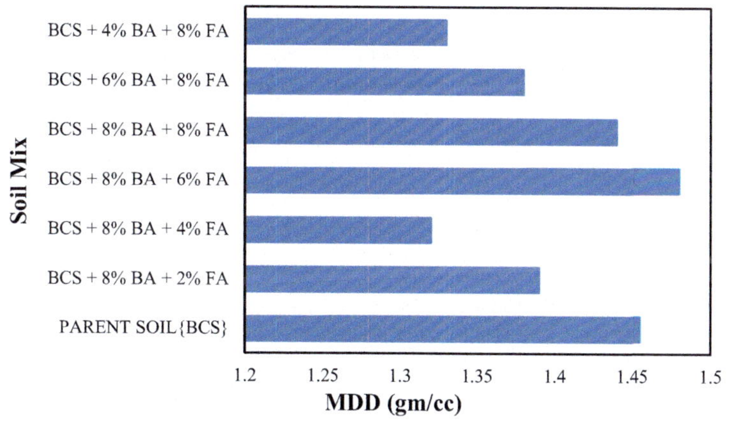

Fig. 10 MDD limit for various combinations of BCS, FA, and BA

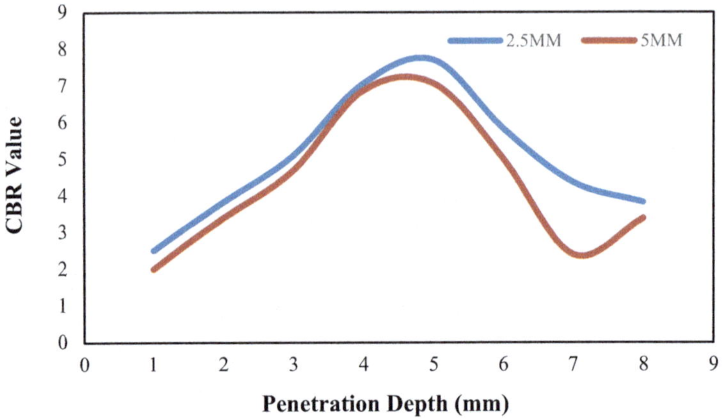

Fig. 11 CBR for various combinations of BCS, FA, and BA

5 Conclusion

The study concluded that the addition of fly ash and bagasse ash significantly improves the geotechnical properties of black cotton soil. Key findings include:

- The liquid limit decreased to 66% with 8% fly ash and 2% bagasse ash.
- The plastic limit was lowest at 25% with 6% fly ash and 8% bagasse ash.
- The optimum moisture content (OMC) decreased to 27%, showing an improvement over the parent soil.
- The maximum dry density (MDD) reduced to 1.48 g/cc at the combination of 6% fly ash and 8% bagasse ash.
- The California Bearing Ratio (CBR) values improved, with peak values observed at mid-range penetration points, indicating enhanced load-bearing capacity.
- The specific gravity decreased to 2.21 with 8% bagasse ash and 2% fly ash, showing the effectiveness of the additives in modifying soil properties.

These results indicate that fly ash and bagasse ash are effective in stabilising black cotton soil, making it more suitable for construction purposes. The research highlights the practical applications of using these industrial by-products in geotechnical engineering, contributing to sustainable construction practices and waste management.

References

1. Bharti, G., Shukla, B. K., Kumar, A., Nagaraju, M. B., Bansal, V., & Varadharajan, S. (2023). Impact of copper and steel slag blend in strengthening mechanical properties of black cotton soil. In *AIP conference proceedings* (Vol. 2800). AIP Publishing.

2. Arya, P., Patel, S. B., Bharti, G., Shukla, B. K., & Hurukadli, P. (2023). Impact of using a blend of bagasse ash and polyester fiber in black cotton soil for improvement of mechanical and geotechnical properties of soil. *Materials Today: Proceedings, 78*, 738–743.
3. Shukla, B. K., Gupta, A., Gowda, S., & Srivastav, Y. (2023). Constructing a greener future: A comprehensive review on the sustainable use of fly ash in the construction industry and beyond. *Materials Today: Proceedings, 93*(3), 257–264.
4. Varadharajan, S., Kirthanashri, S. V., Dwivedi, A., Irfan, D., Shukla, B. K., Kumar, A., Bishetti, P., & Bharti, G. (2023). Use of fly ash and metakaolin to manufacture sustainable concrete. In *AIP conference proceedings* (Vol. 2800). AIP Publishing.
5. Bharti, G., Hurukadli, P., Shukla, B. K., Sihag, P., Jagudi, S., & Tripathi, A. (2023). Environmental impact analysis and utilization of copper slag for stabilising black cotton soil. *Materials Today: Proceedings*. https://doi.org/10.1016/j.matpr.2023.04.305
6. Bhuvaneshwari, S., Robinson, R. G., & Gandhi, S. R. (2005). Stabilization of expansive soil using fly ash. In *Fly ash utilization Programme (FAUP), TIAFC.*
7. Yadu, L. K., Tripathi, R. K., & Singh, D. V. (2011). Comparison of fly ash and rice husk ash stabilized black cotton soil. *International Journal of Earth Sciences and Engineering, 4*(6 SPL), 42–45.
8. Tiwari, A. (2014). Experimental study on stabilization of black cotton soil by fly ash, coconut coir fibre & crushed glass. *International Journal of Emerging Technology and Advanced Engineering, 2*(3), 28.
9. Ashraf, A., & Joseph, M. (2014). Soil stabilization using raw plastic bottles. *Proceedings of Indian Geotechnical Conference, 2*(3), 304.
10. Singh, B., & Mali, S. (2013). Soil stabilization using glass fibers. *International Journal of Scientific & Engineering Research, 5*(12), 32.
11. Choudhary, A. K. (2012). A study on CBR behaviour of plastic waste strip reinforced soil. *Emirates Journal for Engineering Research, 2*(1), 118.
12. Joseph, M., & Muhammed Haneef, F. (2014). Effect of plastic granules on the properties of soil. *International Journal of Engineering Research and Applications, 4*(4), 160.
13. Michel, B., & Aderinle. (2014). Clay soil stabilization using powdered glass. *Journal of Engineering Science and Technology, 9*(1), 541.
14. Sivakumar Babu, G. L. (2014). Studies of black cotton soil admixed with plastic waste. *International Journal of Indian Institute of Science, 9*(2), 154.
15. Suresh, K. (2009). Experimental study on stabilization of black cotton soil with stone dust and fibers. *International Journal of Civil Engineering Society, 1*(3), 82.
16. Bureau of Indian Standards. (1985). *IS 2720 (Part V): Determination of liquid & plastic limits.*
17. Bureau of Indian Standards. (1980). *IS 2720 (Part III): Determination of specific gravity.*
18. Bureau of Indian Standards. (1977). *IS 2720 (Part XL): Determination of free swell index.*
19. Bureau of Indian Standards. (1980). *IS 2720 (Part VIII): Determination of moisture content & dry density.*
20. Bureau of Indian Standards. (1975). *IS 2720 (Part IV): Determination of grain size.*
21. Bureau of Indian Standards. (1973). *IS 2720 (Part X): Determination of unconfined compressive strength.*
22. Bharti, G., Shukla, B. K., Kumar, A., & Hurukadli, P. N. (2023). Stabilization of black cotton soil using steel slag. In *AIP conference proceedings* (Vol. 2800). AIP Publishing.
23. Gowda, S., Kunjar, V., Gupta, A., Kavitha, G., Shukla, B. K., & Sihag, P. (2024). Prediction of the subgrade soil California bearing ratio using machine learning and neuro-fuzzy inference system techniques: A sustainable approach in urban infrastructure development. *Urban Science, 8*(1), 4.
24. Bharti, G., Shukla, B. K., Parashar, B., Kumar, A., & Varadharajan, S. (2023). Experimental study on physico-mechanical properties of expansive soil by using wheat straw fiber. In *AIP conference proceedings* (Vol. 2800). AIP Publishing.
25. Bansal, V., & Shukla, B. K. (2022). Application of urea formaldehyde raisin in pavement subgrade stabilization at Hindon barrage. In *Journal of physics: Conference series* (Vol. 2267, p. 012150). IOP Publishing.

26. Sales, A., & Lima, S. A. (2010). Use of Brazilian sugarcane bagasse ash in concrete as sand replacement. *Waste Management, 30*(6), 1114–1122.
27. Faria, K., Gurgel, R., & Holanda, J. (2012). Recycling of sugarcane bagasse ash waste in the production of clay bricks. *Journal of Environmental Management, 101*, 7–12.
28. Kharade, A. S. (2014). Waste product bagasse ash from sugar industry can be used as stabilizing material for expansive soils. *International Journal of Research in Engineering and Technology, 3*, 506–512.
29. Tiwari, A. K., & Mahiyar, H. (2014). Experimental study on stabilization of black cotton soil by fly ash, coconut coir fiber, and crushed glass. *International Journal of Emerging Technology and Advanced Engineering, 4*(11).
30. Chittaranjan, M., Vijay, M., & Keerthi, D. (2011). Agricultural wastes as soil stabilizers. *International Journal of Earth Sciences and Engineering, 4*(6), 50–51.
31. Kommu, S., Asadi, S. S., & Prasad, A. V. S. (2018). Leaching behavior and strength characteristics of black cotton soil stabilized with fly ash. *Materials Today: Proceedings, 5*(9), 17974–17981.
32. Ratna, P. S., Prasad, D. S. V., & Raju, G. V. R. P. (2016). Performance of Recron 3s fibre with lime in expansive soil stabilization. *IOSR Journal of Mechanical and Civil Engineering, 13*(6).
33. Shanthi, K. K., & Sajida, M. K. (2009). Behaviour of black cotton soil reinforced with sisal fibre. In *10th National Conference on technological trends (CVTT09).*
34. Verma, S., Khan, R., Khan, S., & Khan, R. (2015). Stress strain behaviour analysis of polypropylene fibre reinforced soil. *International Journal of Research in Emerging Science and Technology, 2*.
35. Osinubi, K. J., Ijimdiya, T. S., & Nmadu, I. (2009). Lime stabilization of black cotton soil using bagasse ash as admixture. *Advanced Materials Research, 62-64*, 3–10.
36. Suresh, K., Padmavathi, V., & Sultana, A. (2009). *Experimental study on stabilization of black cotton soil with stone dust and fibers* (pp. 502–506). Indian Geotechnical Society (IGS).
37. Saini, H., Khatti, J., & Acharya, B. (2019). Stabilization of black cotton soil by using sugarcane bagasse ash. *International Journal of Scientific Research and Review, 7*(1), 128–132.
38. Kiran, R. G., & Kiran, L. (2013). Analysis of strength characteristics of black cotton soil using bagasse ash and additives as stabilizer. *International Journal of Engineering Research and Technology (IJERT), 2*(7).
39. Obeta, I. N., Ikeagwuani, C. C., Attama, C. M., & Okafor, J. (2019). Stability and durability of sawdust ash-lime stabilised black cotton soil. *Nigerian Journal of Technology, 38*(1), 75–85.
40. Salim, R. W., Ndambuki, J. M., & Adedokun, D. A. (2014). Improving the bearing strength of sandy loam soil compressed earth block bricks using sugarcane bagasse ash. *Sustainability, 6*, 3686–3696.
41. Sowmiya, K. S., Akila, J., Rashmi, N. A., Karthika, P., & Monisha, R. (2018). Bearing capacity improvement of black cotton soil using LDPE, polyester and plastic straws. *SSRG International Journal of Civil Engineering - Special Issue.*
42. Selonen, S., Dolar, A., Jemec Kokalj, A., Skalar, T., Parramon Dolcet, L., Hurley, R., & van Gestel, C. A. M. (2020). Exploring the impacts of plastics in soil—The effects of polyester textile fibers on soil invertebrates. *Science of the Total Environment, 700*, 134451.
43. Seshasai, V., Ramakrishna, M., & Sasi Kumar, B. V. (2016). Stabilization of expansive soil using fibers. *International Journal of Technology Research and Engineering, 4*(4), 694–699.
44. Shukla, B. K., Bharti, G., Sharma, P. K., Sharma, M., Rawat, S., Maurya, N., Srivastava, R., & Srivastav, Y. (2024). Sustainable construction practices with recycled and waste materials for a circular economy. *Asian Journal of Civil Engineering., 25*, 5255. https://doi.org/10.1007/s42107-024-01111-y
45. Yilmaz, Y., & Karatas, K. (2011). Effect of polypropylene fiber on the strength characteristics of lightly cemented clayey soil mixtures. In *Proceedings of geo-frontiers 2011: Advances in geotechnical engineering* (pp. 707–716).
46. Ibrahim, A. N. H., Ahmad, A. S., Moh'd Akhir, N., & Borhan, M. N. (2016). Performance evaluation of stone mastic asphalt (SMA) using geopolymer as an asphalt modifier. Jordan. *Journal of Civil Engineering, 10*(4).

Index

S. Pathak et al. (eds.), *Intelligent Infrastructure and Smart Materials*,
https://doi.org/10.1007/978-3-031-92421-7

The manufacturer's authorised representative in the EU is Springer Nature Customer Service Centre GmbH, Europaplatz 3, 69115 Heidelberg, Germany. If you have any concerns regarding our products, please contact ProductSafety@springernature.com

Printed and bound by CPI Group (UK) Ltd, Croydon, CR0 4YY

30/09/2026

02225571-0012